Towards a New Neuromorphology

Rudolf Nieuwenhuys · Luis Puelles

Towards a New Neuromorphology

Springer

Rudolf Nieuwenhuys
Netherlands Institute for Neuroscience
Royal NL Academy of Arts and Sciences
Amsterdam
The Netherlands
RudolfN@planet.nl

Luis Puelles
Department of Human Anatomy
and Psychobiology
University of Murcia
Murcia
Spain
puelles@um.es

ISBN 978-3-319-25692-4 ISBN 978-3-319-25693-1 (eBook)
DOI 10.1007/978-3-319-25693-1

Library of Congress Control Number: 2015955601

Springer Cham Heidelberg New York Dordrecht London

Printed on acid-free paper

Springer International Publishing AG Switzerland is part of Springer Science+Business Media
(www.springer.com)

Preface

This book owes its origin to a remarkable coincidence. In October 2012 Luis Puelles (LP) and I (RN) decided to organize and realize a highly selective two-person conference. Its aims were twofold: (1) to discuss the 'state of the art' in developmental and comparative neuromorphology, our fields of interest, and (2) to look into the possibility of preparing an elementary overview of recent developments in these fields. At the same time, one of us (RN) was invited to participate in a neuroscience course for a group of talented students in the Max Planck Institute for Human Cognitive and Brain Sciences in Leipzig, involving a series of introductory lectures as well as dissections of human brains. We (LP and RN) immediately interpreted this invitation as a unique opportunity to realize our aims. An outline of the lectures to be delivered was forwarded to Leipzig and found approval by Prof. Dr. Robert Turner and Dr. Stephan Geyer, the organizers of the course.

The lectures, illustrated with Powerpoint slides and delivered during March 19–21, 2013, were well received, and Prof. Turner spontaneously proposed that his Institute would take the initiative to publish them in their entirety. This offer, which perfectly matched our original intentions (see above), was gratefully accepted, but the transformation of the lecture notes and transcripts (see below) into the final text (the present course) turned out to be much more complex and laborious than anticipated, hence the relatively long time interval between the presentation of these lectures and their appearance in print. Afterwards, the mediation of Professor Turner and his Institute in the publication of the book appeared to be unnecessary, because Springer-Verlag was found willing to publish the work.

The authors wish to thank Professor Turner and Dr. Geyer for their generous support, Professor L. H. Bannister for critically reading the entire text, and for making many suggestions for improvements with respect to content as well as language, Professors L. Medina and J.L. Ferran for reading and offering valuable advice for changes in the parts of section 10 dealing with the concept of homology, Mrs. Sonja Schall, Juliane Dinse, Nancy Muschall, Aline Peter and Christina Schröder for preparing a complete transcript of the videorecorded lectures, Dr. Jenneke Kruisbrink for help with the collection of literature and Mr. Ton Put for help with the illustrations.

In concluding this preamble, the invaluable moral and practical support of Mrs. Suzanne Bakker M Sc. in all phases of this project is gratefully acknowledged and finally, we extend our most sincere thanks to our publisher, Springer-Verlag, and their staff—especially Dr. Andrea Pillmann, Dr. Jutta Lindenborn and Mr. Alexander Subbareddy—for their kind help during the production of this book.

Rudolf Nieuwenhuys
Luis Puelles

Contents

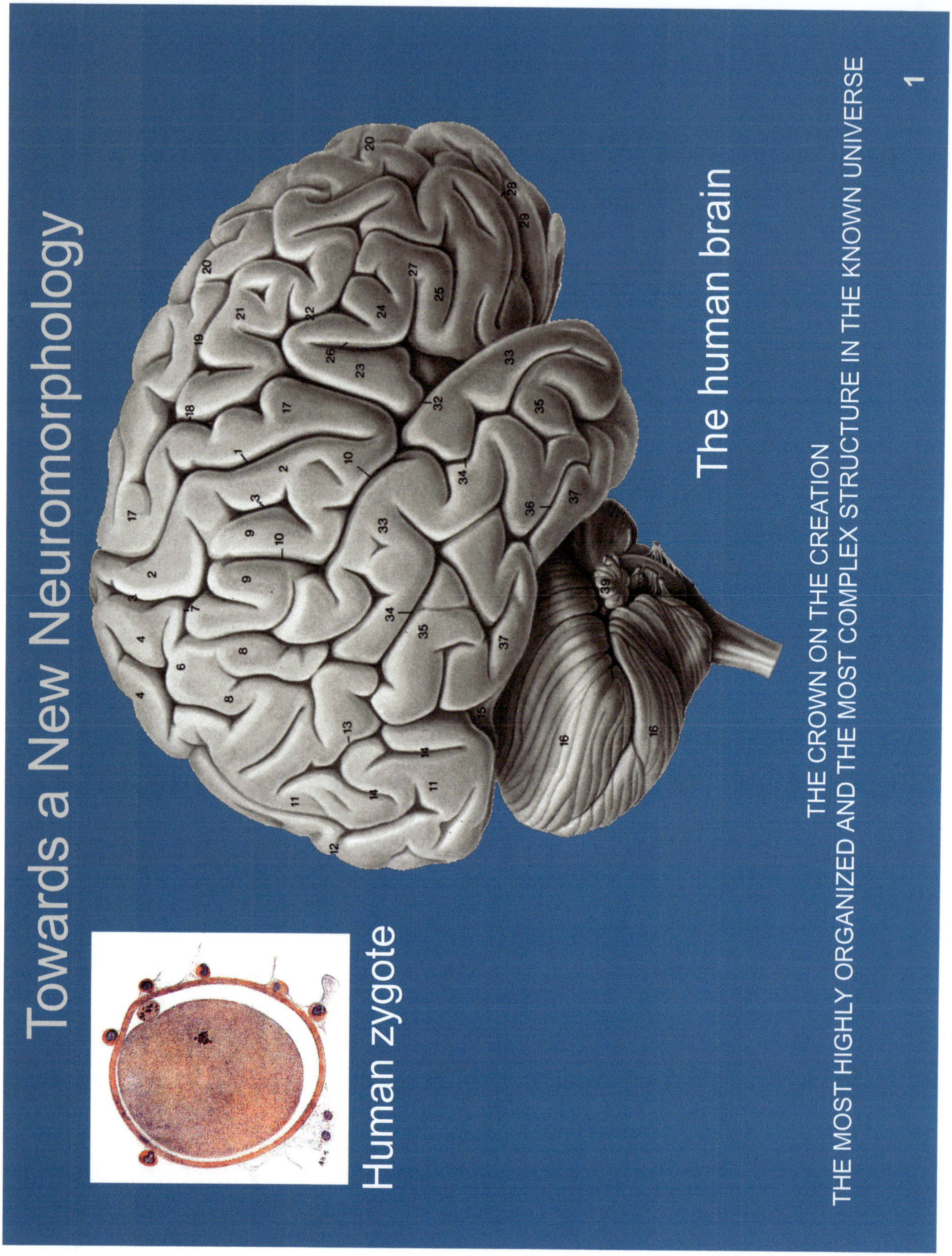
Towards a New Neuromorphology
Human zygote
The human brain
THE CROWN ON THE CREATION
THE MOST HIGHLY ORGANIZED AND THE MOST COMPLEX STRUCTURE IN THE KNOWN UNIVERSE
1

Prologue

Plate 1

Behold the human brain!! For some, it is the crown of the creation. For many others, the most highly organized and the most complex structure in the known universe. It is quite remarkable that all this has developed from a single fertilized oocyte, just as the entire organism has. The telencephalon or cerebrum is very large. Its surface is strongly folded, and directly, under this surface, there is a highly complex zone of grey matter, known as the cerebral cortex. Due to the folding, only one-third of the cortex is exposed to the free surface of the cerebrum. The surface of the second large structure—the cerebellum—is also folded even more deeply than that of the cerebrum. This brain part also contains a superficial cortical zone, of which only one-tenth borders the free surface. The minute granular neurons of the cerebellum represent up to 75 % of all brain neurons.

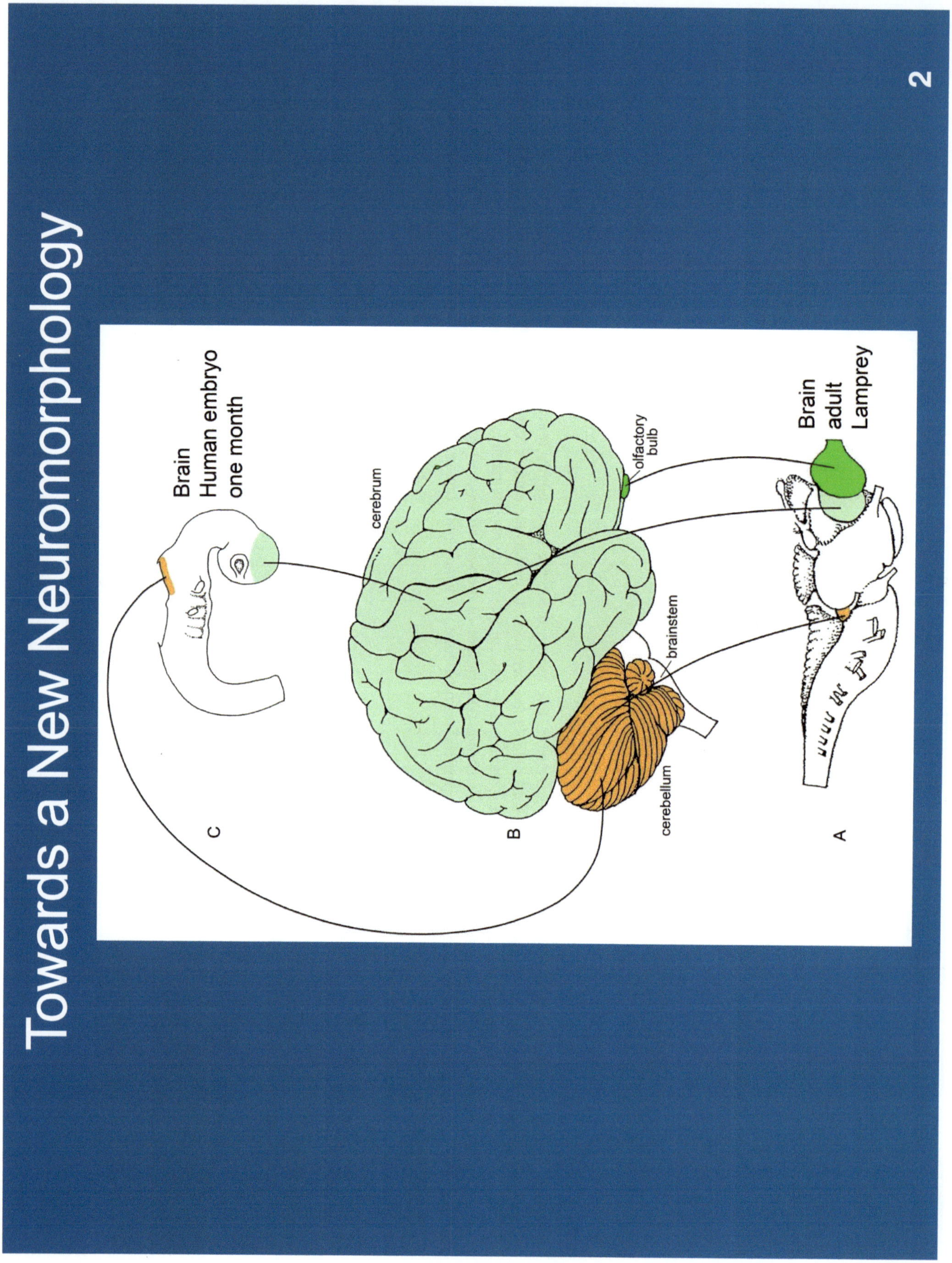
Towards a New Neuromorphology
2
Brain
Human embryo
one month
C
B
A
cerebrum
olfactory
bulb
brainstem
cerebellum
Brain
adult
Lamprey

Plate 2

If we compare the adult human brain with that of an embryo of about one month's gestation, it appears that the cerebrum and the cerebellum, which represent as we have just seen the dominant structures of the adult brain, develop from relatively small primordia or anlagen. And if we compare the adult human brain with that of an adult lamprey, which is the simplest extant vertebrate, we see that in this animal the cerebral hemispheres and the cerebellum are likewise very small brain parts. Note that in the lamprey, the olfactory bulbs, the primary centres for the processing of olfactory stimuli, exceed the cerebral hemispheres in size and that these relations have dramatically changed in the human brain.

It probably is a fair estimate that all over the world some hundred thousand people are working on the human brain, not so much on the human brain as a whole, but rather on partial problems—highly complex problems to be sure—and with the aid of very complex techniques. It could be greatly beneficial if all these workers had a common idea about how this brain is fundamentally structured. What is the real essence of this ultra-complex structure? How far does it compare with other vertebrate brains? It is the central aim of this course to highlight the elementary principles of the ontogenetic and evolutionary development of the human brain.

Towards a New Neuromorphology

1. From the classic to modern neuromorphology, an introductory overview
2. The neural plate
3. From neural plate to neural tube
4. The neural tube
5. The longitudinal axis of the brain
6. The (complex) rostral part of the brain
7. The natural coordinate system of the CNS
8. The fundamental morphological units of the CNS
9. The structural pattern, 'Bauplan' or 'blueprint' of the CNS
10. The development of grisea
11. The "wiring" of the CNS
12. General summary, conclusions and perspective

3

Plan of the Course

Plate 3

As you can see, the course comprises twelve sections. It begins with an introductory overview of the matter to be covered. This is doubtless the most relaxing part of the whole course, because all of what we will be telling you therein will later come back in more detail. The course ends, not surprisingly, with a brief section entitled: General summary, conclusions and perspective. Throughout the course, numerous smaller and larger *excursions* will be made to domains/aspects/subjects deserving especial attention.

Towards a New Neuromorphology

1. From the classic to modern neuromorphology, an introductory overview

1.1 Neural plate – Neural groove – Neural tube

An enormous increase of insight into the molecular regulatory processes and gene expression patterns during early development

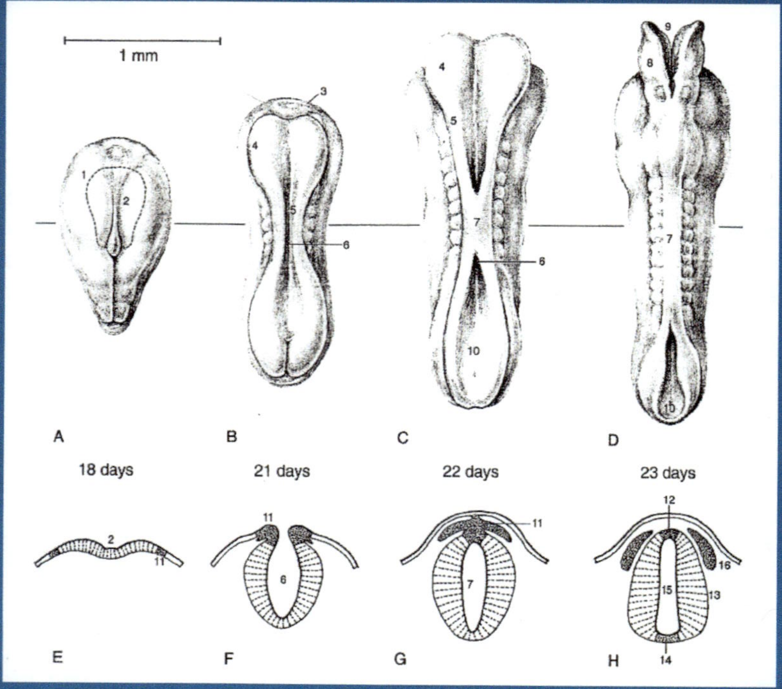

Dorsal aspects and cross sections of some human embryos

Expression patterns of some genes in the anterior neural plate of a chick embryo (Sánchez-Arrones et al. 2009)

1 From the Classic to Modern Neuromorphology, an Introductory Overview

1.1 Neural Plate—Neural Groove—Neural Tube

Plate 4

Most of you will know that the human brain begins as a tiny epidermal plate, called the 'neural plate', whose lateral edges rise upward, approach each other and finally fuse, so that the originally flat neural plate is transformed into a hollow neural tube with a fused roof plate (left panel). Note that this closing process does not occur synchronously over the length of the developing central nervous system (CNS). It begins in the future cervical (neck) region and proceeds from there anteriorly and posteriorly, so that the initially large anterior and posterior neural openings or 'neuropores' become progressively smaller and finally disappear. Experimental fate-mapping has shown that the rostralmost closing point lies at the locus of the future anterior commissure.

It should be emphasized that our knowledge concerning this early ontogenesis of the CNS just outlined, has enormously increased during the last decades, due to the development of molecular neurobiology. The right-hand panel symbolizes this progress. It shows the expression patterns of some developmental regulatory genes in the anterior neural plate of a chick embryo illustrating prospective telencephalic- (pink) and eye fields (green).

R. Nieuwenhuys and L. Puelles, *Towards a New Neuromorphology*,
DOI 10.1007/978-3-319-25693-1_1

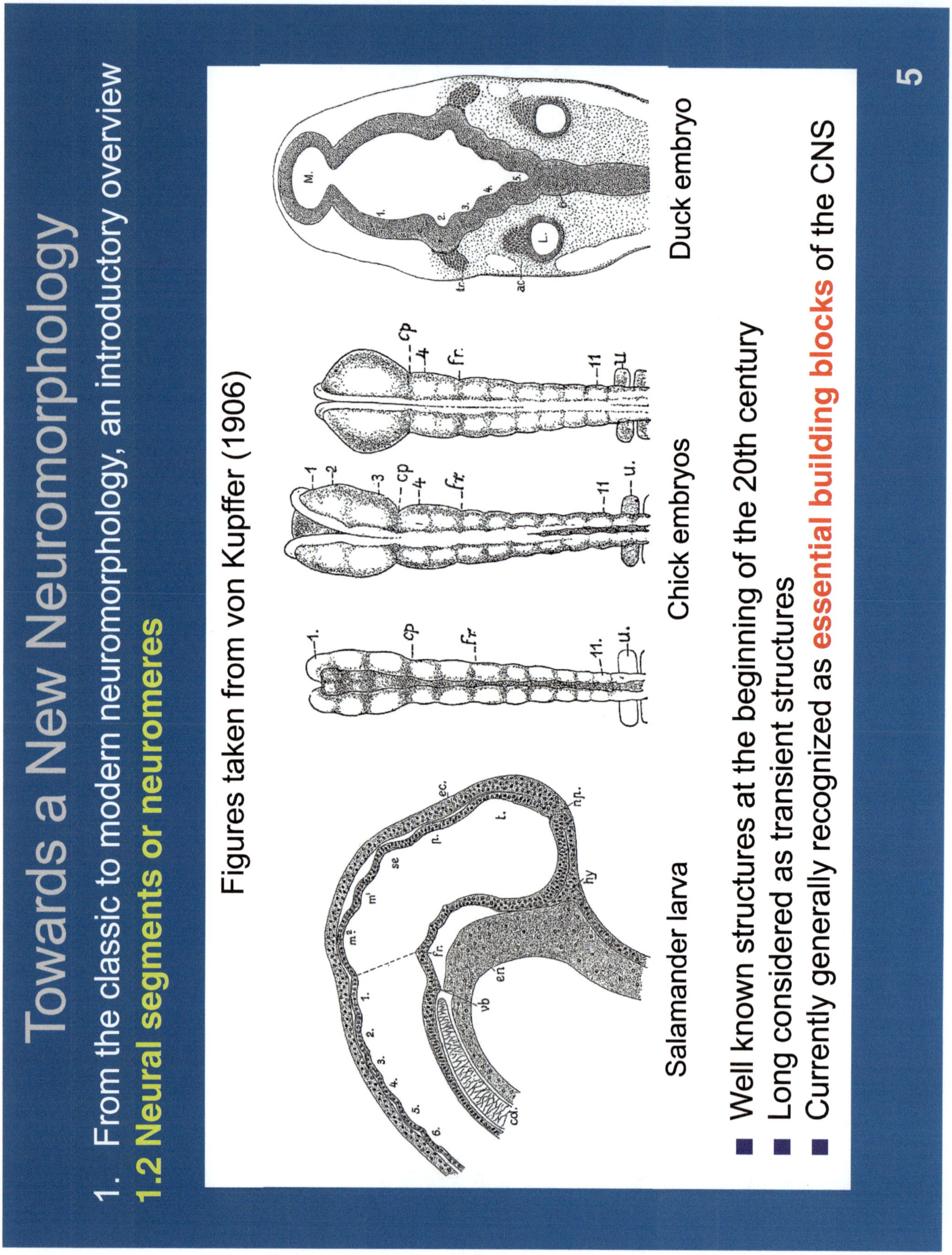
Towards a New Neuromorphology
1. From the classic to modern neuromorphology, an introductory overview
1.2 Neural segments or neuromeres
Figures taken from von Kupffer (1906)
Salamander larva
Chick embryos
Duck embryo
Well known structures at the beginning of the 20th century
Long considered as transient structures
Currently generally recognized as essential building blocks of the CNS
5

1.2 Neural Segments or Neuromeres

Plate 5

The first point we want to emphasize is that the brain—not only the human brain, but rather the brain of all vertebrates—is *segmented*. The neural segments are known as *neuromeres*.

The left hand panel of plate 5 shows a median section through the brain of an early embryo of the terrestrial salamander, *Salamandra*. At mesencephalic levels, the neuromeres can be clearly recognized as the bulges m1 and m2, and the same holds true for the rhombencephalic neuromeres (known as rhombomeres) 1–6.

The centre of the panel shows reconstructions of the neural tube of chick embryos at 24, 25 and 26 h of incubation. Some of the neuromeres are numbered (1, 2, 3, 11).

Finally, on the right, a horizontal section through the hindbrain of a duck embryo of three days incubation shows rhombomeres (1–6) clearly visible as local bulges of the neural wall (note, M stands for mesencephalon). Recent work has demonstratrated the existence of additional, ‘hidden’, cryptorhombomeres, entities that can be delimited only molecularly (Cambonero and Puelles 2000; Marin et al. 2008; Tomas-Roca et al. 2014).

All of the pictures just shown are taken from the outstanding contribution of Von Kupffer (1906) to O. Hertwigs’ Handbuch der vergleichenden und experimentellen Entwickelungslehre der Wirbeltiere, which was capably complemented by Ziehen’s (1906) chapter on the development of the mammalian CNS. These works illustrate that neuromeres were well-known structures at the beginning of the 20th century. It is highly remarkable that, during most of the remainder of that century, the existence and/or relevance of the phenomenon neuromerism was played down by numerous authors, among whom Neal (1918), Streeter (1933) and de Lange (1936) may be especially mentioned. It was argued that in vertebrates only the mesoderm is truly segmented and that it is the *somitomeres*, which impose their segmental pattern to the developing neural tube. Part of the explanation for the neglect of the neuromeres may well have been that during most of the 20th century, comparative neuroanatomists were primarily interested in the relation between structure and function. The noted comparative neuroanatomist Charles Judson Herrick (1908) claimed that the only essential units in the brain are *functional* units. Because it was not clear what the functional significance of the neural segments could be, they were neglected. As will be discussed later on during this course, the segments are now generally considered to represent the fundamental building blocks of the CNS. It will also be shown that the segments play a prominent role in the development of the functional circuitry.

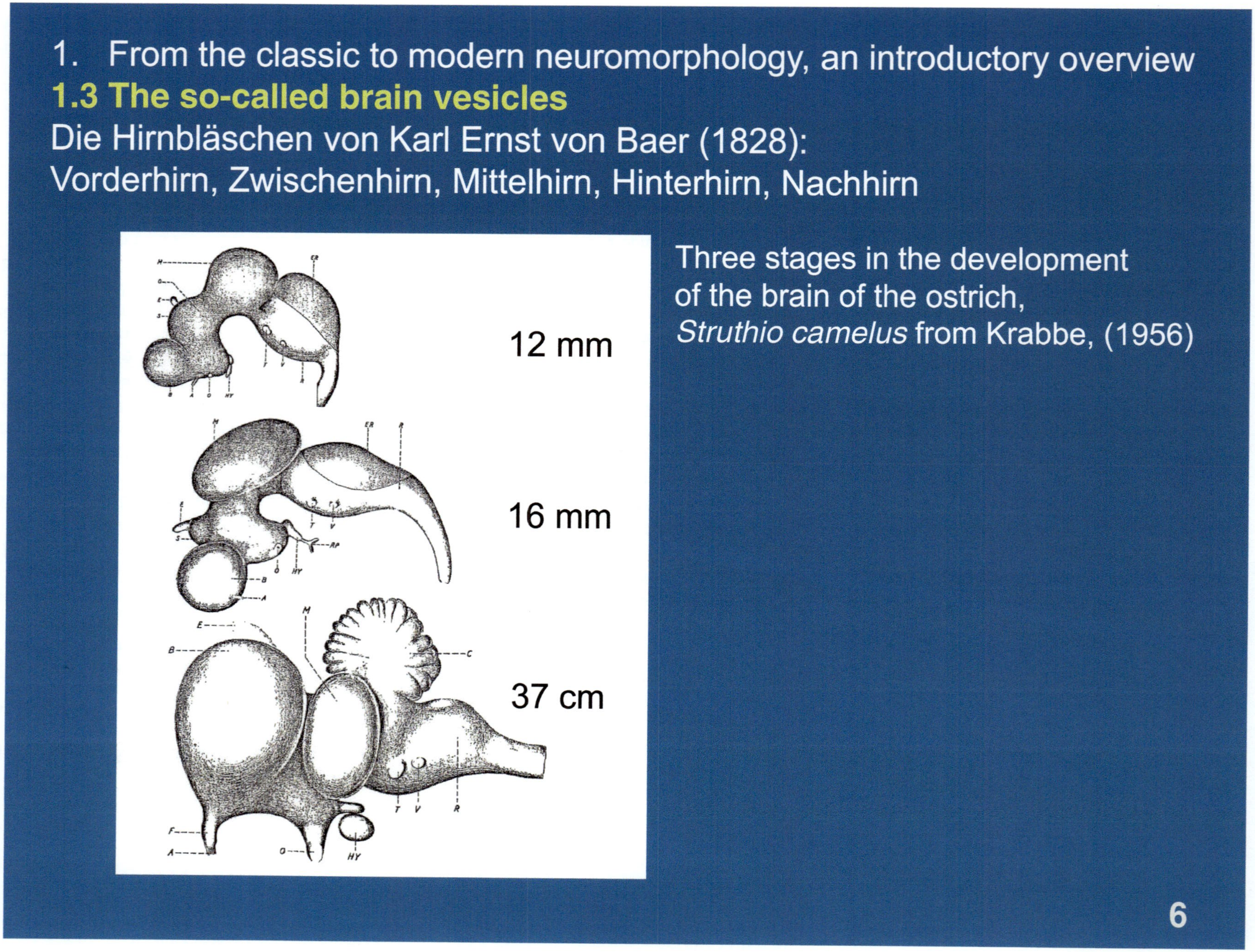
1. From the classic to modern neuromorphology, an introductory overview
1.3 The so-called brain vesicles
Die Hirnbläschen von Karl Ernst von Baer (1828):
Vorderhirn, Zwischenhirn, Mittelhirn, Hinterhirn, Nachhirn
12 mm
16 mm
37 cm
Three stages in the development
of the brain of the ostrich,
Struthio camelus from Krabbe, (1956)
6

1.3 The So-Called Brain Vesicles

Plate 6

The second point to be stressed in this general introduction is that during its development bulges form in the wall of the vertebrate brain, known as brain vesicles. These structures were already known by Karl Ernst Von Baer (1828), the founder of embryology.

Plate 6 shows three stages in the development of he brain of the ostrich, *Struthio camelus*, taken from the relatively unknown work of the Danish neuroembryologist Knud Krabbe (1956). It will be seen that most of the brain parts in these embryos have a clearly vesicular appearance.

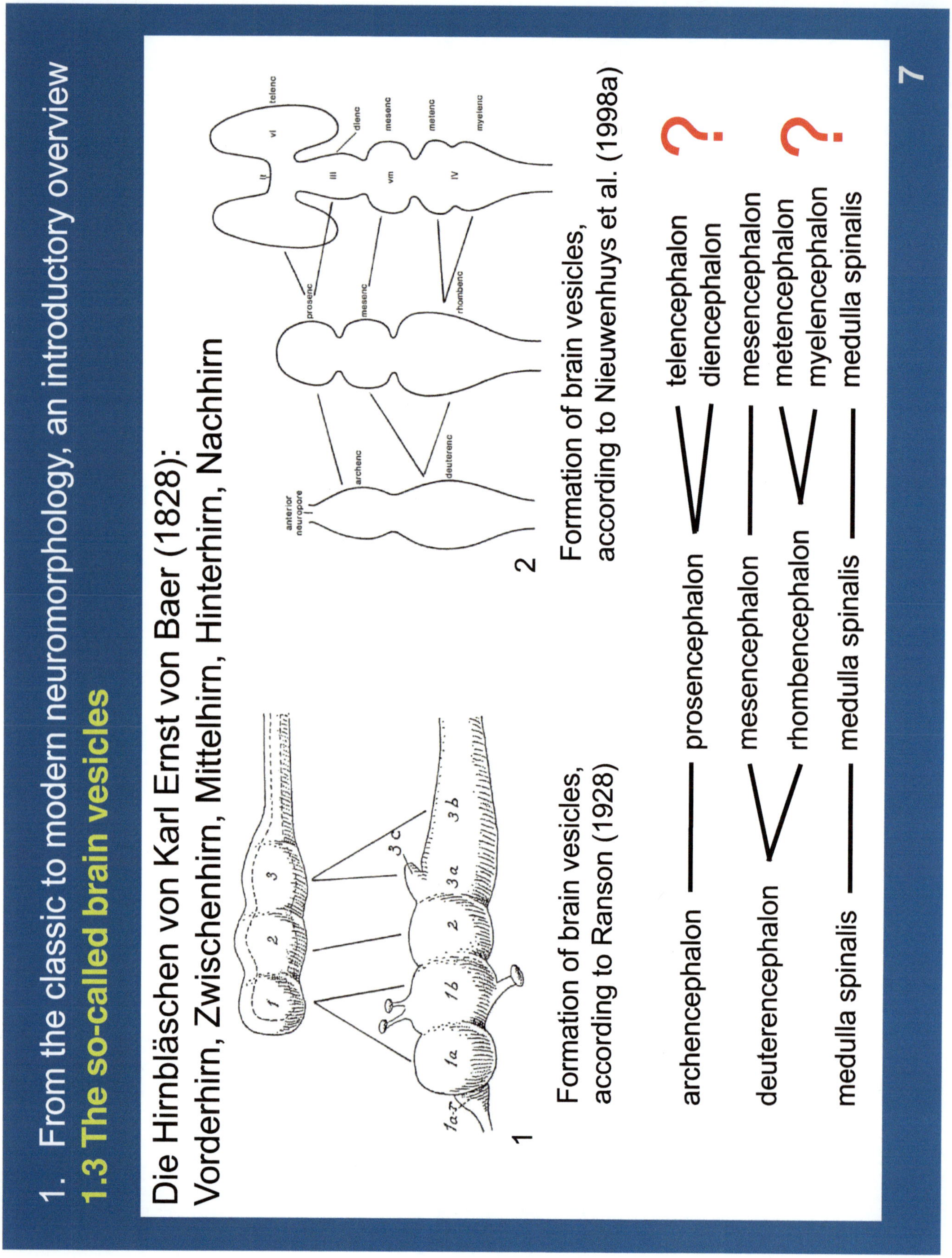
1. From the classic to modern neuromorphology, an introductory overview
1.3 The so-called brain vesicles
Die Hirnbläschen von Karl Ernst von Baer (1828):
Vorderhirn, Zwischenhirn, Mittelhirn, Hinterhirn, Nachhirn
1
1a
1b
2
3a
3b
3c
1
2
3
2
anterior neuropore
archenc
deuterenc
prosenc
mesenc
rhombenc
telenc
dienc
mesenc
metenc
myelenc
Formation of brain vesicles, according to Ranson (1928)
Formation of brain vesicles, according to Nieuwenhuys et al. (1998a)
archencephalon
deuterencephalon
medulla spinalis
prosencephalon
mesencephalon
rhombencephalon
medulla spinalis
telencephalon
diencephalon
mesencephalon
metencephalon
myelencephalon
medulla spinalis
?
?
7

Plate 7

The classic views on the morphogenesis of the brain, derived primarily from von Baer (1828) and von Kupffer (1906) may be epitomized as follows:

1. Early in development the rostral part of the neural plate becomes wider than the caudal part. The former gives rise to the brain, whereas the spinal cord develops from the latter.
2. Prior to the closure of the anterior neuropore, the rostral part of the primordial brain shows a transversely oriented ventral fold, the *plica encephali ventralis*, also known as the 'Kopfbeuge' (cephalic flexure) of von Kupffer (1906) (not shown). This fold, situated near the rostral end of the notochord, is considered to roughly mark the boundary between two principal brain regions, the *archencephalon* and the *deuterencephalon* (right panel).
3. Soon after the formation of the neural tube, its rostral, primordial brain part shows three dilatations (left and right panels). These so-called *primary brain vesicles* are known in sequence from rostral to caudal as the forebrain or *prosencephalon*, the midbrain or *mesencephalon* and the hindbrain or *rhombencephalon* (Plate 7, left panel, 1,2,3 respectively). The prosencephalon develops from the archencephalon, whereas the mesencephalon and the rhombencephalon are derivatives of the deuterencephalon. (It is of note that recent studies on differential molecular profiles suggest that the midbrain is also archencephalic, so that the arch/deuterencephalic boundary would lie just caudal to the mesencephalon at a region known as the isthmus instead, see below.)
4. Ultimately, in the classic view the prosencephalon and the rhombencephalon show each a differentiation into two parts each. The prosencephalon becomes subdivided into a rostral endbrain or *telencephalon* and a caudal between-brain or *diencephalon*, whereas it was held that the rhombencephalon becomes subdivided into a rostral *metencephalon* and a caudal *myelencephalon* (Plate 7, right panel). These four subdivisions form with the mesencephalon, which remains undivided, the five *secondary brain vesicles*, which, as we have seen, were already known to Von Baer (1828).

The morphogenetic processes just described are pictorially summarized in Plate 7. The red question marks indicate that recently doubt has been cast on the classic 'five-vesicle-concept'.

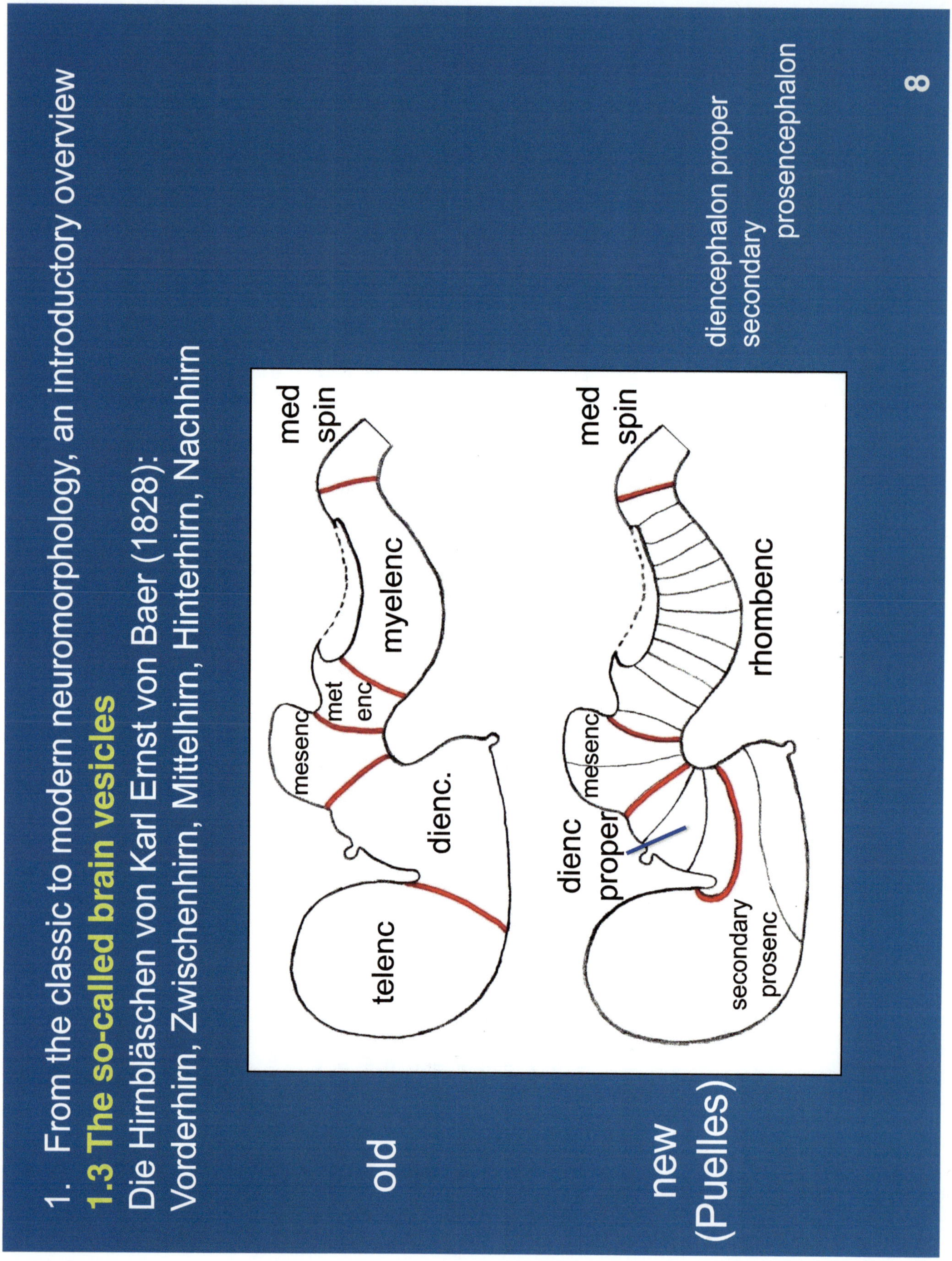
1. From the classic to modern neuromorphology, an introductory overview
1.3 The so-called brain vesicles
Die Hirnbläschen von Karl Ernst von Baer (1828):
Vorderhirn, Zwischenhirn, Mittelhirn, Hinterhirn, Nachhirn
old
new
(Puelles)
telenc
dienc.
mesenc
met
enc
myelenc
med
spin
dienc
proper
mesenc
rhombenc
secondary
prosenc
med
spin
diencephalon proper
secondary
prosencephalon
8

Plate 8

The upper picture of Plate 8 shows the classic subdivision of the brain into the five secondary brain vesicles: telencephalon, diencephalon, mesencephalon, metencephalon and myelencephalon. However, we know now that it is not the brain vesicles, but rather the neural segments or neuromeres which form the basic building blocks of the brain. In various works, one of us (L. P.) compared the neuromeric organization based on structural and molecular data with the classic subdivision of the brain. His findings are summarized in the lower picture of Plate 8. It appears that the boundaries of the three primary brain vesicles perfectly match the neuromeric boundaries. In mammals the boundary between the metencephalon and the myelencephalon roughly marks in mammals the caudal border of the pons but is not essential in non-mammalian vertebrates. As will be discussed in Sect. 8, the pontine bulge is a secondary formation, formed by cells originating from the myelencephalic rhombic lip. The apparent sulcal border between the telencephalon and the diencephalon has no relation whatsoever with the neuromeric organization of the brain. The telencephalon and the hypothalamic part of the classic diencephalon form together a morphological entity, which is composed of two neuromeres. The remainder of the classic diencephalon: pretectum, epithalamus plus dorsal thalamus and ventral thalamus form another entity, composed of three neuromeres. One of us (L. P.) proposes to designate these new entities as *secondary prosencephalon* and *diencephalon proper*, respectively. It will take some time before these terms uniformly reach the neuroanatomy textbooks, though they are implemented in the reference atlases of the Allen Developing Mouse Brain Atlas (www.developingmouse.brain-map.org).

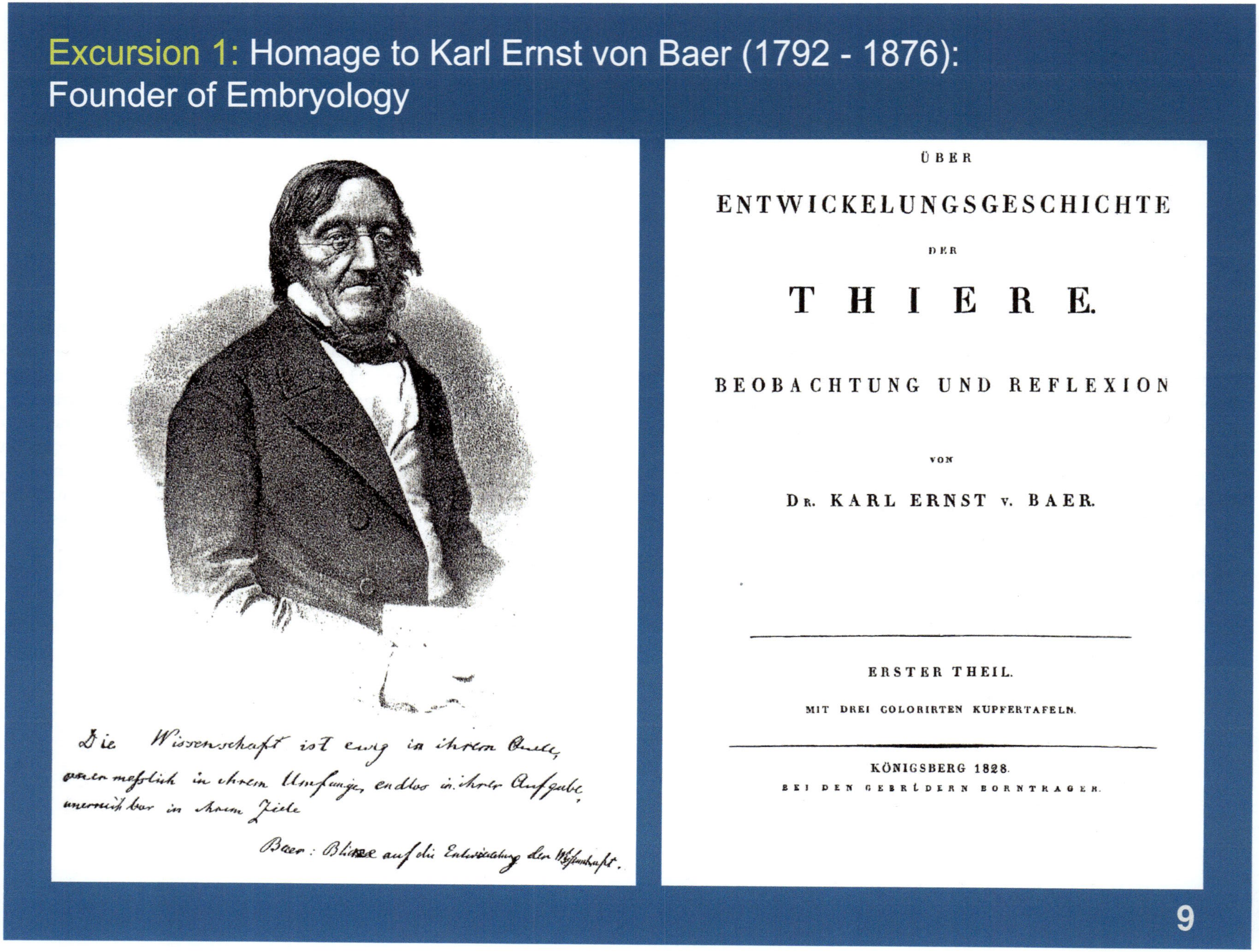
Excursion 1: Homage to Karl Ernst von Baer (1792 - 1876):
Founder of Embryology
Die Wissenschaft ist ewig in ihrem Quelle, unermeßlich in ihrem Umfange, endlos in ihrer Aufgabe, unerreichbar in ihrem Ziele
Baer: Blicke auf die Entwickelung der Wissenschaft.
ÜBER
ENTWICKELUNGSGESCHICHTE
DER
THIERE.
BEOBACHTUNG UND REFLEXION
VON
DR. KARL ERNST v. BAER.
ERSTER THEIL.
MIT DREI COLORIRTEN KUPFERTAFELN.
KÖNIGSBERG 1828.
BEI DEN GEBRÜDERN BORNTRÄGER.
9

Excursion 1: Karl Ernst Von Baer

Plate 9

Meanwhile, it is high time for our first excursion, a brief tribute to Karl Ernst Von Baer (1792–1876). It has already been mentioned that this Estonian scientist has to be regarded as the founder of embryology. His *opus magnum:* "On the development of animals" is a real classic, and it has an absolutely superb subtitle: "Observation and reflection". The title page of this work is shown on the right side of the plate. The left side shows a portrait of Von Baer, and in his own handwriting, his views on the development of science, *in facsimile.* In translation it reads; "Science is eternal in its source, immeasurable in its dimensions, endless in its assignment, unattainable in its purpose". This should be kept in mind, particularly when you read in your newspaper that a group of 'scientists' has declared itself determined to solve the mysteries of the human brain within ten years (or less).

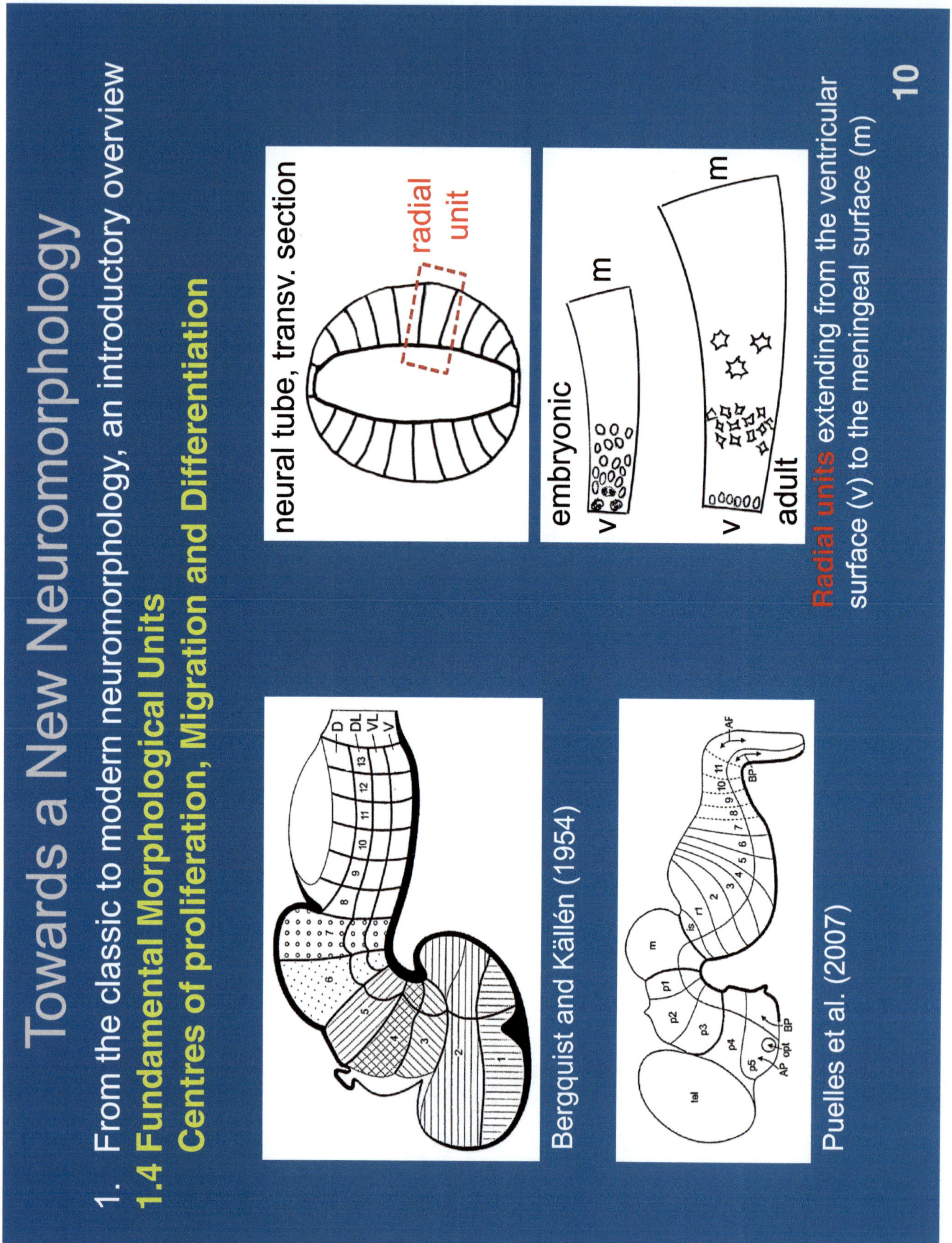
Towards a New Neuromorphology
1. From the classic to modern neuromorphology, an introductory overview
1.4 Fundamental Morphological Units
Centres of proliferation, Migration and Differentiation
Bergquist and Källén (1954)
Puelles et al. (2007)
neural tube, transv. section
radial unit
embryonic
v
m
adult
v
m
Radial units extending from the ventricular surface (v) to the meningeal surface (m)
10

1.4 The Fundamental Morphological Units of the CNS

Plate 10

We have emphasized already the importance of neural segments or neuromeres.

On the left, two diagrammatic representations of the segmental organization of the vertebrate brain are depicted. The upper is taken from a paper of Harry Bergquist and Bengt Källén, published in 1954, under the modest title: "Notes on the early histogenesis and morphogenesis of the central nervous system in vertebrates". This publication, its fundamental significance notwithstanding, was completely neglected at the time of its appearance. It has already been mentioned that neural segments were not *bien reçu* at that time.

The lower picture is taken from the more recent work of Puelles et al. (2007). Both upper and lower pictures show that the vertebrate brain is not only segmented but also divisible into longitudinal zones, which follow the overall curvature of the brain. The intersecting boundaries of the neuromeres and the longitudinal zones delimit quadrangular areas of the brain wal, which represent early thin progenitor domains or histogenetic areas that later transform into three-dimensional, radially oriented structures.

The upper right-hand picture represents a diagrammatic transverse section through the neural tube. It shows that the radial structures, just mentioned, extend from the ventricular surface to the meningeal surface of the tube. These radial structures form the *fundamental morphological units of the CNS*. The principal histogenetic processes, cellular proliferation, -migration and -differentation, occur specifically within the confines of these units, controlled by characteristic sets of developmental regulatory genes.

The lower right-hand picture shows two of these radial units, the one taken from an embryonic, immature stage, the other from an adult brain (m, meningeal surface; v, ventricular surface).

Towards a New Neuromorphology

1. From the classic to modern neuromorphology, an introductory overview

1.5 Morphogenesis: Topology versus Topography

Topology is the geometry of distortion, the branch of mathematics, which investigates the properties that remain unchanged when geometric configurations are subjected to one-to-one continuous transformations. On these transformations it is allowed to change distances, to bend, stretch or twist the configurations.

However, it is forbidden to 'cut' to 'tear', or to 'fuse'.

In other words: Topology is the geometry of invariant neighbourhood relationships irrespective of distortion

11

1.5 Topology Versus Topography in (Neuro) Morphology

Plate 11

In what follows, we will explain that for a correct morphological interpretation of (neural) structures, it is essential to take their *relative* or *topological*- rather than their *topographical position* into consideration.

In anatomy, it is customary to employ for descriptive purposes an orthogonal, Cartesian coordinate system. In this system three axes or directions are distinguished, namely rostral-caudal or anterior-posterior, dorsal-ventral and medial-lateral. This coordinate system and its related terminology are also frequently used in the naming of structures within the CNS: lateral pallium, nucleus dorsomedialis anterior, tractus olfactorius lateralis and so on. Such names are appropriate for the description of structures identified in one particular embryonic stage of a given species, or in the adult stage of one species, but they may be totally inadequate to designate corresponding parts or entities in the brains of different stages of one and the same species, or in the adult brains of two or more different species. The adjective *topological* used above refers to the branch of mathematics, known as topology.

Plate 11 briefly explains what topology is. Read it carefully. Note that the configurations a, b, c and d, can all be continuously transformed into each other, and are hence topologically equivalent. However, none of these configurations can be topologically transformed into the open curve shown in e. Analogously, the early closed neural tube is subject to numerous local shape modifications during its growth, but the primary relative positions of the parts and the corresponding boundary relationships persist—sometimes very deformed—into the adult brain shape. Morphological topologic analysis starts accordingly with the closed neural tube, since the transformation of the neural plate into the neural tube involves the non-topologic fusion of the roof plate. Leaving this aspect out, the topologic comparison of neural plate and neural tube mappings remains of prime importance.

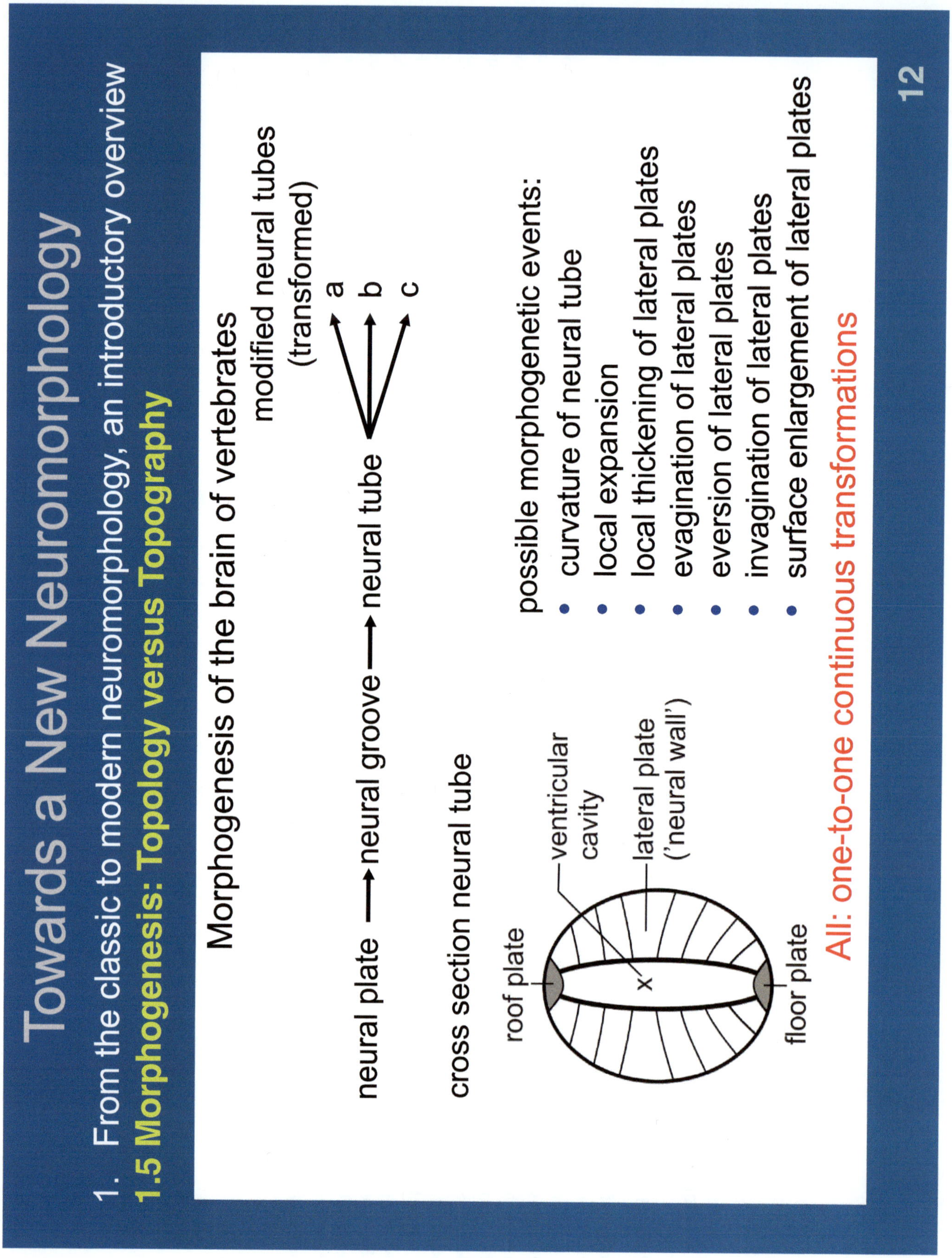
Towards a New Neuromorphology
1. From the classic to modern neuromorphology, an introductory overview
1.5 Morphogenesis: Topology versus Topography
Morphogenesis of the brain of vertebrates
neural plate → neural groove → neural tube
modified neural tubes (transformed)
a
b
c
cross section neural tube
roof plate
ventricular cavity
lateral plate ('neural wall')
x
floor plate
possible morphogenetic events:
• curvature of neural tube
• local expansion
• local thickening of lateral plates
• evagination of lateral plates
• eversion of lateral plates
• invagination of lateral plates
• surface enlargement of lateral plates
All: one-to-one continuous transformations
12

Plate 12

Plate 12 shows that the early morphogenesis of the brain of all vertebrates may be epitomized as a topological transformation: neural plate → neural groove → neural tube.

A diagrammatic cross section of the neural tube is shown. It comprises a thin roof plate, a thin floor plate, and bilateral lateral plates. Together, these structures surround the slit-like ventricular cavity. The lateral plates may also be designated as 'neural walls' because most nerve cells or neurons develop within these structures. However, in some places the floor and roof plates also produce neurons.

The adult brains of all vertebrates may be characterized as modified or transformed neural tubes, but the events occurring during morphogenesis may differ among the various parts of the developing brain, as well as among the different groups of vertebrates.

The most common morphogenetic events in the brains of vertebrates are indicated in the right (see later for further descriptions)., As we shall see, most brain parts are the product of combinations of two or more of these events.

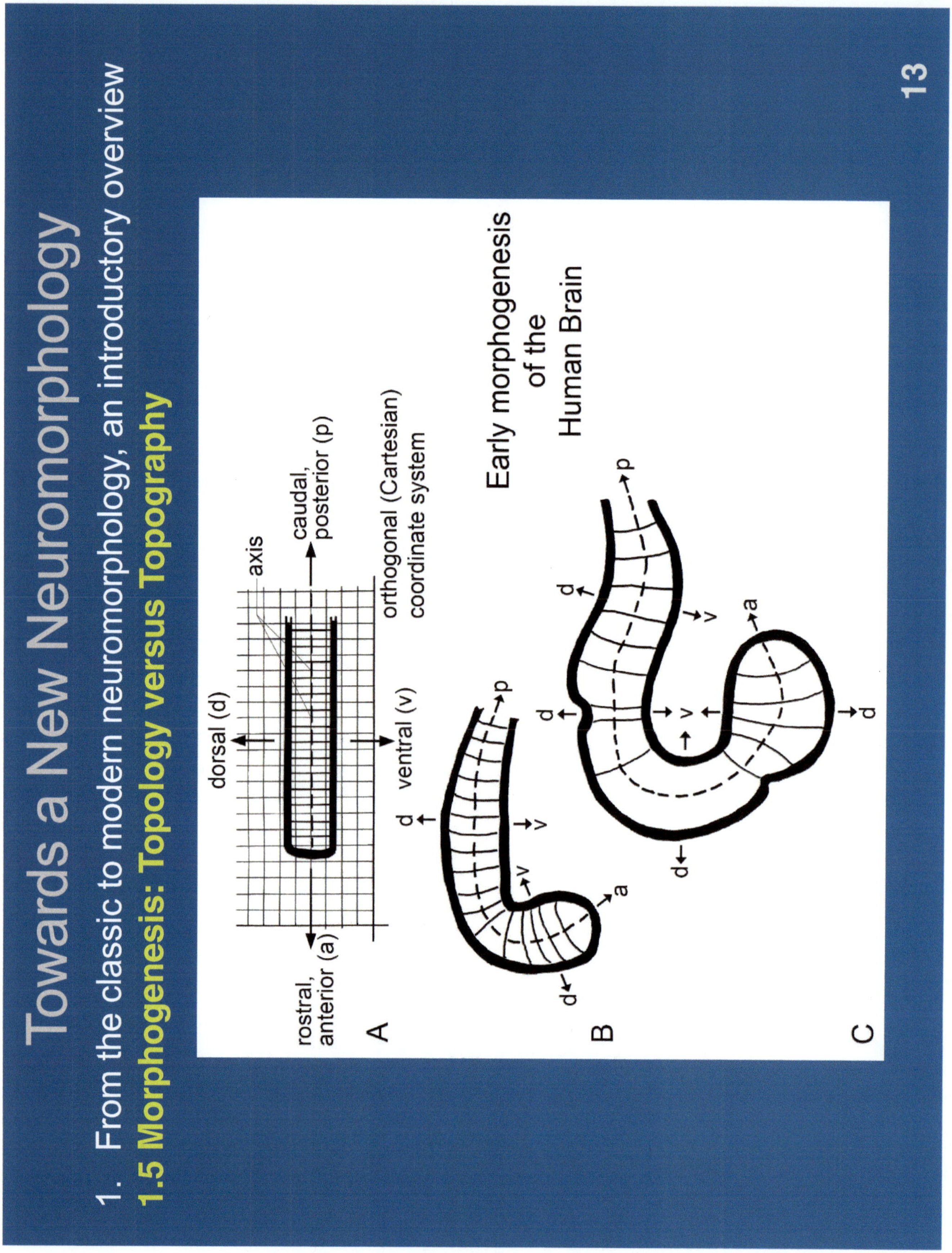
Towards a New Neuromorphology
1. From the classic to modern neuromorphology, an introductory overview
1.5 Morphogenesis: Topology versus Topography
A
rostral, anterior (a)
dorsal (d)
ventral (v)
axis
caudal, posterior (p)
orthogonal (Cartesian) coordinate system
B
C
Early morphogenesis of the Human Brain
13

Plate 13

Plate13 shows a comparison between the systems of topology and (Cartesian) topography applied to the early morphogenesis of the human brain. The figures show diagrammatic midsagittal or median sections through a simple neural tube stage, embedded in an orthogonal or Cartesian coordinate system (A), the brain of an embryo of about one month (B), and the brain of an embryo of about five weeks (C).

The neural tube depicted in A is strongly 'idealized'. Such perfectly straight neural tubes have never been observed. It is known that in the human, the bending of the tube, shown in B, actually becomes manifest before the closure of the tube.

The 'idealized' neural tube fits perfectly in the Cartesian coordinate system. We can attribute an *axis* to the tube, represented by a straight line along the virtual centre of its ventricular cavity. The anterior or rostral end of the tube lies opposite its posterior or caudal end. The upper or dorsal wall of the tube extends over its entire length, oriented parallel to the lower or ventral wall. There are multiple neural segments whose boundaries are all vertically oriented (perpendicular to the axis) and strictly parallel to each other.

The embryonic brain shown in B is curved, and that shown in C is even more strongly so. However, it is customary in morphology to indicate directional and positional relations within developing (and adult!) brains ***as if*** these brains were still straight and stretched tube-like structures. Or, to phrase it in a more sophisticated way: bringing these developing brains back to the tube-stage *by subjecting them to one-to-one continuous topological transformations.*

Generalizing the notions just presented, it may be said that, in contrast to invariant topological (intrinsic) positional relationships, the *topographical* terms dorsal and ventral (or rostral and caudal) indicate developmentally variable directions relative to arbitrary external references derived from a Cartesian coordinate system. The intrinsic *morphological* or *topological* directions dorsal and ventral (or rostral and caudal) may vary widely in apparent topography, as shown in figures B and C.

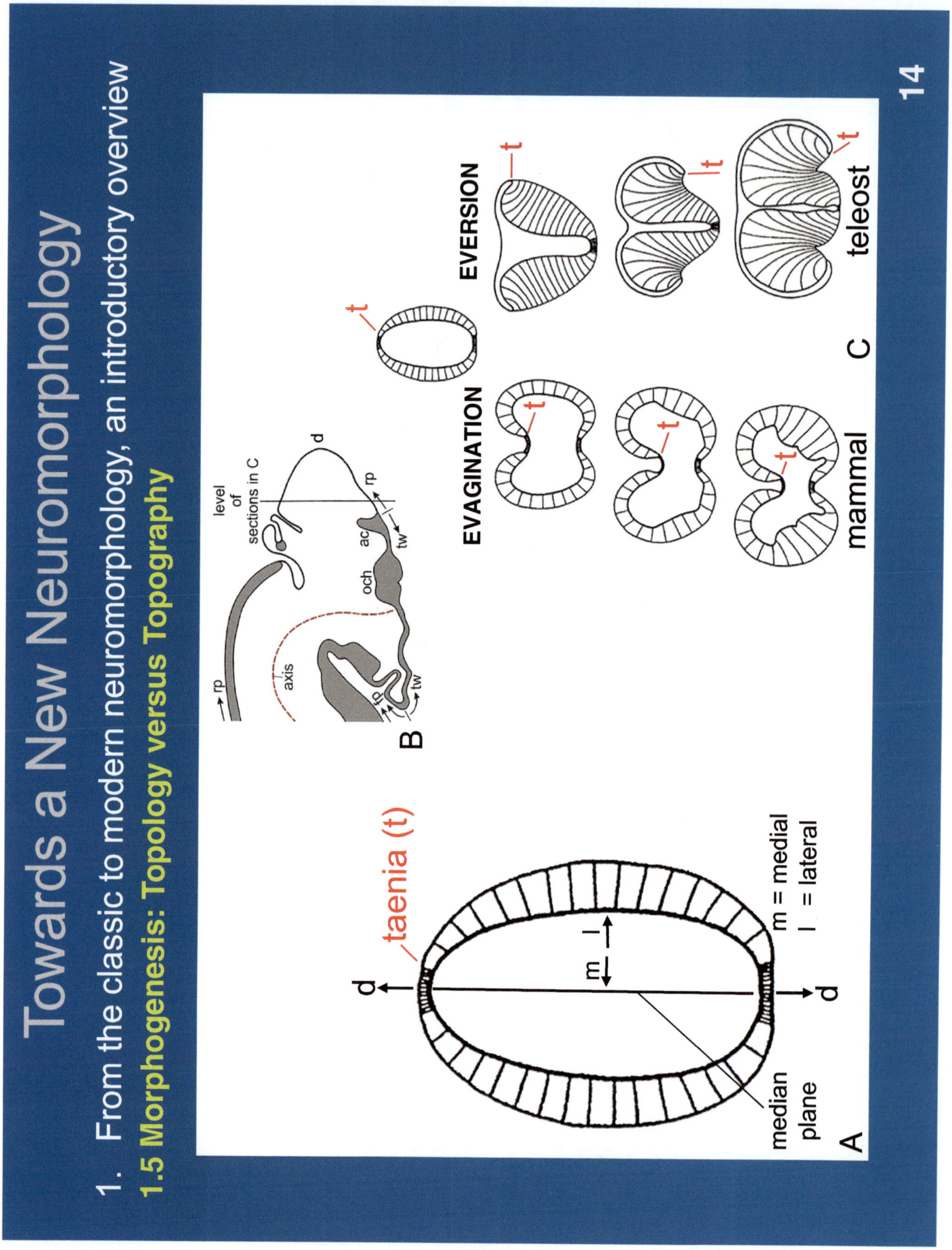
Towards a New Neuromorphology
1. From the classic to modern neuromorphology, an introductory overview
1.5 Morphogenesis: Topology versus Topography
taenia (t)
d
m
l
d
median plane
m = medial
l = lateral
A
rp
axis
fp
tw
och
ac
tw
rp
level of sections in C
d
B
EVAGINATION
EVERSION
t
mammal
C
teleost
14

Plate 14

Developing and adult brains are generally bilaterally symmetrical, meaning that they can be divided into two halves, which mirror each other, by a median or midsagittal plane (Plate 14, Fig. A). The third positional component of the Cartesian coordinate system used in descriptive anatomy is oriented perpendicular to the median plane, describing the distance from this landmark. Within the context of this system, the descriptor *medial* means: situated close to or directed towards the median plane, whereas *lateral* means the opposite: situated far from or directed away from the median plane. These relations are illustrated in Fig. A, which shows—once again—a diagrammatic cross section through the neural tube. What has been said earlier about the meaning of morphologically or topologically dorsal and ventral, and rostral and caudal, also holds true for the designations morphologically or topologically medial and lateral. Here too, these designations indicate invariant directional and positional relations described in the developing or adult CNS after morphogenetic transformation of the tube-like configuration.

The importance of this matter may be illustrated and elucidated by a brief sketch of the morphogenesis of the telencephalon in two groups of vertebrates: mammals and teleost fishes (Plate 14, Fig. C). In both of these groups (and in all vertebrates), the anlage of the telencephalon initially shows a simple, tube-like configuration. However, due to the ventral curvature of the rostral part of the brain (see red axial line in Plate 14, Fig. B), the lateral plates are at telencephalic levels dorsally as well as ventrally (topographically) interconnected by the, topologically always dorsal roof plate. This remarkable feature is illustrated in Plate 14, Fig. B, which shows a median section through the rostral part of the brain of a teleost fish, and will be fully explained in Sect. 6. (Abbreviations in the insert: ac, anterior commissure; fp, floor plate; och, optic chiasma; rp, roof plate; tw, terminal wall) It is, for what follows, important to note that the bilateral lines of attachment between the choroidal roof plate and the lateral plates at the topographically dorsal side are known as the taeniae (t).

Evagination or bulging-out of the lateral plates (Plate 14, Fig. C) is the dominant morphogenetic event in the developing telencephalon of mammals (and of almost all other groups of vertebrates). This process leads to the formation of hollow cerebral hemispheres, whose walls surround the lateral ventricles. The illustrations show that, later on the evagination is complicated by a local thickening of the lateral walls. The roof plate remains a narrow structure and the taeniae maintain their relative positions.

In contrast, the morphogenesis of the teleostean telencephalon is characterized by an eversion. During this process, the lateral plates diverge with a consequent widening of the roof plate. The lateral plates thicken and subsequently recurve laterally. The choroidal roof plate widens progressively and ultimately becomes transformed into a thin membrane, which largely surrounds the thickened solid hemispheres. The figures show clearly that the lines of choroidal attachment, the taeniae, shift laterally and ultimately attain a ventrolateral topographic position, though they remain topologically dorsal. Note that the roof plate is not always structurally choroidal, as happens at the other roof location illustrated in these schemata, where the septal anterior commissure develops; this locus seems topographically ventral, but is topologically dorsal (see Plate 14, Fig. B).

Due to their radically different morphogeneses, uncritical comparison of the cell masses in evaginated forebrains with those in everted forebrains has led to endless debates and controversies (see: Nieuwenhuys 2009a, b, 2011a). It will be clear that, for a correct comparison, the evaginated and everted telencephalic walls have to be transformed ‘back’ towards their initial configuration. The taeniae occupy in both evaginated and everted forebrains a morphologically or topologically dorsomedial position.

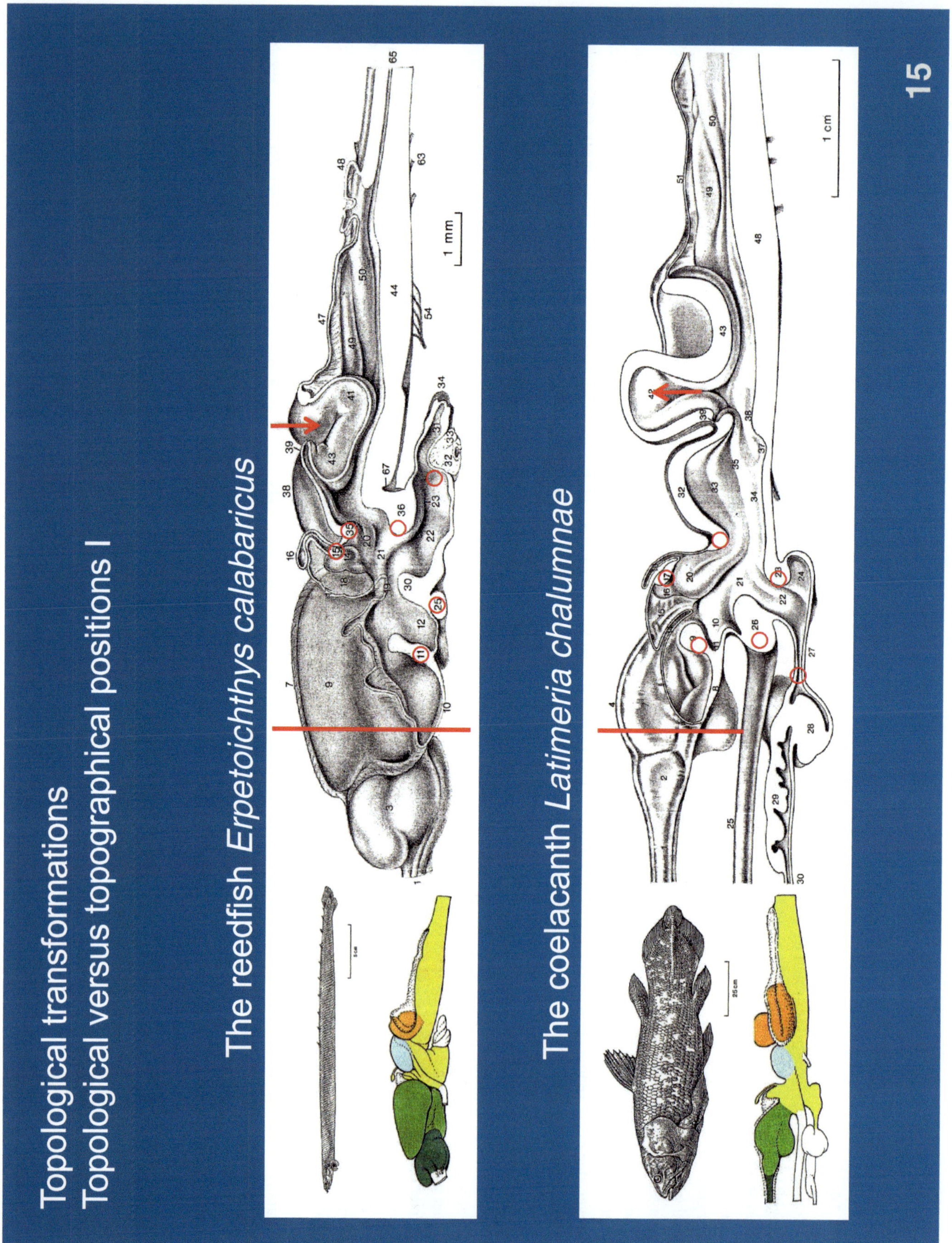
Topological transformations
Topological versus topographical positions I
The reedfish Erpetoichthys calabaricus
The coelacanth Latimeria chalumnae
1 mm
1 cm
15

Plate 15

We are going to conclude this rather long section on topology versus topography in neuromorphology with a comparison of some structures in the brain of two fishes, the reedfish *Erpetoichthys calabaricus* and the coelacanth *Latimeria chalumnae* (Nieuwenhuys 1998d, e) All fish are remarkable, and some are more remarkable than others, but coelacanths are the most remarkable. These fishes were long thought to have become extinct 60 million years ago, but in 1938 a strange fish caught off the coast of southeast Africa proved to be a surviving coelacanth and was named *Latimeria chalumnae*. Coelacanths are large fish with a length of about 1.5 m. They have fleshy-lobed paired fins, which are believed to 'foreshadow' the extremities of terrestrial vertebrates.

Plate 15 shows holotypes of the two selected fishes with lateral views of their brains and, to the right, medial views of their bisected brains. Note the configuration of three structures in the brains of these fishes: the cerebellum (red arrows), the diencephalon in continuity with the hypothalamus (red circles at equivalent landmarks), and the telencephalic hemispheres (cross sectioned at the red lines).

In many vertebrates, the cerebellum develops from a dorsally directed evagination of the rostral part of the rhombencephalon. The cerebellum of the coelacanth fully answers to this general description. However, in the reedfish, the cerebellum is, quite remarkably and quite exceptionally, the product of a ventrally directed invagination into the rhombencephalic or fourth ventricle.

Six structures, namely the commissura anterior, the commissura habenulae, the commissura posterior, the tuberculum posterius, the infundibulum and the chiasma opticum, can be easily recognized in median sections through the forebrain in all vertebrates and, hence may be considered as landmarks in this brain region. They are marked with red circles in the right-hand pictures of the plate and can be seen to be brought by differential morphogenesis into variant topographical positions, while maintaining invariant topological (relative) positions.

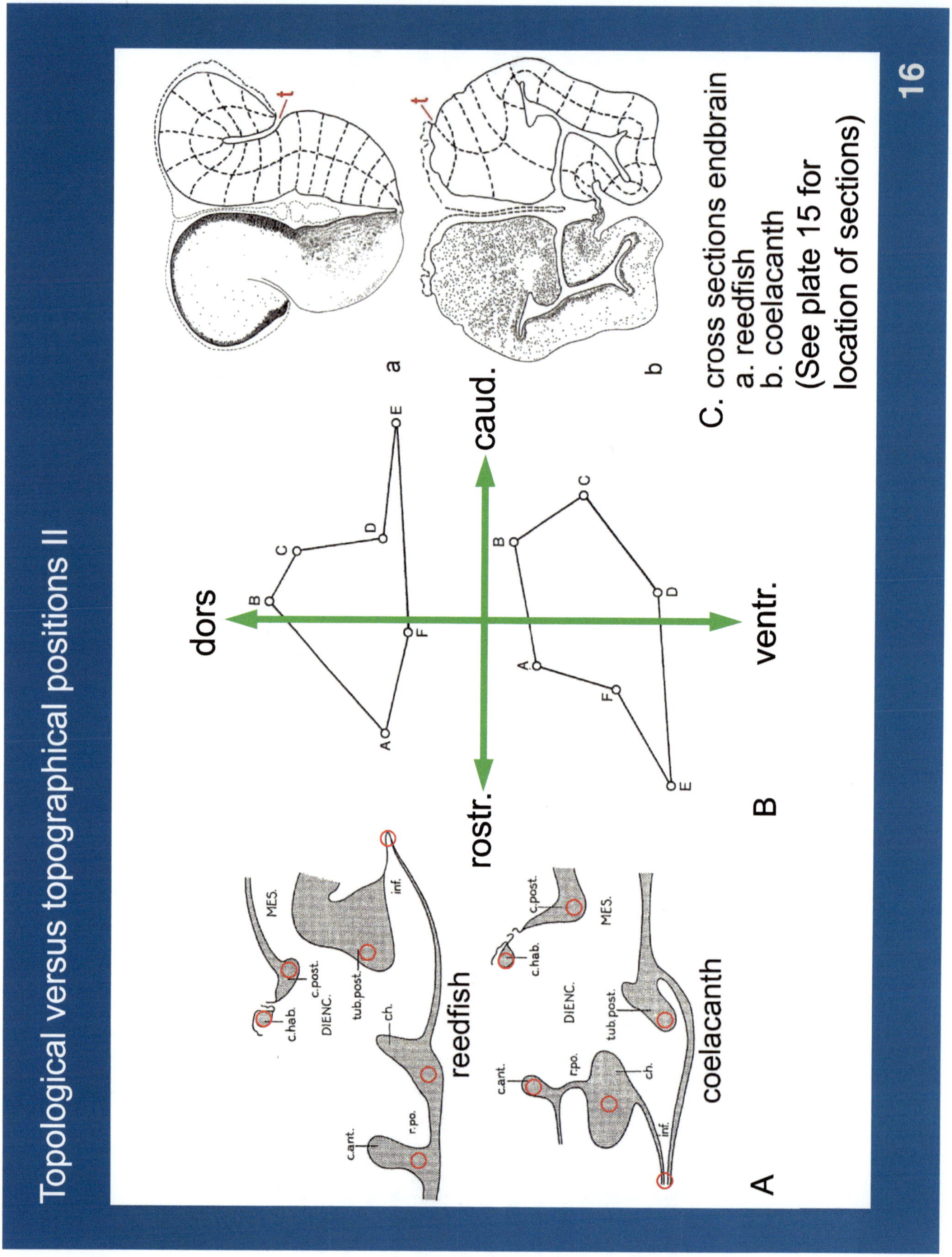
Topological versus topographical positions II
16
A
c.ant.
r.po.
ch.
tub.post.
DIENC.
c.hab.
c.post.
MES.
inf.
reedfish
coelacanth
B
dors
ventr.
rostr.
caud.
A
B
C
D
E
F
a
b
t
C. cross sections endbrain
a. reedfish
b. coelacanth
(See plate 15 for
location of sections)

Plate 16

The positions of the six structures shown in Plate 15 are indicated once again in Plate 16, Fig. A. This shows median sections through the prosencephalon of these two species, in which the structures are again marked with red circles. In Plate 16, Fig. B, they are diagrammatically represented as nodes interconnected hexagonally. Reference to the Cartesian coordinate system (green) shows immediately that the six structures selected do not occupy fixed topographical positions with respect to each other. Note, for instance, that the infundibulum (E) occupies in the reedfish a topogaphically caudal-, but in the coelacanth a topographically rostral position.

The vertical red lines in the right hand figures of Plate 15, indicate the levels of the transverse sections through the cerebral hemispheres of the reedfish and the coelacanth, shown in Plate 16, Fig. C.

It will be clear that the hemispheres of the reedfish (Plate 16, Fig. C: a) are of the everted type. It is remarkable, however that, unlike the teleosts, hemispheric eversion in the reedfish (see Plate 14) is not accompanied by a thickening of the lateral walls.

The telencephalic hemispheres of the coelacanth (Plate 16, Fig. C: b), finally, show a highly remarkable hybrid configuration. Its thin-walled basal parts are clearly evaginated. They surround narrow but extensive lateral ventricles. The walls of the dorsal parts of the hemispheres are, on the other hand, greatly thickened, and show a slight tendency towards eversion. The latter feature appears from the widening of the choroidal roof plate, and the consequent lateral shift of the taeniae (Nieuwenhuys 1965, 2009b).

1.6 Summary of the Introductory Overview

1. During the last decades molecular neurobiological studies have greatly deepened our insights into the regulation of the development of the CNS.
2. Neural segments or neuromeres are essential building blocks of the CNS.
3. The validity of the subdivision of the early embryonic brain into three primary brain vesicles, prosencephalon, mesencephalon and rhombencephalon has been corroborated, but this does not apply to the five secondary brain vesicles.
4. The prosencephalon is subdivisible into a diencephalon proper and a secondary prosencephalon, or hypothalamo-telencephalic complex (Puelles).
5. Intersection of transverse (neuromeric) and longitudinal zones leads in the embryonic CNS to the formation of radially oriented fundamental morphological units, which represent centres of proliferation, migration and differentiation.
6. A topological approach is *conditio sine qua non* for an adequate practising of developmental and comparative neuromorphology.

Towards a New Neuromorphology

2. Neural plate

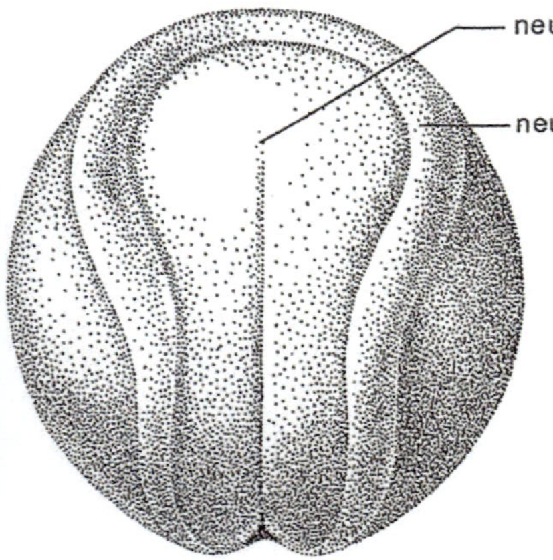

A. Dorsal view of a frog embryo (early neurula stage) (Nieuwenhuys 1998a)

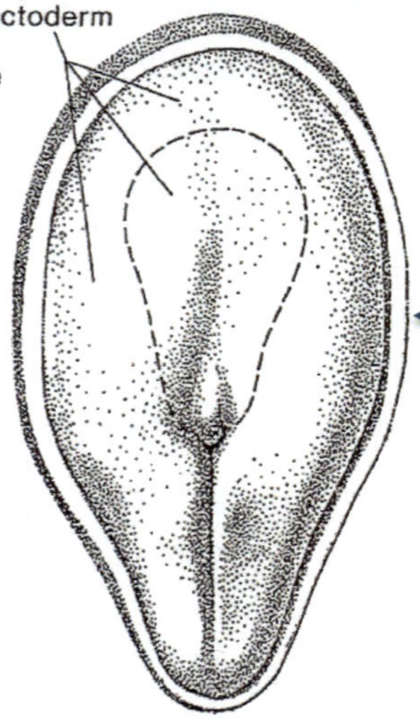

B. Dorsal view of a human embryo (16-day' presomite stage) (Nieuwenhuys 1998a)

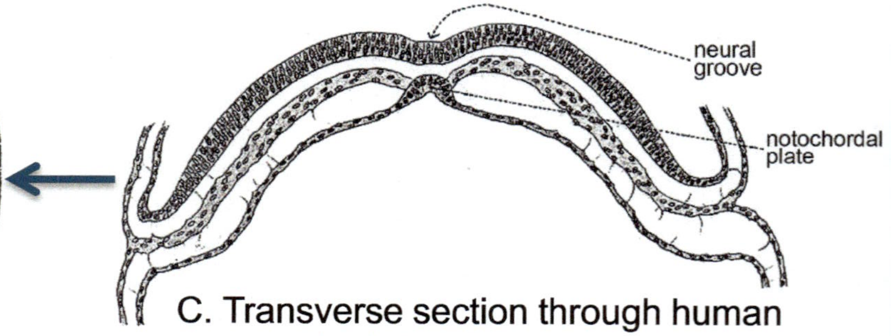

C. Transverse section through human embryonic disk (16-day' presomite stage) (Hamilton et al. 1947)

D. Structure of neural plate

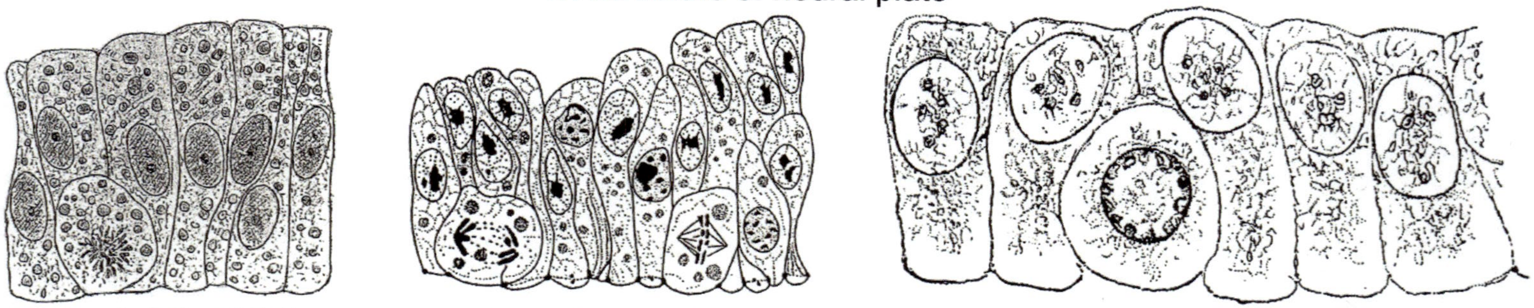

shark (His 1889)

chick (Sauer 1935b)

rabbit (His 1889)

2 The Neural Plate

2.1 Structure

Plate 17

Plate 17 presents dorsal views of early frog and human embryos (Fig. A, B). In the frog embryo, the neural plate is surrounded by a distinct ridge (neural fold). In the human embryo, such a demarcation is lacking and the neural plate can only be arbitrarily delimited morphologically from the remainder of the external germ layer, though a precise neural versus non-neural boundary is visible with molecular markers. The apparent lack of a border can also be observed in the transverse section through the human embryonic disk, shown in Plate 17, Fig. C.

The three microscopic images, which form the lower half of the plate (Plate 17, Fig. D), show that the neural plate consists of columnar epithelial cells. In all of the three pictures one or more cells in mitosis are visible. The neural plate is thus formed by a *proliferating* epithelium.

Before going on with the structural organization of the neural plate, we have to insert some remarks on the elementary principles of molecular developmental (neuro)biology.

R. Nieuwenhuys and L. Puelles, *Towards a New Neuromorphology*,
DOI 10.1007/978-3-319-25693-1_2

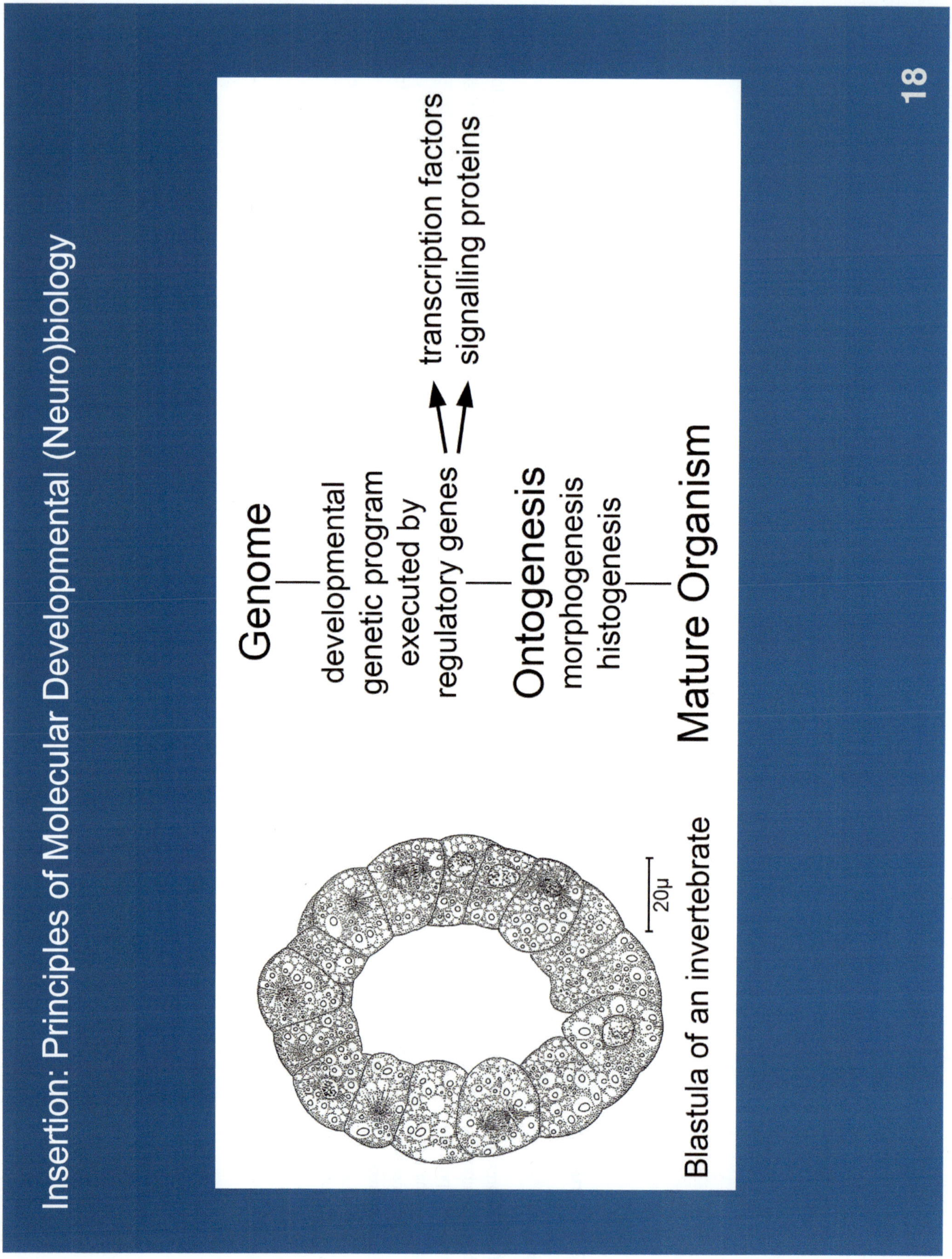
Insertion: Principles of Molecular Developmental (Neuro)biology
18
Blastula of an invertebrate
20µ
Genome
developmental
genetic program
executed by
regulatory genes
transcription factors
signalling proteins
Ontogenesis
morphogenesis
histogenesis
Mature Organism

Insertion: Principles of molecular developmental (neuro)biology

Plate 18

The left half of Plate 18 shows an early embryo (blastula) of an invertebrate. It consists of a single layer of epithelial cells (several of them in mitosis), which surround a central, fluid-filled blastocoel.

Much of the development occurs intracellularly and is governed by intracellular regulatory networks (see below). Other developmental events are initiated by the secretion or action by contact of signalling proteins, by which some embryonic cells influence the genomic readout of embryonic cells, either by direct contact, or by diffusion of signals via the fluid in the blastocoel (or along the plane of the epithelium).

The right half of Plate 18 presents a diagram of the processes involved in the development of an organism. It indicates that a mature organism is the product of morphogenetic (shaping) and histogenetic (structure generating) processes, and that both of these are governed by the genome. The genome contains all of the information needed for constructing an entire organism and an entire nervous system. It does so by unfolding a developmental genetic program, which is executed by a special set of genes, called regulatory genes or master control genes. These genes initiate the production of two special types of proteins, *transcription factors* and *signalling proteins*.

The transcription factors produced in a cell act on the genome within the nucleus of the same cell, where they influence in various combinations other regulatory genes by activating or repressing them.

The signalling proteins are released by the parent cells, and indirectly influence the activity of regulatory genes in other cells of the developing organism (via tissular diffusion, attachment to specific membrane receptors, and triggering of intracytoplasmatic secondary messenger systems). Cell-cell interaction via contact implies direct activity of a signalling protein inserted in the membrane of one cell upon a receptor in another cell. The end result of the regionally diverse genomic cascades is the patterned activation and regulation of genes providing for structural proteins and 'house-holding' metabolic enzymes, thus guiding step-wise growth and differentiation.

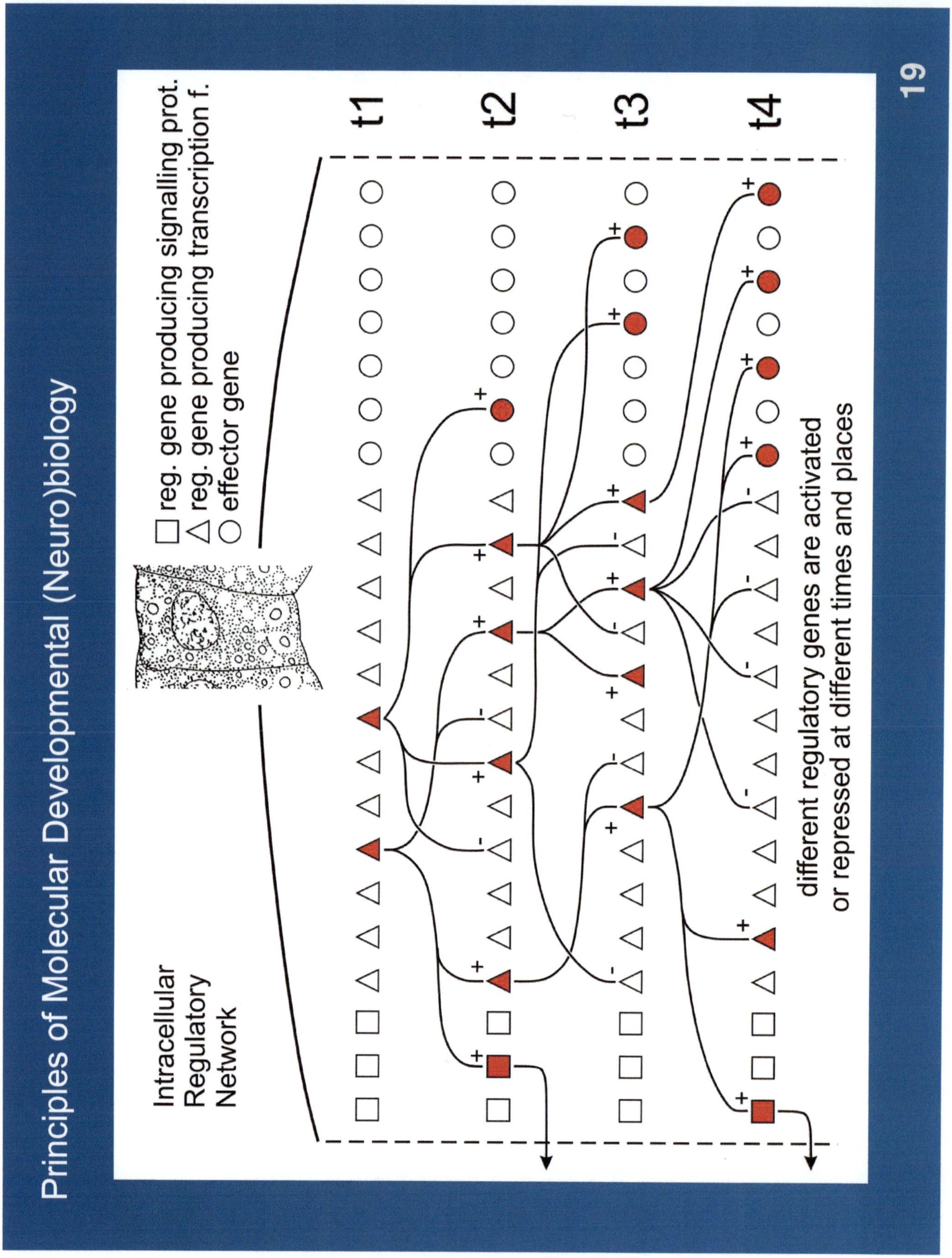
Principles of Molecular Developmental (Neuro)biology
19
Intracellular Regulatory Network
reg. gene producing signalling prot.
reg. gene producing transcription f.
effector gene
t1
t2
t3
t4
+
-
different regulatory genes are activated or repressed at different times and places

Plate 19

Plate 19 presents in more detail the genomic processes just outlined. It shows, still very diagrammatically, the pattern of activity of some developmental regulatory genes at four successive time segments in a single developing cell. Some of the activated regulatory genes produce signalling proteins, which leave the cell. An increasing number of transcription factors is involved in the activation of *effector genes* which are responsible for the production of structural proteins, such as those involved in the construction of microtubules or receptor complexes.

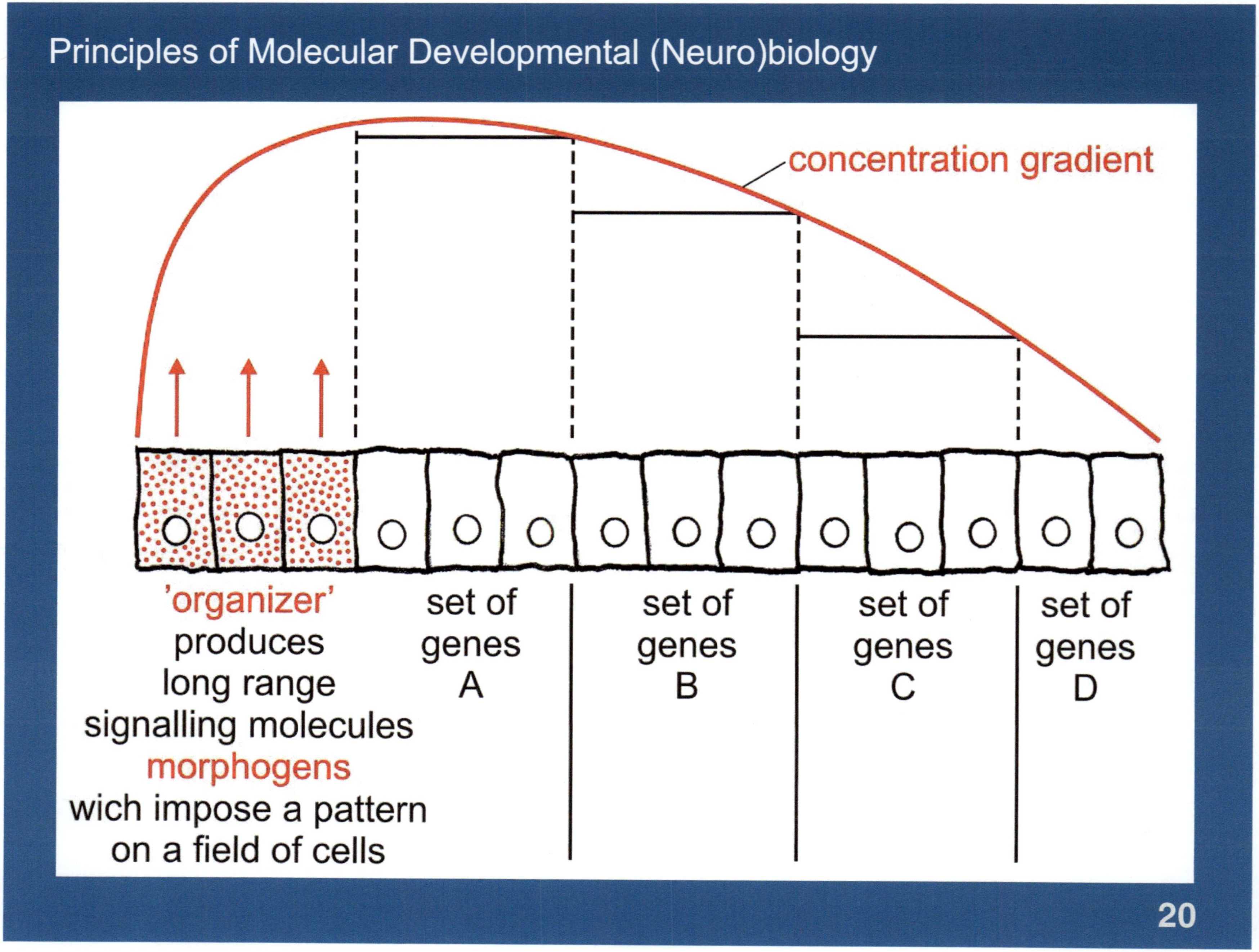
Principles of Molecular Developmental (Neuro)biology
concentration gradient
'organizer'
produces
long range
signalling molecules
morphogens
wich impose a pattern
on a field of cells
set of
genes
A
set of
genes
B
set of
genes
C
set of
genes
D
20

Plate 20

An organizer is a group of cells which produce long range signalling molecules or *morphogens*. These morphogens orchestrate gradientially the activation of specific sets of regulatory genes in different responding cells depending on the local concentration of the signal (genomic selector mechanism), and therewith influence in a generic way the differentiation of sets of specific cells in adjacent groups of embryonic cells at different distances from the organizer.

Plate 20 shows some cells of an *organizer*, releasing signalling molecules. These molecules form a gradient of decreasing concentration. The different concentrations of morphogens elicit the activation of different sets of regulatory genes, and therewith the differentiation of different sets of cells in the target field of embryonic cells. [For details, we refer to Davidsons' (2006) excellent book: "The regulatory genome".]

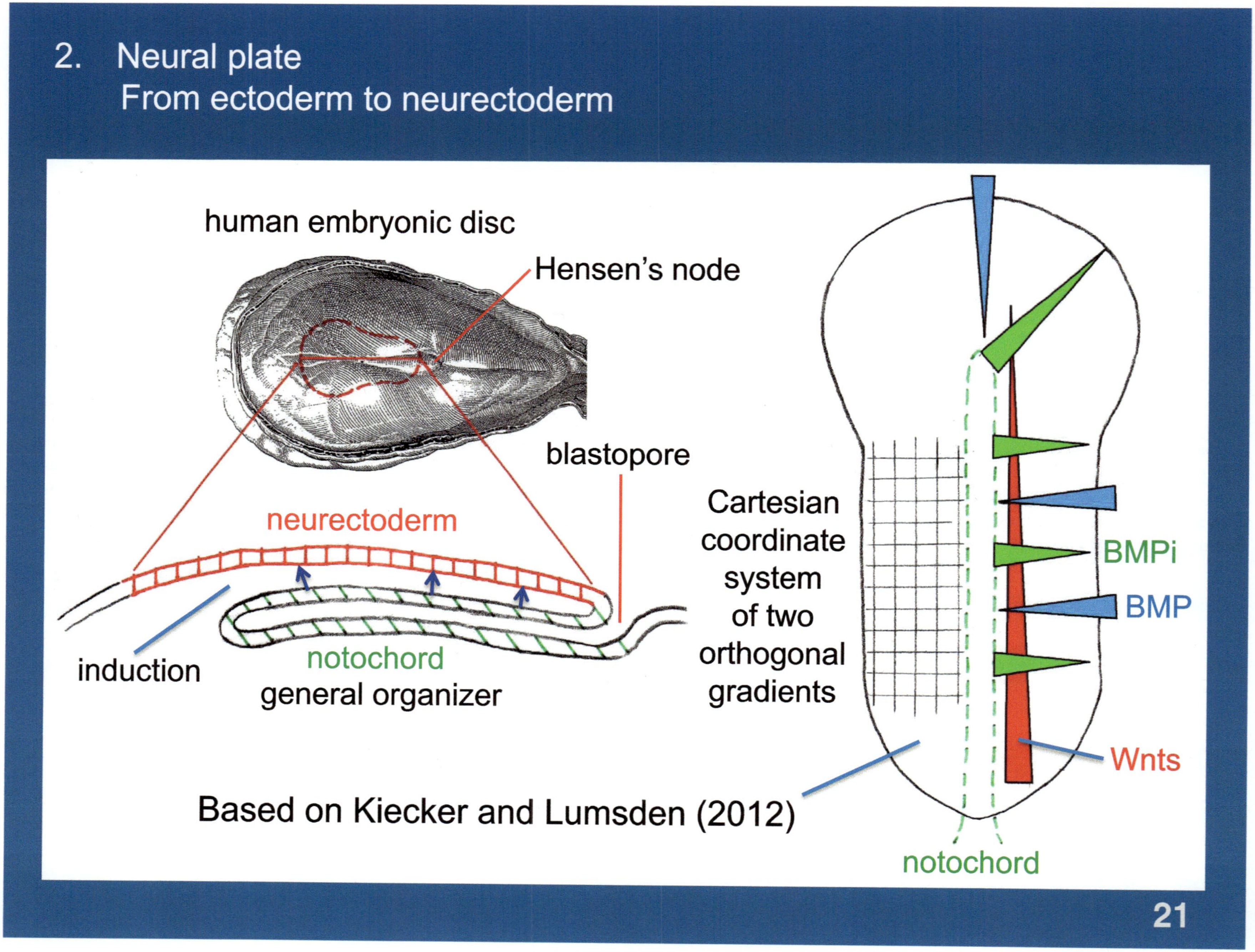
2. Neural plate
From ectoderm to neurectoderm
human embryonic disc
Hensen's node
blastopore
neurectoderm
induction
notochord
general organizer
Cartesian coordinate system of two orthogonal gradients
BMPi
BMP
Wnts
notochord
Based on Kiecker and Lumsden (2012)
21

2.2 From Ectoderm to Neurectoderm

Plate 21

Back to the neural plate!

The left-hand half of Plate 21, shows a dorsal view of a human embryonic disk. It will be seen that, just behind the posterior end of the future neurectoderm, there is a small area of strong proliferation, known as *Hensen's node*, which lies at the rostral end of the primitive line. In the centre of this proliferation and along the primitive line, an invagination of the epiblast occurs, related to gastrulation. Cells born at the node are deposited underneath the plane of the ectoderm, and the node, which is a dynamic epiblastic configuration with changing cellular populations (like a wave), then recedes caudally. After some time, we detect a rostrally directed median process lying underneath the ectoderm rostral to Hensen's node, which is known as the *notochord* or axial mesoderm (lower left). The rostral tip of the notochord represents the locus where the node started its activity. Once in place, the notochord represents a general organizer for the entire body, secreting a diffusing morphogen, sonic hedgehog (SHH), which among other functions helps to induce the formation of the neural plate from the overlying ectoderm (so-called *vertical* induction, as opposed to earlier *planar* induction generated by prospective notochordal cells before they leave the node). The induction process just outlined, was first deduced by Spemann and Mangold (1924), on the basis of ingenious ablation and grafting experiments in amphibian embryos. The notochord is designated as a general organizer, because it sequentially influences the formation of the entire neural plate. During neurulation (closure of the neural tube), the notochord further induces the differentiation of the floor plate, and both entities build an important axial ventralizing organizer, which antagonizes the dorsalizing influence of the roof plate. As we shall see, during later development of the CNS, several secondary or local organizers become active, whose position may be transverse as opposed to longitudinal.

The right-hand half of Plate 21 shows how the initial regional differentiation of the neural primordium is induced in a dose-dependent manner by gradients of three signalling proteins. The ectoderm that surrounds the neural plate expresses dorsalizing bone morphogenetic protein (BMP: *blue wedges*), while the notochord and the floor plate release ventralizing BMP inhibitors (*green wedges*). In addition to these gradients along the topological DV dimension, there is a longitudinal, postero-anterior gradient of wingless type proteins (Wnts) activity (*red wedge*), the exact source of which is unknown. Anti-Wnt-acting substances are produced and released at the rostral end of the neural plate (not shown). As indicated in the picture, the resulting, orthogonal gradients of signalling-protein activity establish a Cartesian coordinate system of positional information across the neural plate, which is reflected in variations in molecular profile (combinations of active and repressed genes). Thus, each point on the neural plate is differentially specified by particular concentrations of the three (or more) signalling proteins and resulting molecular signalling cascades (for details, see: Kiecker and Lumsden 2012).

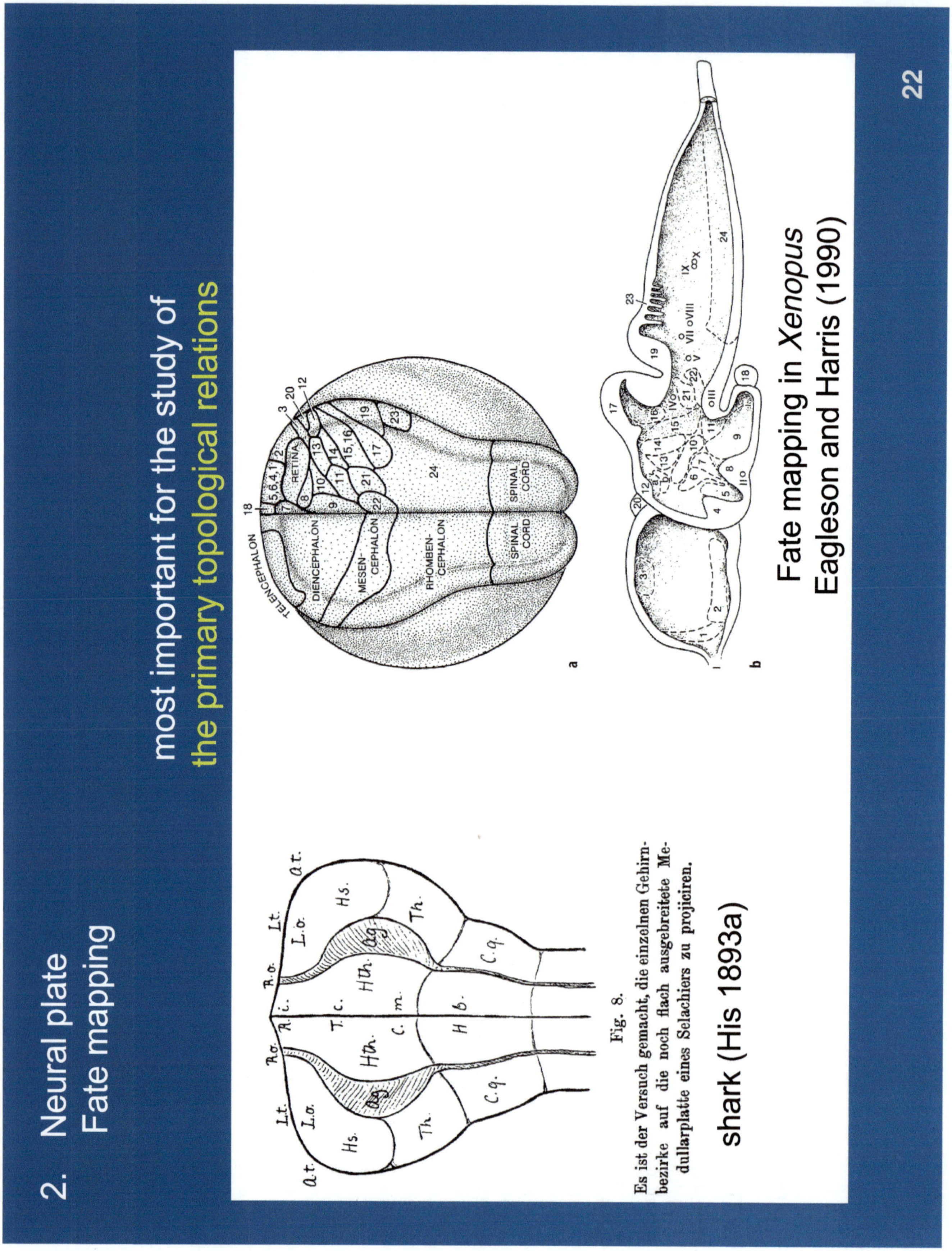
2. Neural plate
Fate mapping
most important for the study of
the primary topological relations
Fig. 8.
Es ist der Versuch gemacht, die einzelnen Gehirnbezirke auf die noch flach ausgebreitete Medullarplatte eines Selachiers zu projiciren.
shark (His 1893a)
TELENCEPHALON
DIENCEPHALON
MESEN-CEPHALON
RHOMBEN-CEPHALON
SPINAL CORD
RETINA
a
b
Fate mapping in *Xenopus*
Eagleson and Harris (1990)
22

2.3 Regionalization and Fate Mapping

Plate 22

This group of pictures shows that the morphological identities of the various regions in the developing CNS are progressively determined as a function of their *primary topological positions*, that are the positions, which their primordia occupy in the neural plate.

The left figure of Plate 22 shows a pioneering attempt by the noted neuroembryologist Wilhelm His (1831–1904) to project the various parts of the mature brain on the neural plate, in this case of a shark embryo. He indicates the supposed positions of the primordia of, among other structures, the cerebral hemispheres (Hs), hypothalamus (Hth), eye (Ag), thalamus (Th), mesencephalic tegmentum (Hb), and corpora quadrigemina (Cq).

The right figure of Plate 22 shows the results of an experimental mapping study of Eagleson and Harris (1990) of the neural plate of the clawed frog *Xenopus laevis*, in which the positions of the primordia of a number of brain regions were determined. In these experiments, small amounts of fluorescent dyes were applied as vital markers to the neural plate, and the fates of the patches of labeled cells were followed.

Figure a shows the presumptive brain areas as identified in the right half of the neural plate in a larval *Xenopus* brain; Figure b shows the anatomical position of the brain areas mapped in a, viewed in sagittal section. The main areas mapped include: (1) area olfactoria; (3) primordium hippocampi; (6) magnocellular preoptic nucleus; (10) ventromedial thalamic nucleus; (12a, b) dorsal and ventral habenular nuclei; (13–15) anterior, central and posterior thalamic nuclei; (17) mesencephalic tectum); (18) pituitary gland; (19) cerebellum; (24) medulla oblongata.

The left side of Figure a shows the authors' interpretation of main brain borders and the telencephalic field. Note that the 'diencephalon' would be divided nowadays into 'diencephalon proper' and 'secondary prosencephalon'; the latter enclosing the hypothalamus and the telencephalon (see in the right half of Figure a the possible oblique boundary between them passing caudal to the retinal field and ending peripherally between areas 3 and 20).

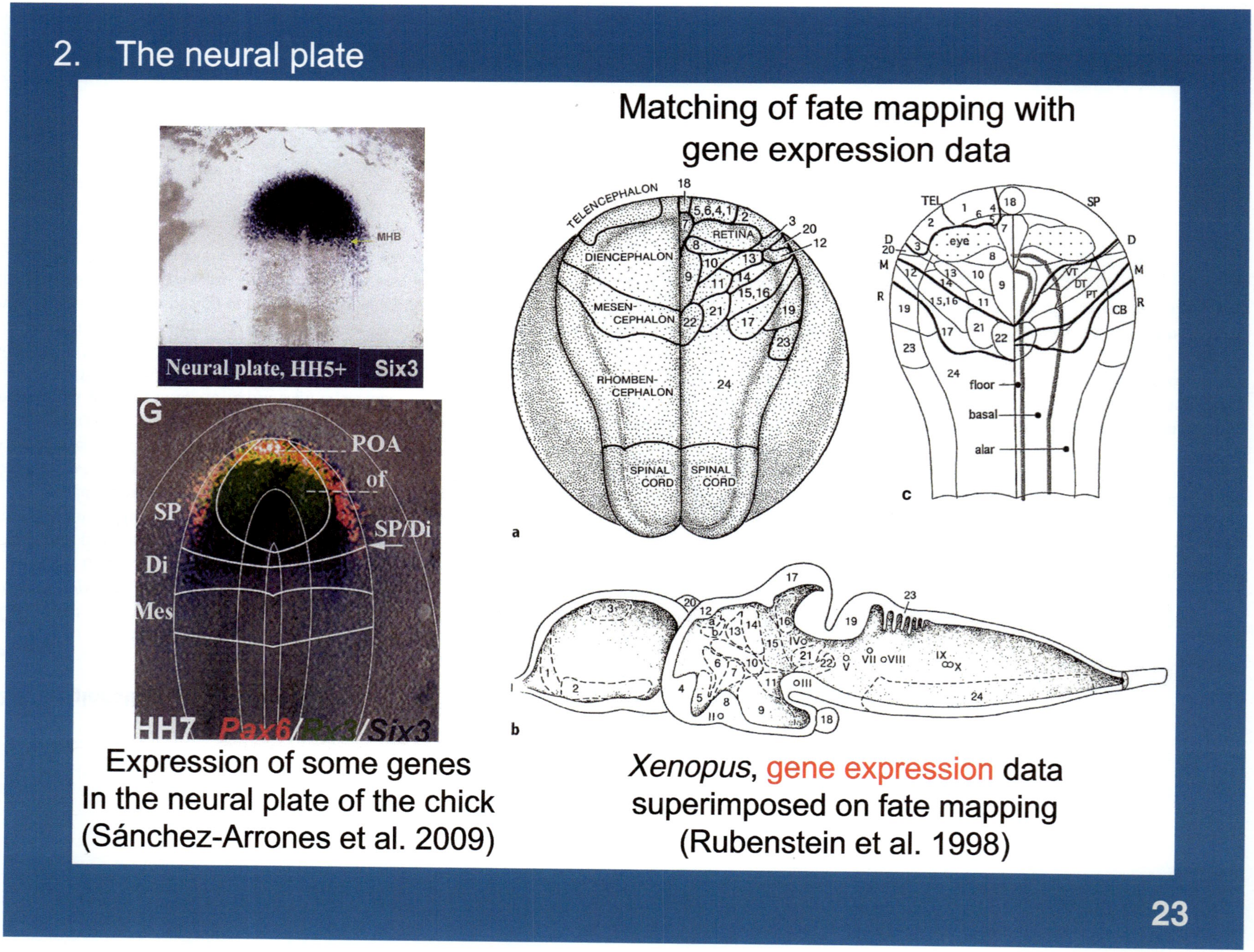
2. The neural plate
Neural plate, HH5+
Six3
MHB
G
POA
of
SP
SP/Di
Di
Mes
HH7 Pax6/Rx3/Six3
Expression of some genes
In the neural plate of the chick
(Sánchez-Arrones et al. 2009)
Matching of fate mapping with
gene expression data
TELENCEPHALON
DIENCEPHALON
MESEN-CEPHALON
RHOMBEN-CEPHALON
SPINAL CORD
SPINAL CORD
RETINA
a
TEL
eye
SP
D
M
R
VT
DT
PT
CB
floor
basal
alar
c
b
Xenopus, gene expression data
superimposed on fate mapping
(Rubenstein et al. 1998)
23

Plate 23

During recent decades, our insight into the regionalization of the brain has greatly increased by studies on the expression patterns of developmental regulatory genes and their products. Such studies have shown that the organization of the brain may be reduced to a set of basic molecularly defined entities that are common to all vertebrates. A few examples of such gene expression patterns in the neural plate of the chick, derived from Sánchez-Arrones et al. (2009), are shown in the left half of Plate 23.

The upper left figure shows that in the Hamburger-Hamilton (HH) stage 5 + (end of neural induction), the expression of the *Six3* gene is confined to the prospective forebrain and midbrain, and is excluded from the prospective floor plate and more caudal brain regions.

The lower left figure shows the expression patterns of *Pax6* (red), *Rx2* (green) and *Six3* (blue) in the chicken neural plate at HH7. The boundaries of future brain regions, as derived from experimental fate mappings are indicated. *Rx2* is expressed in the eye fields (ef); *Pax6* is mainly expressed at this stage in the telencephalic field of the secondary prosencephalon (SP), whereas the receding expression of *Six3* still extends from the SP over the lateral (topologically dorsal) part of the diencephalon proper (Di).

The right half of Plate 23 shows the results of comparisons of fate mapping with gene expression data in the CNS of *Xenopus*. The results of the experimental mapping study of Eagleson and Harris (1990), already discussed above (Plate 22), are compared firstly with the left half of c showing the accentuated boundaries of the primordia of the telencephalon, conventional diencephalon and mesencephalon, and secondly with the right half of c depicting instead the subdivision of the *Xenopus* neural plate suggested by Rubenstein et al. (1998), based on the expression patterns of various genes (note the transverse topology of the three diencephalic prosomeres: PT, DT, VT). There is a fair match between the mappings of Eagleson and Harris (1990) and Rubenstein et al. (1998), but direct comparison is hampered by the fact that the former authors adhere to the classic subdivision of the brain, whereas the latter use the new subdivision proposed by Puelles (see Plate 8). Thus, in the right half of c, Sp and D stand for secondary prosencephalon and diencephalon proper, respectively. In addition, thick and thin grey lines indicate the fate-mapped positions of the longitudinal boundaries that separate the prospective floor, basal, and alar plates.

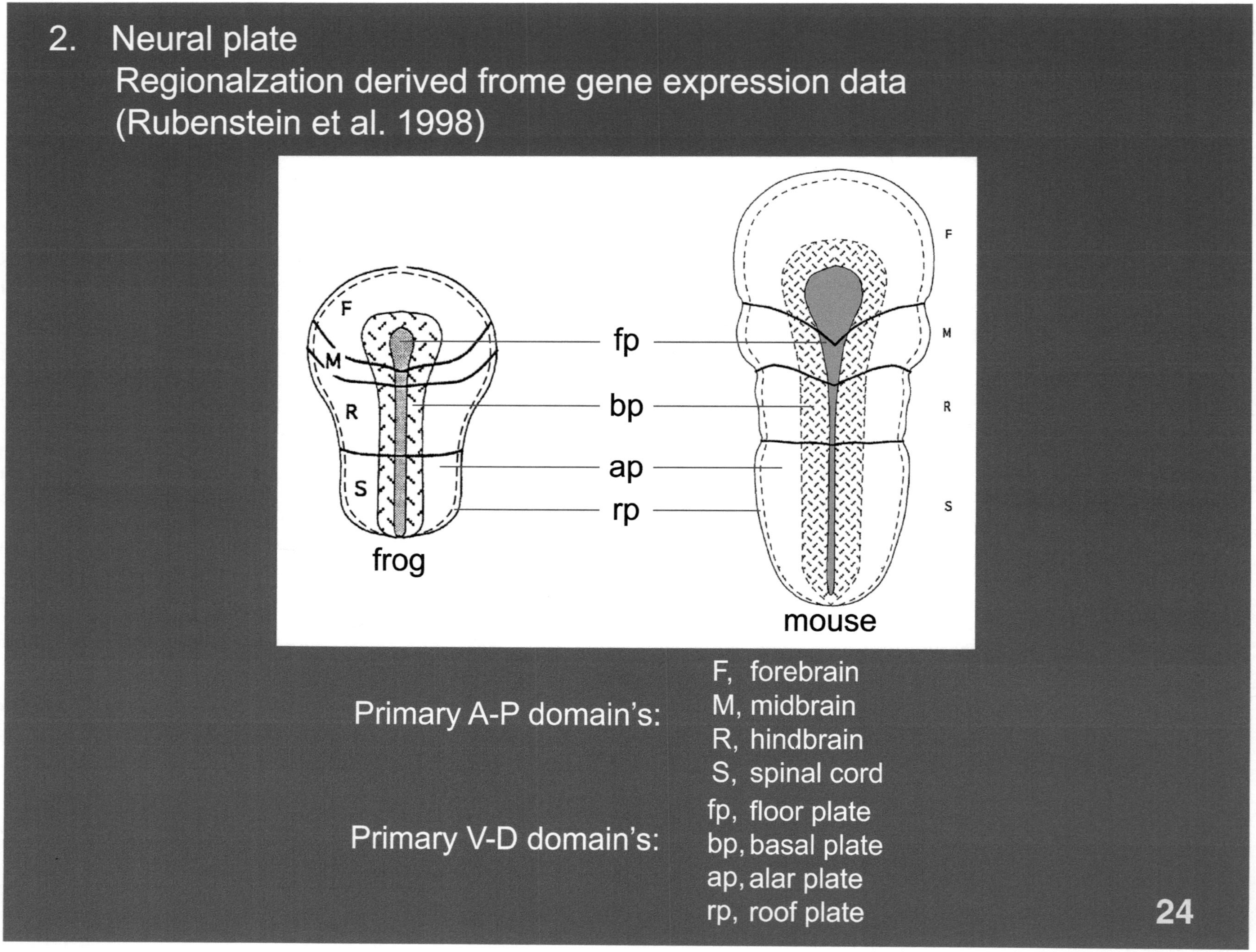
2. Neural plate
Regionalzation derived frome gene expression data
(Rubenstein et al. 1998)
F
M
R
S
fp
bp
ap
rp
frog
mouse
Primary A-P domain's:
F, forebrain
M, midbrain
R, hindbrain
S, spinal cord
Primary V-D domain's:
fp, floor plate
bp, basal plate
ap, alar plate
rp, roof plate
24

Plate 24

Gene expression data, reviewed by Rubenstein et al. (1998), allow an early regionalization of the neural plates of the frog and mouse into four primary antero-posterior and four primary ventro-dorsal domains. The positions and names of these primary domains are indicated in the plate.

Towards a New Neuromorphology

3. From neural plate to neural tube: 1. Neurulation

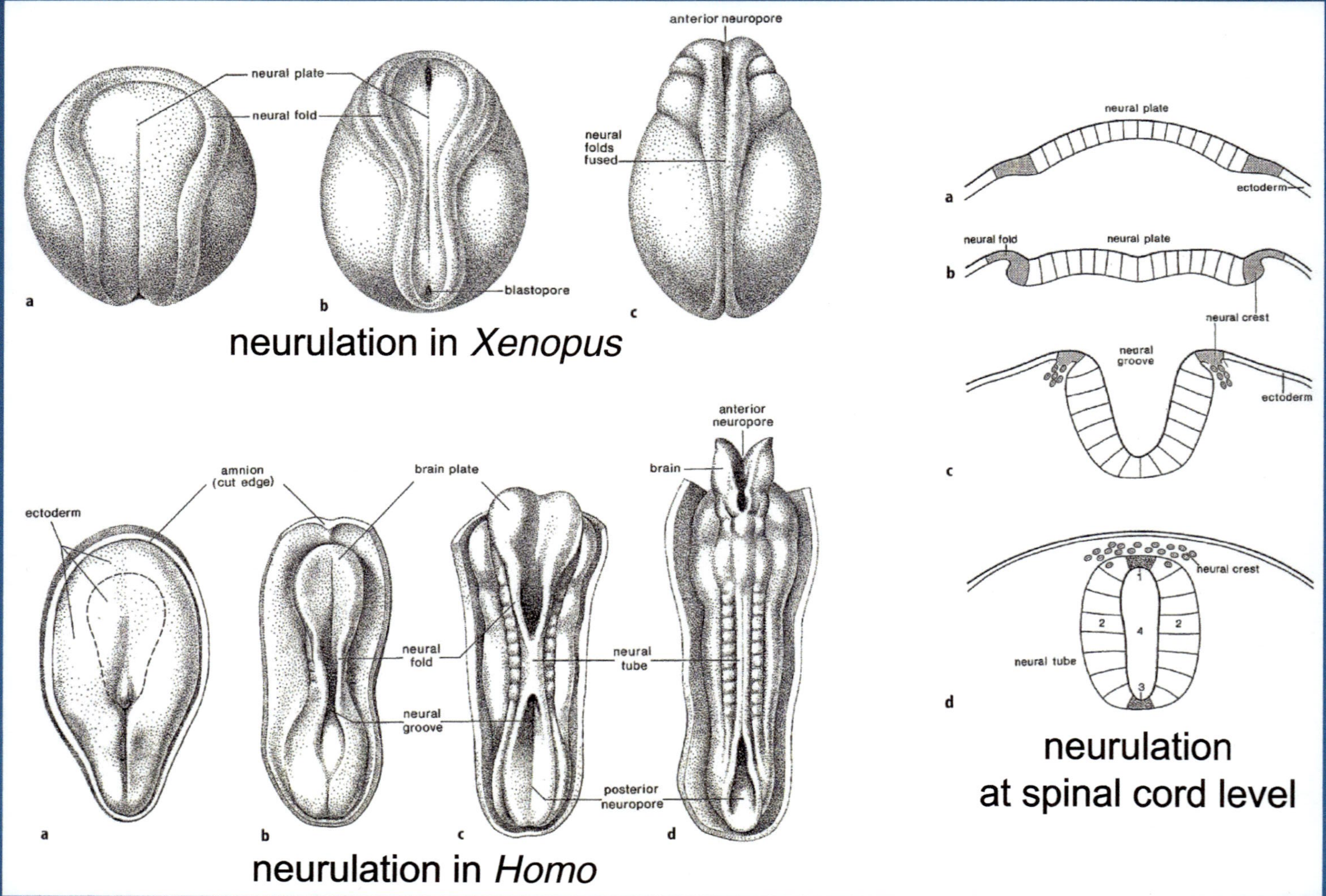

3 From Neural Plate to Neural Tube

3.1 Neurulation

Plate 25

The transformation of the neural plate into the neural tube is known as *neurulation*. During this process, the bilateral outer borders of the neural plate fuse together point-to-point along the median line, forming the roof plate. The neural crest population is released into the mesenchyme at the same time, and the skin also fuses, covering the whole.

The left part of Plate 25 illustrates this process in the frog and the human.

The upper panel shows dorsal views of an early neurula (a), a mid neurula (b), and a late neurula stage (c) of the frog. In a the neural plate is still flat; in b the neural folds approach each other, and in **c** they have fused and the neural tube is largely closed, leaving transiently open neuropores at both ends.

The lower panel shows dorsal views of reconstructions of human embryos.

(a) 16-day presomite stage; the approximate extent of the neural plate is marked by a dashed line.
(b) Early somite and neural groove stage of a 20-day embryo.
(c) Seven-somite embryo of about the 22nd day, showing beginning of closure of the neural tube.
(d) Ten-somite embryo of about the 23rd day with advanced closure of the neural tube and the rostral and caudal neuropores.

The right part of Plate 25 presents a series of diagrammatic transverse sections showing the formation of the neural tube at the level of the spinal cord. a Early neural plate stage; b late neural plate stage with formation of neural folds; c neural groove stage with incipiently separate neural crests, and d early neural tube stage with interstitial migrating neural crest elements and closed skin overhead. The names of the various parts of the early neural tube have been presented already in the general introduction (Plate 12): (1) roof plate, (2) lateral plates, (3) floor plate, and (4) ventricular cavity.

The figures show that the ectoderm along the lateral edges of the neural plate forms bilaterally a band-like strip, the primordial neural crest. This strip separates the primordial neural ectoderm from the primordial general body ectoderm. As the edges of the neural groove approach each other, they carry with them the adjoining primordial general body ectoderm. It is of note that, when the process of fusion occurs, not only the neural ectoderm but also the body ectoderm fuses in the median plane. As a result of this double fusion, the neural ectoderm detaches completely from the body ectoderm to form a submerged neural tube. This central displacement or 'internation' is a major developmental event. It enables the CNS to expand almost unlimitedly by surface growth and later to become protected by skeletal elements of ectomesodermal origin.

With the closure of the neural tube, the cells of the bilateral neural crests (they do not reach the rostral end of the plate) separate off and move into the space between the dorsal part of the neural tube and the overlying ectoderm. A certain proportion of these neural crest cells aggregates to form the sensory ganglia of cranial nerves and spinal nerves.

R. Nieuwenhuys and L. Puelles, *Towards a New Neuromorphology*,
DOI 10.1007/978-3-319-25693-1_3

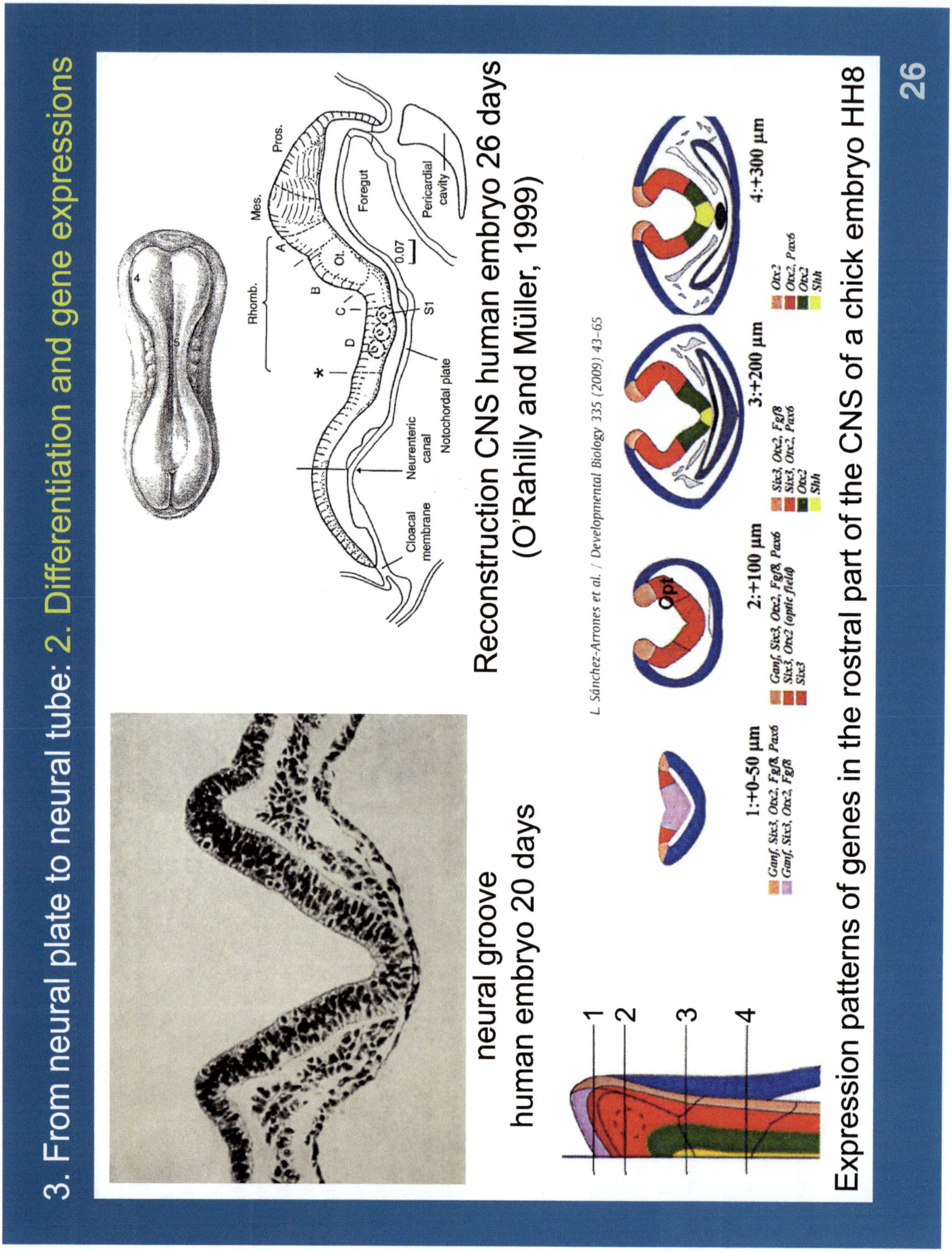
3. From neural plate to neural tube: 2. Differentiation and gene expressions
neural groove
human embryo 20 days
1
2
3
4
Rhomb.
Mes.
Pros.
A
B
C
D
Ot.
Foregut
Pericardial cavity
0.07
S1
Notochordal plate
Neurenteric canal
Cloacal membrane
Reconstruction CNS human embryo 26 days
(O'Rahilly and Müller, 1999)
L. Sánchez-Arrones et al. / Developmental Biology 335 (2009) 43–65
Opt
1:+0-50 µm
Ganf, Six3, Otx2, Fgf8, Pax6
Ganf, Six3, Otx2, Fgf8
2:+100 µm
Ganf, Six3, Otx2, Fgf8, Pax6
Six3, Otx2 (optic field)
Six3
3:+200 µm
Six3, Otx2, Fgf8
Six3, Otx2, Pax6
Otx2
Shh
4:+300 µm
Otx2
Otx2, Pax6
Otx2
Shh
Expression patterns of genes in the rostral part of the CNS of a chick embryo HH8
26

3.2 Differentiation and Gene Expressions

Plate 26

Plate 26 presents some aspects of the morphological, histological and molecular differentiation at the neural groove stage.

The upper left figure of Plate 26 shows a transverse section of a human embryo of about 21 days taken through the prospective rhombencephalic region. The walls of the neural groove are formed by the thick, pseudostratified epithelium. Numerous mitoses can be observed in the inner, future ventricular zone of the neural walls.

The upper right half of Plate 26 shows a dorsal view of a human embryo of about 26 days postfertilization (above) and below it a reconstruction of a median view of an embryo of the same age, derived from the 'Detailed atlas of developmental stages of the human brain', by O'Rahilly and Muller (1999). It will be seen that the authors were able to delineate in this neural groove stage the primordia of several brain parts, among which are the prosencephalon, the mesencephalon, and the four primary rhombomeres A, B, C and D. The asterisk indicates the future cerebrospinal junction.

The lower part of Plate 26 shows diagrammatic transverse sections through a neural-groove-stage (HH8) of the chick. The positions of the sections depicted, is indicated on the map of the unfolded neural groove to the left. The subdivision in color–coded longitudinal zones (floor plate: yellow; basal plate: green; alar plate, including prospective telencephalon: red; telencephalic subpallial and pallial roof plate: purple; roof plate plus neural crest from pallium caudalwards: pink) is based on the expression patterns of the indicated combinations of marker genes, also color-coded. The extraneural ectoderm is marked with blue. These pictures are derived from Sanchez-Arrones et al. (2009).

We next conclude Sects. 2 (neural plate) and 3 (From neural plate to neural tube) with a brief summary of the main points.

3.3 Summary of Neural Plate and Neurulation

1. The neural plate consists of a monolayered or pseudostratified epithelial sheet (although each cell spans the entire wall, the cell nuclei occupy diverse radial positions, giving the false impression of multiple layers).
2. General longitudinal and rostral/caudal organizers specify the neurectoderm by the secretion of morphogens, and endow the neural plate with an overall, grid-like pattern along its topologic anteroposterior and ventrodorsal axes.
3. Neurulation is a major morphogenetic event by which the primordial flat and open CNS emancipates itself completely from the surrounding ectoderm, and transforms via intrinsic cell shape changes into a closed neural tube.
4. Gene expression studies have shown that the specification of the principal longitudinal zones (floor plate, basal plate, alar plate, roof plate), as well as the principal preneuromeric transverse domains of the CNS (forebrain, midbrain, hindbrain, spinal cord) precedes the closure of the neural tube.
5. Fate mapping experiments supplemented by gene expression studies render it possible to transform the neural plate into a two-dimensional representation ('blueprint') of the (highly complex) topological relations of anatomic parts in the mature CNS. Although the early neural tube is molecularly patterned into longitudinal and transverse domains, it consists histologically of a proliferative, pseudostratified, highly columnar epithelium destined to form the future thickness of the brain wall (radial dimension).

Towards a New Neuromorphology

4. The neural tube

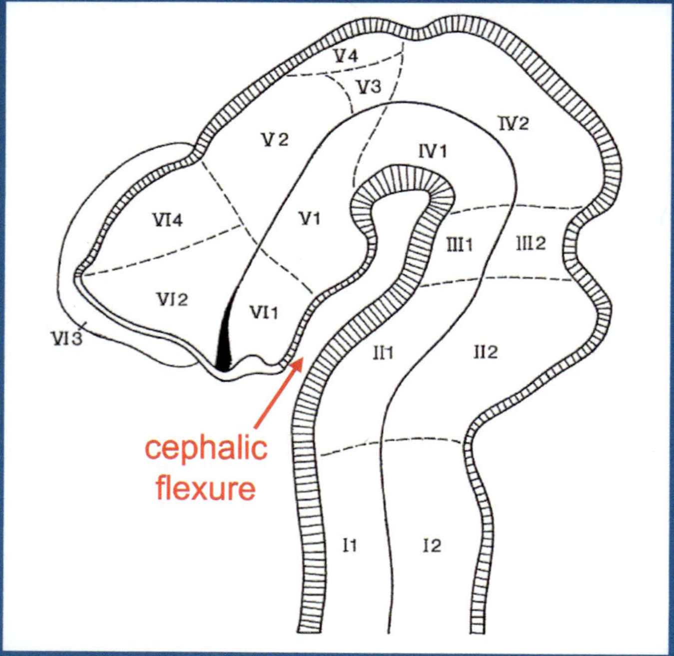

Brain subdivisions of an one-month old human embryo (His 1893b)

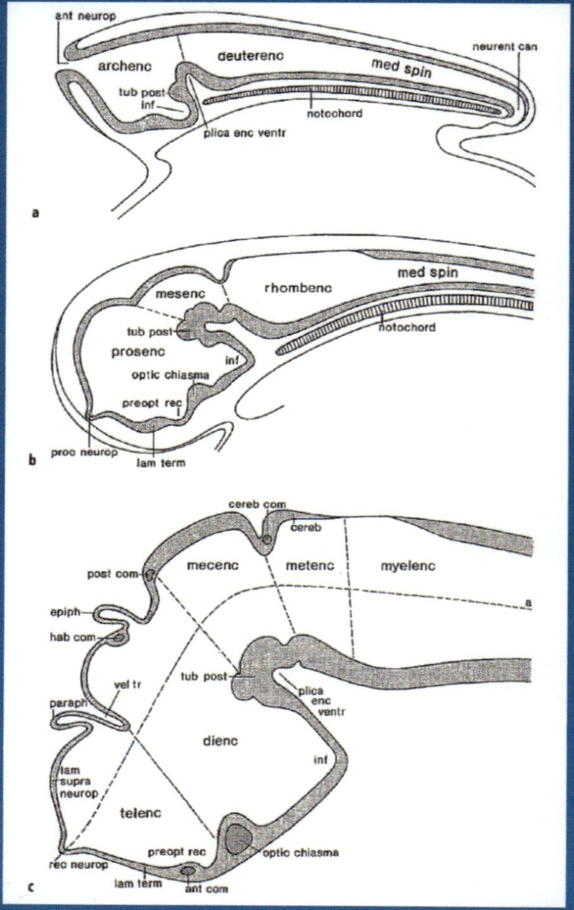

Median sections through the embryonic brain, (a) two vesicle stage; (b) three and (c) five veside stage (von Kupffer 1906)

4 The Neural Tube

4.1 Historical Introduction-1

Plate 27

Let us begin our discussion of the neural tube and its regional differentiation by looking at some historic pictures, one from His (1893b), and three from Karl Von Kupffer (1906).

The picture of His (Plate 27, left side) shows a median section through the brain of a human embryo about one month old. The brain axis is already strongly bent at this stage, particularly at the illustrated hairpin-like cephalic flexure. A longitudinal groove visible internally in the lateral walls of the brain generally divides them into a morphologically dorsal zone and a morphologically ventral zone. His proposed to designate this longitudinal ventricular groove as the *sulcus limitans* and the bounded zones as alar and basal plates, respectively. He indicated that all motor nuclei in the brain are situated ventral to this sulcus and are basal, whereas all sensory nuclei are situated dorsal to it and are alar.

The main divisions of the brain are depicted as areas roughly oriented topologically perpendicular to the sulcus limitans. The Roman numerals: I, II. IV, V and VI, indicate the five secondary brain vesicles contemplated by this author in caudal—rostral order: myelencephalon, metencephalon, mesencephalon, diencephalon and telencephalon, respectively, as already discussed (Plate 7). It is of note that His considered the narrowed sector III, which is intercalated between metencephalon and mesencephalon, as a separate division of the brain, which he named: *isthmus rhombencephali*. We now include this region in the hindbrain. Note also that he gave the hypothalamus its name and, following Forel (1877), defined it as restricted to the basal forebrain territory, ascribing parts of it to both diencephalon (V1) and telencephalon (VI1).

The three pictures, which form the right-hand panel of Plate 30 are, as already mentioned, derived from Von Kupffer (1906). They show his idea of the two-vesicle—(a); three-vesicle—(b), and five-vesicle (c) stages in the development of the brain of a'generalized' vertebrate. In figure c, von Kupffer has indicated the hypothetic course of the longitudinal axis of the brain (a virtual line along the centre of the ventricular cavity). Note that the rostral end of this axis ends at the neuroporic recess, which marks the last point of closure of the anterior neuropore (at the roof plate). In contrast the axis drawn by His (1893b) ends rostrally at the optic recess (alar/basal boundary). Both authors trace the floor plate practically up to the optic chiasma. All these points are controversial and have been amply discussed in the literature (see e.g. Kingsbury 1922). Recent gene expression studies (Puelles et al. 2012; Puelles 2013), which will be extensively discussed in Sects. 5 and 6, have yielded new insights into the course of the axis and the morphological relations at the rostral end of the neural tube.

Before entering into the discussion of the further differentiation of the brains of vertebrates, we should examine the brain of the lancelet, *Branchiostoma lanceolatum*, generally known as amphioxus. Amphioxus is a member of the Cephalochordata, which form a small (23 species) separate phylum of the Animal Kingdom. It has long been thought that the cephalochordates are the closest living relatives of the vertebrates (better: Craniata), but this honour has been recently passed on to the Urochordata or Tunicates, another phylum of acraniate chordates. [A later excursion of this course will be devoted to the CNS of a highly remarkable larval tunicate, the so-called ascidian tadpole.]

R. Nieuwenhuys and L. Puelles, *Towards a New Neuromorphology*,
DOI 10.1007/978-3-319-25693-1_4

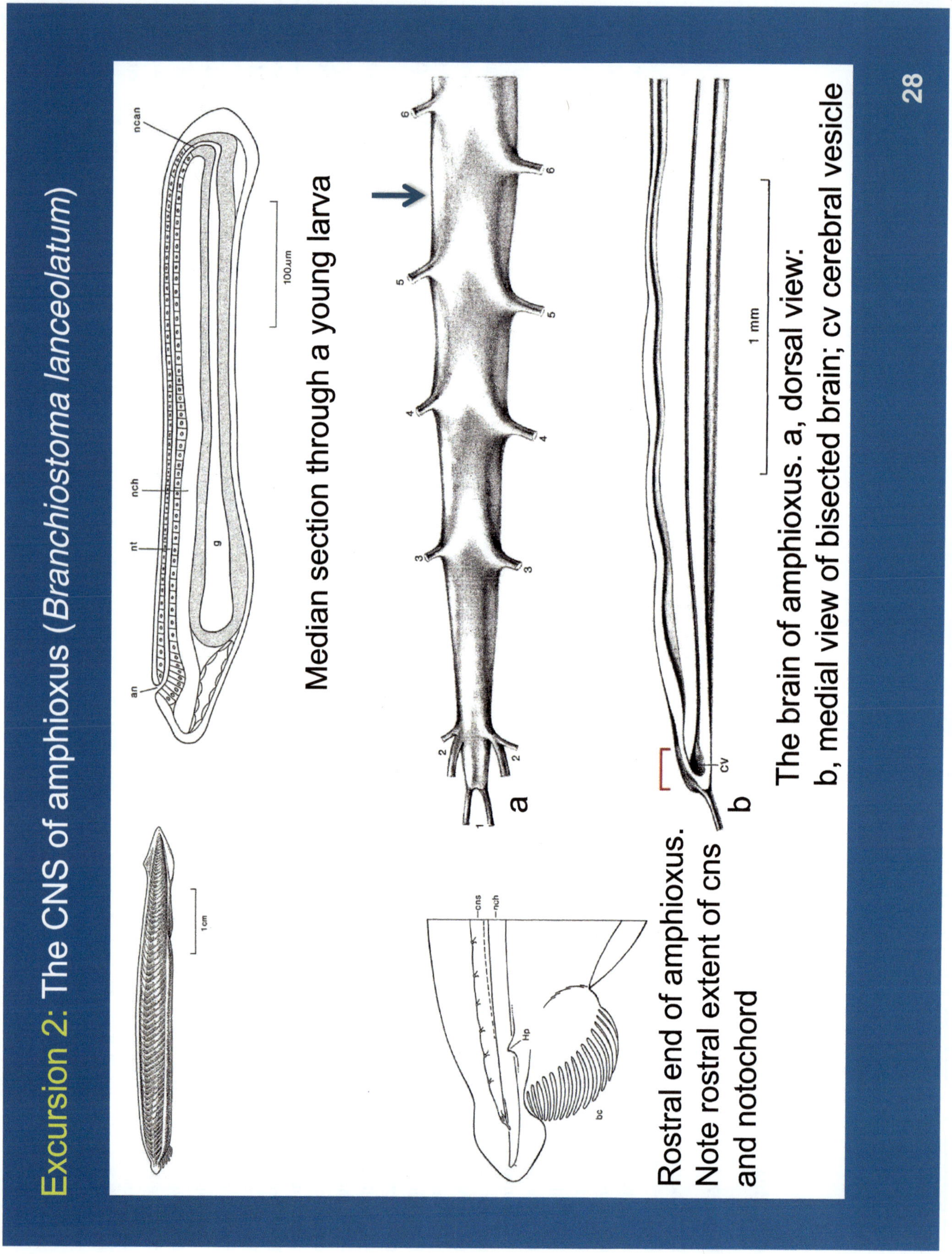
Excursion 2: The CNS of amphioxus (Branchiostoma lanceolatum)
1cm
an
nt
nch
g
ncan
100µm
Median section through a young larva
cns
nch
Hp
bc
Rostral end of amphioxus. Note rostral extent of cns and notochord
1
2
3
4
5
6
a
cv
b
1 mm
The brain of amphioxus. a, dorsal view: b, medial view of bisected brain; cv cerebral vesicle
28

Excursion 2 The CNS of amphioxus (Branchiostoma lanceolatum)

Plate 28

Amphioxus is a small, translucent marine animal some 5–10 cm in length (Plate 28, upper left figure). Although amphioxus resembles a fish, it has a much simpler organization. There is no true head, and paired special sense organs are entirely lacking. Its body axis is formed by a well-developed notochord, extending from the very tip of the rostrum to the end of the tail. The mouth is surrounded by tentacles for straining food particles from the water (Plate 28, lower left figure).

Plate 28, upper right picture shows that in a young larval stage the anlage of the CNS is represented by a simple epithelial tube, which is situated directly dorsal to the notochord.

The CNS of amphioxus maintains a simple tubular shape throughout life. The adult CNS attains its largest size in the middle of the body, from where it tapers gradually towards its rostral and caudal ends. Interestingly, the rostral tip of the notochord extends further rostrally than that of the nerve cord (Plate 28, lower left figure).

Although the rostral part of the nerve cord does not show any enlargements (Plate 28, lower right figures; a, dorsal view; b, median section), it differs sufficiently from the remainder in structure and genetic profile to warrant its designation as a forebrain vesicle. There is no telencephalon and there are no evaginated eye vesicles, but several gene markers suggest the presence of hypothalamic and diencephalo-mesencephalic primordia (Albuixech-Crespo et al. 2015). Gene expression studies also suggest that the equivalent of the craniate spinomedullary junction is situated at the level of the blue arrow in the middle right figure.

The most rostral part of the brain of amphioxus is commonly designated as the *cerebral vesicle* because its thin walls surround a small but distinct ventricular widening, reminiscent of the third ventricle (red bracket in lower right figure).

Excursion 2: The CNS of amphioxus (**Branchiostoma lanceolatu)m**

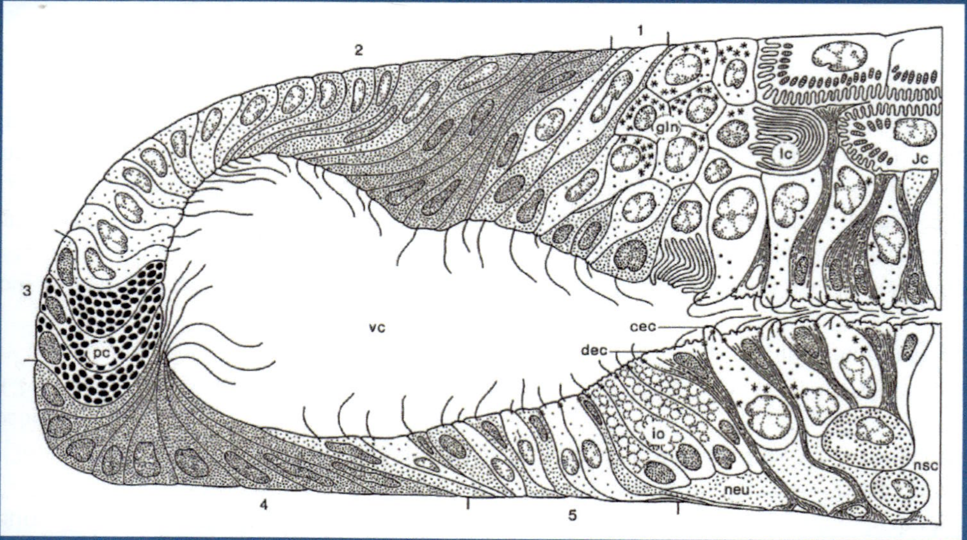

Paramedian section through the cerebral vesicle
Redrawn from Meves (1973)

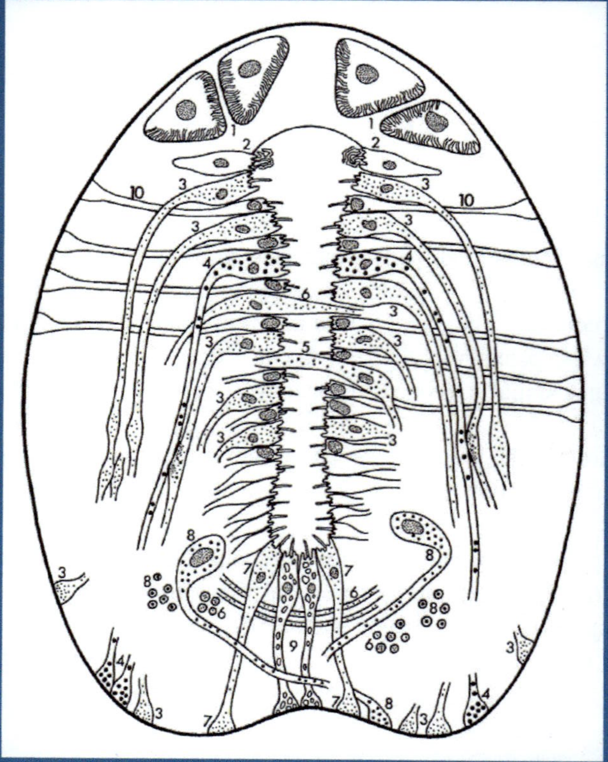

Transverse section through
caudal part of cerebral vesicle
(Obermüller-Wilén 1984)

29

Plate 29

There is a vast literature on the structure and organization of the CNS of amphioxus (for reviews, see: Nieuwenhuys 1998c; Wicht and Lacalli 2005). Even a superficial structural survey of this literature would require a course in itself. Here, we point to the remarkable feature that in the larval, as well as in the adult CNS of amphioxus, there occur all sorts of neural elements, representing 'transitional stages' between epithelial cells and true neurons.

The ultrastructural studies of Meves (1973) (Plate 29, left picture) have shown that the walls of the cerebral vesicle are formed mostly of ciliated columnar epithelial cells. She distinguished five groups of cells passing from the dorsocaudal roof to the ventrocaudal floor, with different structural and ultrastructural characteristics. Prominent among these is the rostrally situated Group 3, the elements of which contain numerous pigment granules. The detailed E.M. studies of Lacalli et al. (1994) and Lacalli (1996) have shown that in larval stages, the pigmented cells constitute, together with adjacent non-pigmented cells in the basal epithelial aspect of the cerebral vesicle, a photoreceptive complex which they designated as the frontal eye. The more anteriorly stituated cells of this complex presumably represent receptor cells, whereas those more posteriorly situated were interpreted as primordial retinal neurons. All of these elements appeared to be provided with axonal processes.

The right-hand picture of Plate 29 presents a semidiagrammatic transverse section through the caudal part of the cerebral vesicle, derived from Obermüller-Wilén (1984). The cell types depicted include: 10, ependymal gliocytes, extending their radially oriented processes throughout the width of the neural wall; 2, lamellar photoreceptor cells, extending their highly modified, lamella-like cilium into the ventricle; 3, catecholaminergic cells, fibres and terminals; 4, neuroendocrine FSH-immunoreactive cells, fibres and terminals, the axons of which elements are contacted by local widenings (*boutons en passants*) of catecholaminergic cells; 8, serotoninergic neurons, the somata of which have migrated outwards into the neuropil and fibre zone of the brain wall. It is noteworthy that not only in the brain, but also in the spinal cord the majority of neurons retain their contact with the ventricular cavity and therewith the epithelial cell-like shape of their somata.

Obermüller-Wilén (1984) observed that at the level of the caudal part of the cerebral vesicle, the basal lamina apposed to the ventral brain surface is contacted by terminal profiles of the peptidergic FSH-cells (4), as well as of catecholaminergic (3, 7) and serotoninergic (8) neurons. She considered it likely that all of these terminals release their products through the basal lamina, from where they may diffuse to the blood space bordering the ventrolateral part of the brain.

Excursion 2: Neurons in the CNS of amphioxus larvae

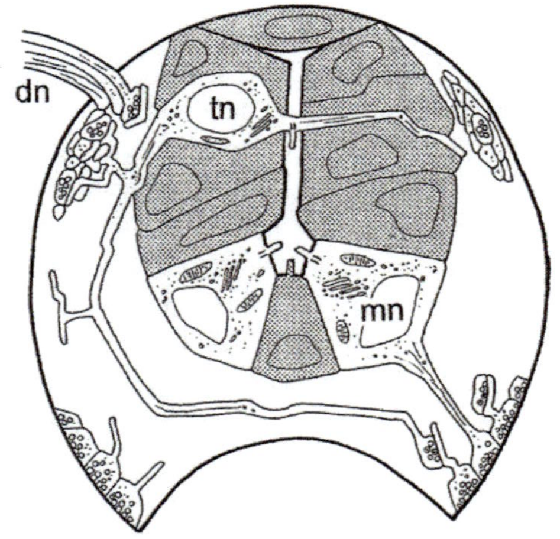

The spinal cord of an early amphioxus larva (transverse section)
dn dorsal nerve
tn interneuron with translumenal process
mn motoneuron
(Wicht and Lacalli 2005)

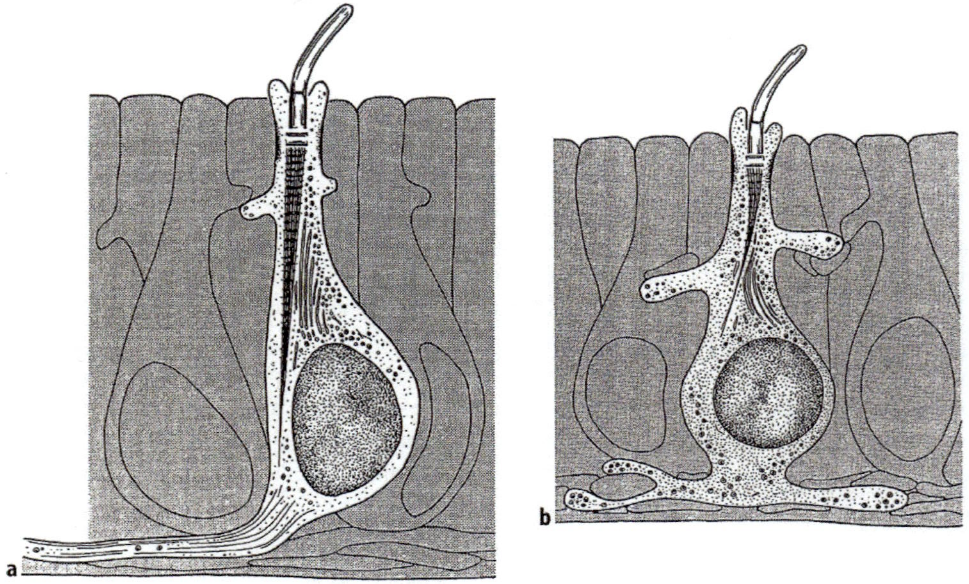

Two main cell types encountered in the rostral part of the cerebral ventricle of an early amphioxus larva (Lacalli 1996)

30

Plate 30

This plate presents some examples of neurons in the larval CNS of amphioxus, embedded in the epithelial lining of the ventricular cavity.

The left-hand figure of Plate 30, derived from Wicht and Lacalli (2005), shows a motoneuron (mn) and a translumenal interneuron (tn) in a transverse section of the anterior cord of an early amphioxus larva. These neurons form part of the locomotory control system, the axons of the motoneurons remarkably forming neuromuscular junctions at the surface of the cord. The translumenal interneuron is provided with two processes, a shorter one and a longer one, projecting from opposite poles of the cell. The shorter process passes, just like that of many other interneurons, across the ventricular cavity to the contralateral side of the cord. The longer process forms an arcuate fibre, which, passing through the most ventral part of the cord, also reaches the contralateral side. Laterally directed branches of the initial part of this fibre contact incoming dorsal nerve roots (dn), whereas its distal part forms a preterminal ending on the axon of a motoneuron.

The right-hand figure of Plate 30 shows two main cell types observed by Lacalli (1996) in the rostral part of the cerebral vesicle of an early amphioxus larva. Both of these cells are provided with a cilium projecting into the ventricular cavity, which suggests a sensory function.

The cell type shown in a, is provided with a typical axon, thus combining a putative sensory with a typical neuronal feature. Such elements are known as *neurosensory cells*. They are very common in invertebrates. In vertebrates they are confined to the olfactory epithelium. The cell in question has, in addition, a few short branches in the subapical zone.

The cell type shown in b, is provided with several subapical and basal processes, but lacks an obvious axon.

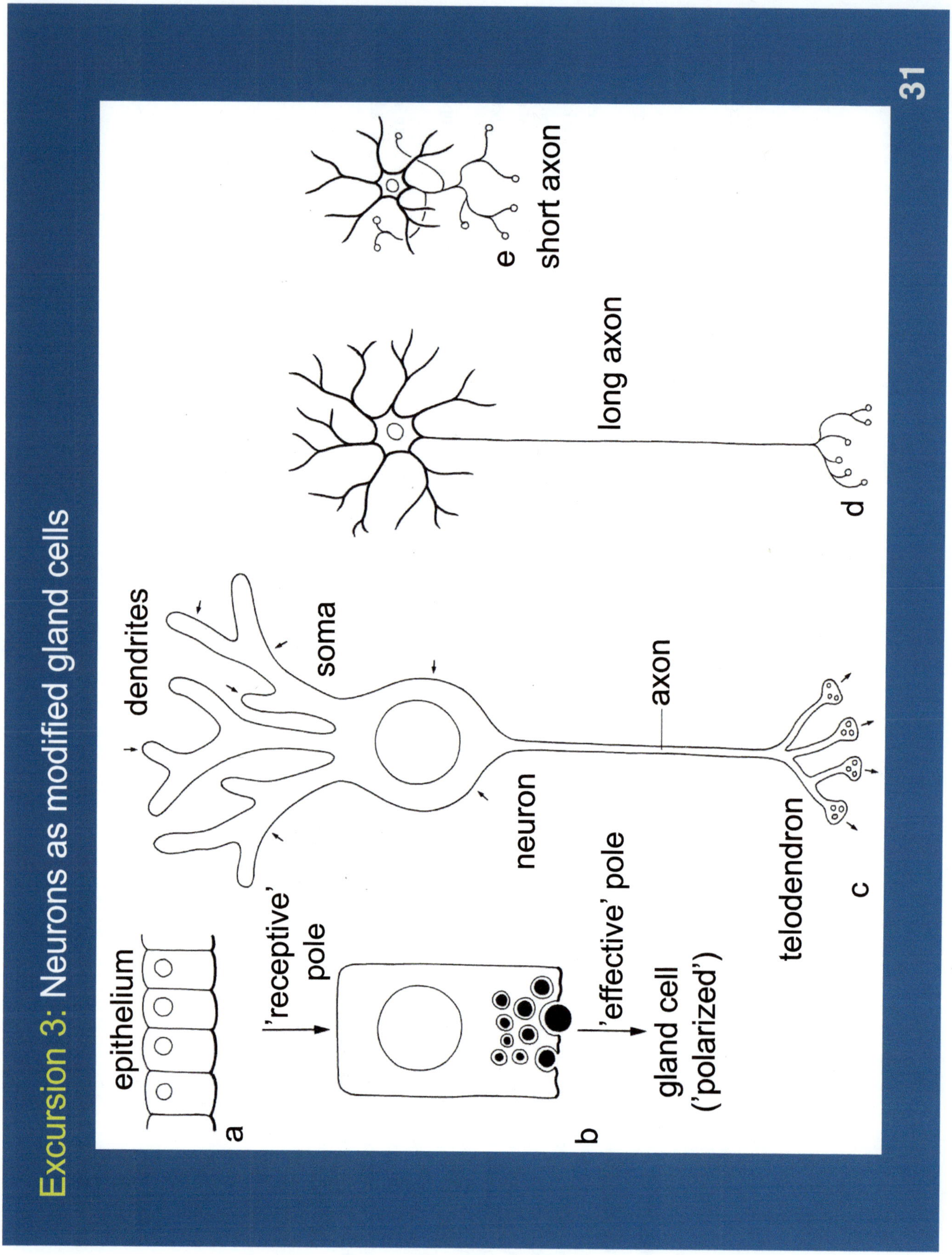
Excursion 3: Neurons as modified gland cells
31
epithelium
a
'receptive'
pole
'effective' pole
gland cell
('polarized')
b
dendrites
soma
neuron
axon
telodendron
c
long axon
d
short axon
e

Excursion 3: Neurons as modified gland cells

Plate 31

We include an additional brief excursion to illustrate that neurons can be envisioned as *modified gland cells.*

Figure a shows some columnar epithelial cells.

Figure b shows a typical gland cell: an epithelial cell specialized in the secretion of a specific substance. Note that the activity of this element is directed or *polarized.* It has a receptive pole where the stimuli activating the cell arrive, and an effective pole where the specific secretory product is released.

Figure c shows a typical neuron. It can be envisioned as a gland cell of which the receptive pole is drawn out into a number of ramifying processes, constituting together a *dendrtic tree*. The effective pole is also greatly stretched and transformed into (i) an intermediate, conductive segment, the *axon*, and (ii) a ramifying terminal segment, the *telodendron*. By this transformation, the effective pole is fractionated into numerous 'microglands', called terminals or 'boutons terminaux'. The latter contain numerous small vesicles in which molecules of specific humoral messengers or *neurotransmitters* of the cell are stored. The terminals form specialized contacts, *synapses*, with other neurons. The presynaptic and postsynaptic membranes are separated by narrow, fluid-filled *synaptic clefts*. (It is important to note that the synaptic clefts are directly continuous with the general extracellular space, and that this space occupies no less than 15–20 % of the total volume of the CNS.) The synaptic vesicles release their contents into the synaptic clefts. The transmitter molecules diffuse to the postsynaptic membrane, where they bind to specific receptor complexes. The neurotransmitter substances are, in accordance with the influence they exert on their target elements, characterized as excitatory or inhibitory transmitters.

The transformations on the receptive side of the hypothetical modified gland cell under discussion (c), have led to an enormous enlargement of the receptive (somal plus dendritic) surface of the element, providing room for numerous afferent, excitatory and inhibitory terminals of other neurons. Neurons are, just like gland cells, typically polarized elements, because the flow of information is, under normal conditions, always directed from the receptive to the effective pole.

The element depicted in d represents a typical long axoned or *Golgi type I neuron*. In *Golgi type II neurons,* also known as *local circuit neurons* (e) the axon is short and ramifies in the vicinity of the soma.

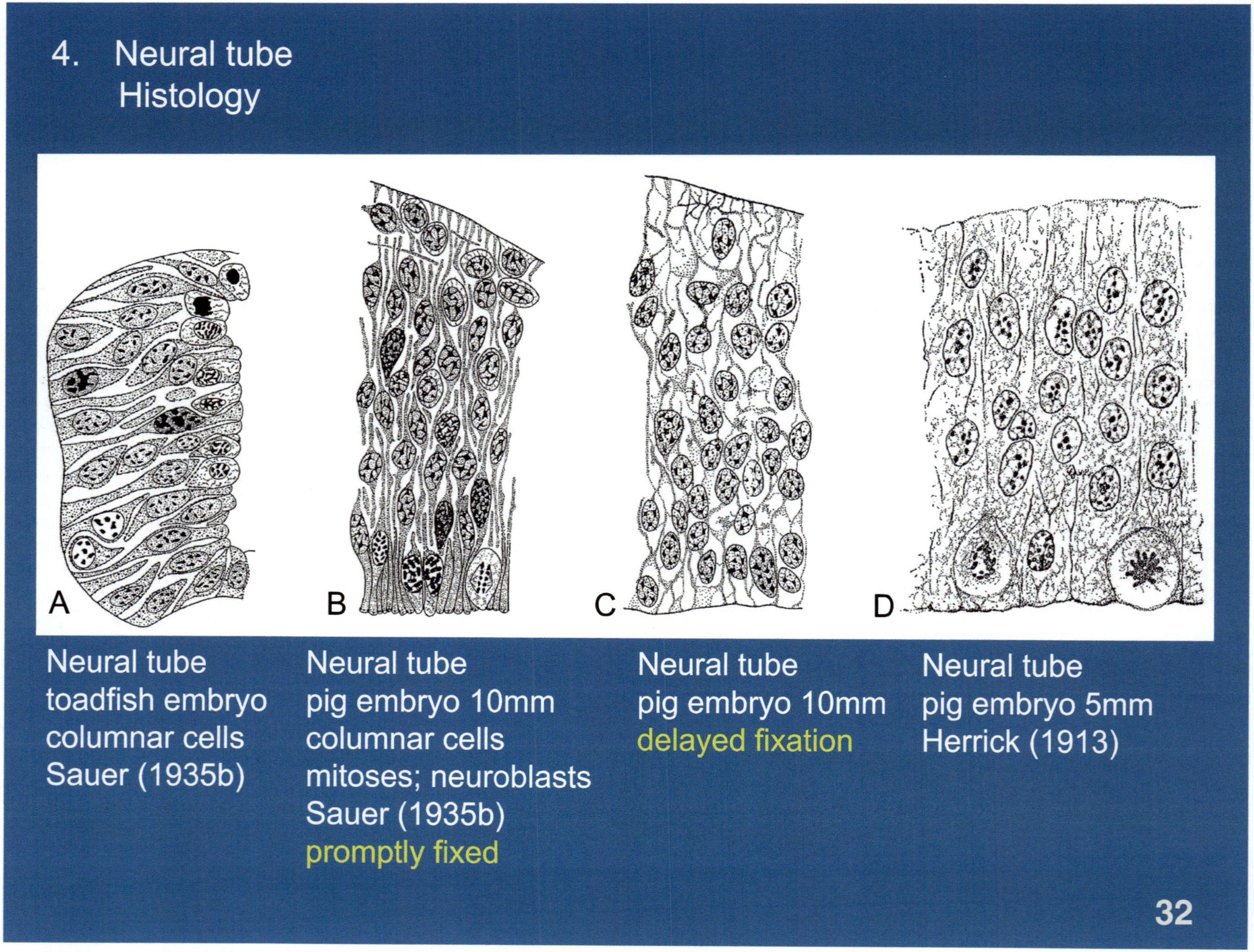
4. Neural tube
Histology
A
B
C
D
Neural tube
toadfish embryo
columnar cells
Sauer (1935b)
Neural tube
pig embryo 10mm
columnar cells
mitoses; neuroblasts
Sauer (1935b)
promptly fixed
Neural tube
pig embryo 10mm
delayed fixation
Neural tube
pig embryo 5mm
Herrick (1913)
32

4.2 Structure of the Neural Tube

Plate 32

Plate 32 presents some diagrams of the cellular structure of the early neural tube of vertebrates, taken from the excellent work of Sauer (1935a). Sauer emphasized that the early neural tube consists largely of very long columnar epithelial cells, extending from the internal, ventricular surface to the external surface of the neural tube. He took vigorous issue against the prevailing view at that time that these columnar cells anastomose to form a syncytial structure.

Figure A shows a transverse hemisection through the neural tube of an embryonic teleost, the toadfish *Opsanus tau.* The columnar form of the cells is clearly visible. Sauer emphasizes that the figure shows the appearance of the cells at one focal level only, and that most of the elements could be traced through their entire length by focusing.

Figure B shows a small part of the alar plate of the neural tube of a 10-mm. pig embryo. Here too, the columnar shape of the individual cells is clearly visible. The rounded, non-columnar cells in the superficial zone of the section represent neuroblasts. (The development and fate of these elements will be discussed later on in this course.)

Figure C shows a section through the neural tube of a pig embryo, entirely similar to that shown in B, except that the fixation in C was delayed for 15 min. It will be seen that in this preparation the cells have lost their individuality and seem to be connected by anastomosing, strand-like structures.

Sauer concluded from these observations that the cells in the undistorted neural tube are independent elements, not parts of a syncytium. The numerous pictures in the literature, seemingly demonstrating the contrary, such as the one reproduced in D, show in Sauer's opinion artifacts, due to delayed fixation.

4. Neural tube
Ventrodorsal and dorsoventral regionalization

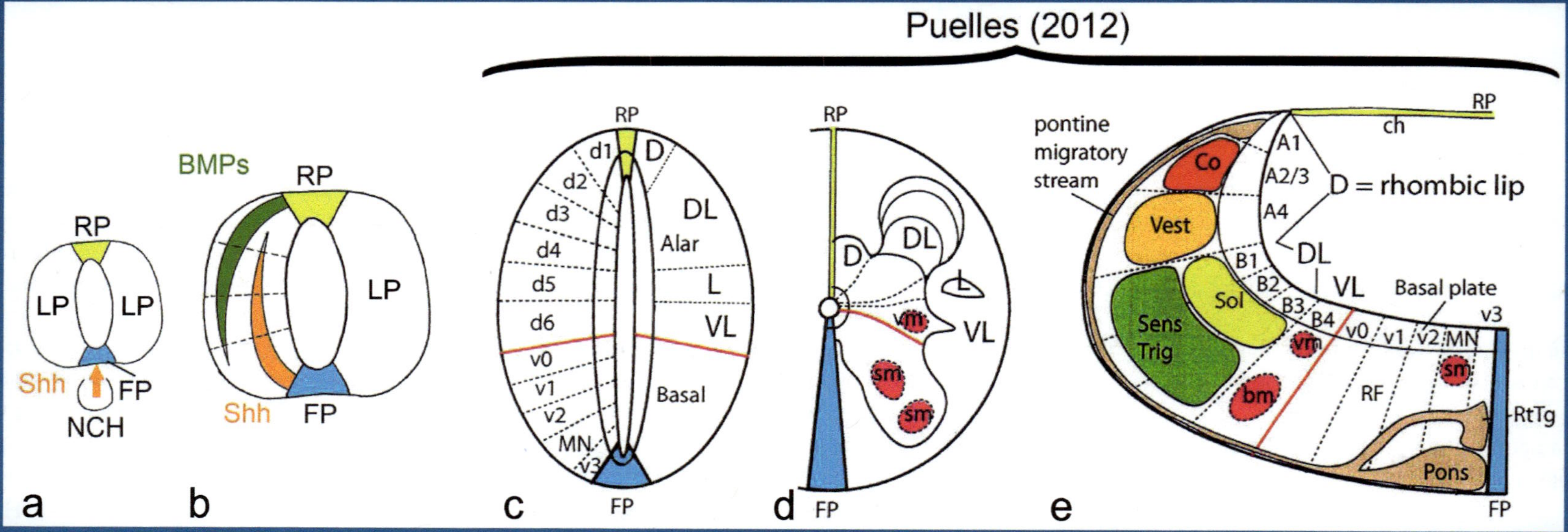

a early neural tube

b later neural tube

c developing spinal cord

d mammalian spinal cord adult

e developing mammalian rhombencephalon

The morphogens secreted by FP and RP form concentration gradients that specify different cell fating in a dose-dependent fashion, via transcriptional activators and repressors acting on different classes of genes (Kiecker and Lumsden 2012)

33

4.3 Ventrodorsal Regionalization: Longitudinal Zones

Plate 33

Plate 33 presents an elementary overview of the formation of longitudinal zones in the neural tube. It is based on review articles by Kiecker and Lumsden (2012) and Puelles (2013), to which we refer for details.

Figure a shows a diagrammatic transverse section of an early neural tube. The lateral plates (LP) are dorsally and ventrally connected by the roof plate (RP) and the floor plate (FP), respectively. The FP is a strip of wedge-shaped glial cells along the ventral midline of the tube. It is indicated in the figure that the notochord (NCH), which underlies the neural tube, expresses *Sonic hedgehog* (*Shh*), a vertebrate ortholog of the *Drosophila* segment polarity gene *hedgehog*. *Shh* encodes for a signalling factor, the protein Sonic Hedgehog (SHH), which is secreted by the notochord and induces the formation of the floor plate. In somewhat later stages, *Shh* is expressed by the floor plate itself, which becomes a secondary source of diffusing SHH.

Figure b shows that the floor plate (FP) and the roof plate (RP) both release morphogens, which form, respectively, ventrodorsally and dorsoventrally directed concentration gradients with antagonistic ventralizing and dorsalizing effects. It is well established that SHH protein represents the morphogen secreted by the FP. The identity of the RP organizer has not been fully determined as yet; there is, however good evidence that the Bone Morphogenetic Proteins (BMPs) are involved. The opposed ventrodorsal and dorsoventral concentration gradients of the morphogens specify the fates of the neuronal precursors in the lateral plates of the neural tube in a dose-dependent fashion, as explained in Plate 21. The different concentrations of the FP and RP morphogens elicit the compartmentalized expression of different activating and repressing regulatory Bone Morphogenetic Proteins, which regulate the local activity of different sets of transcriptional control genes.

We have seen that, according to His (1893b), a longitudinal, ventricular *sulcus limitans* divides the lateral wall of the neural tube into a primarily motor basal plate, and a primarily sensory alar plate (Plate 28, left side). Puelles (2012) introduced a molecular reinterpretation of the His subdivision by specifying the alar–basal boundary as a linear equilibrium between alternative histologic fates, established by interplay of opposite ventralizing and dorsalizing signalling effects originating from the FP and RP. This newly defined boundary between the basal and alar plates is indicated in Figs. c, d and e, with a red line. It is of note that the ventricular end of this boundary does not coincide with the sulcus limitans, but is rather situated more ventrally. Due to this shift, the visceromotor (vm) and branchiomotor (bm) cell groups, which according to the classic interpretation of His are born and remain situated in the basal plate, now turn out to arise indeed within the basal plate, but secondarily come to lie within the alar plate (Fig. e), after performing tangential migration (Windle 1971; Ju et al. 2004).

Figure c shows that the interplay of opposing ventral and dorsal signals triggers the formation of 11 molecular sectors (or microzones), each characterized by a specific combination of transcription factors, which produce given types of (postmitotic) neuroblasts. During further development, the neuroblasts sort out of the proliferative ventricular zone and migrate outward into the growing mantle zone, where the elements originating from the various molecular sectors intermingle and finally settle via mutual adhesive interactions into definitive structures, such as the layered dorsal horn in the spinal cord (Fig. d). The somatic motoneurons (sm), which originate from sector MN, form discrete groups (columns) within the spinal anterior horn, within which they are surrounded by interneurons stemming from the basal sectors v0, v1 and v2. The sympathetic preganglionic visceromotor neurons (vm) originate from the thoracic sector v3, which is situated directly adjacent to the FP. From there, they migrate tangentially upward, to attain their ultimate position in the spinal lateral horn, which forms part of a ventrolateral alar area (VL), situated directly dorsal to the basal/alar border. Elements originating from the sectors d2–d4 participate in the formation of the spinal dorsal horn (DL).

The ventrodorsal and dorsoventral patterning mechanisms just outlined also operate in the developing brain. In the hindbrain or rhombencephalon (Fig. e), the lateral neural walls diverge lateralwards, and the RP is transformed into a wide membrane, the future choroid plexus (ch). The topologically ventrodorsally arranged molecular subdomains correspond by and large to those in the spinal cord. The somatomotor neurons (sm), which will form the nuclei of the XIIth, VIth and IVth nerves, remain close to their site of origin, sector MN. The branchiomotor cells (bm), which will form the cranial motor nuclei of the XIth, Xth, IXth, VIIth and Vth nerves, and the parasympathetic preganglionic visceromotor cells (vm), which will give rise to the dorsal motor vagal nucleus and the salivatory nuclei, migrate tangentially from sector v3 to the ventrolateal alar zone VL. The latter forms the topologically most ventral part of the alar plate. The remaining part of the rhombencephalic alar plate can be divided into a dorsolateral (DL) and a dorsal domain (D). The former gives rise to the sensory trigeminal and solitary nuclei. The latter contains a special germinal zone, which is known as the rhombic lip. Molecularly distinct compartments of this germinal zone give rise to a large number of cell masses, including the cochlear, vestibular and pontine nuclei. The cochlear and vestibular nuclei are the product of short radial migrations. The cells destined to form the pontine nuclei, on the other hand, form a long, superficially situated tangential migratory stream, originating in the rhombic lip area of rhombomeres 6–8, which moves subpially into rhombomeres 3–4, and finally settles there within the most ventral part of the basal plate (Fig. e). The precursor cells of other ‘precerebellar’ nuclei (nuclei which project to the cerebellum), such as the inferior olive and the nucleus reticularis tegmenti pontis (RtTg) also originate from the rhombic lip, attaining their ultimate positions via a long tangential migration. The most rostral part of the rhombic lip contributes substantially to the formation of the cerebellum, by providing the precursors of the adult granule cells, which represent the most numerous cell type in the cerebellum.

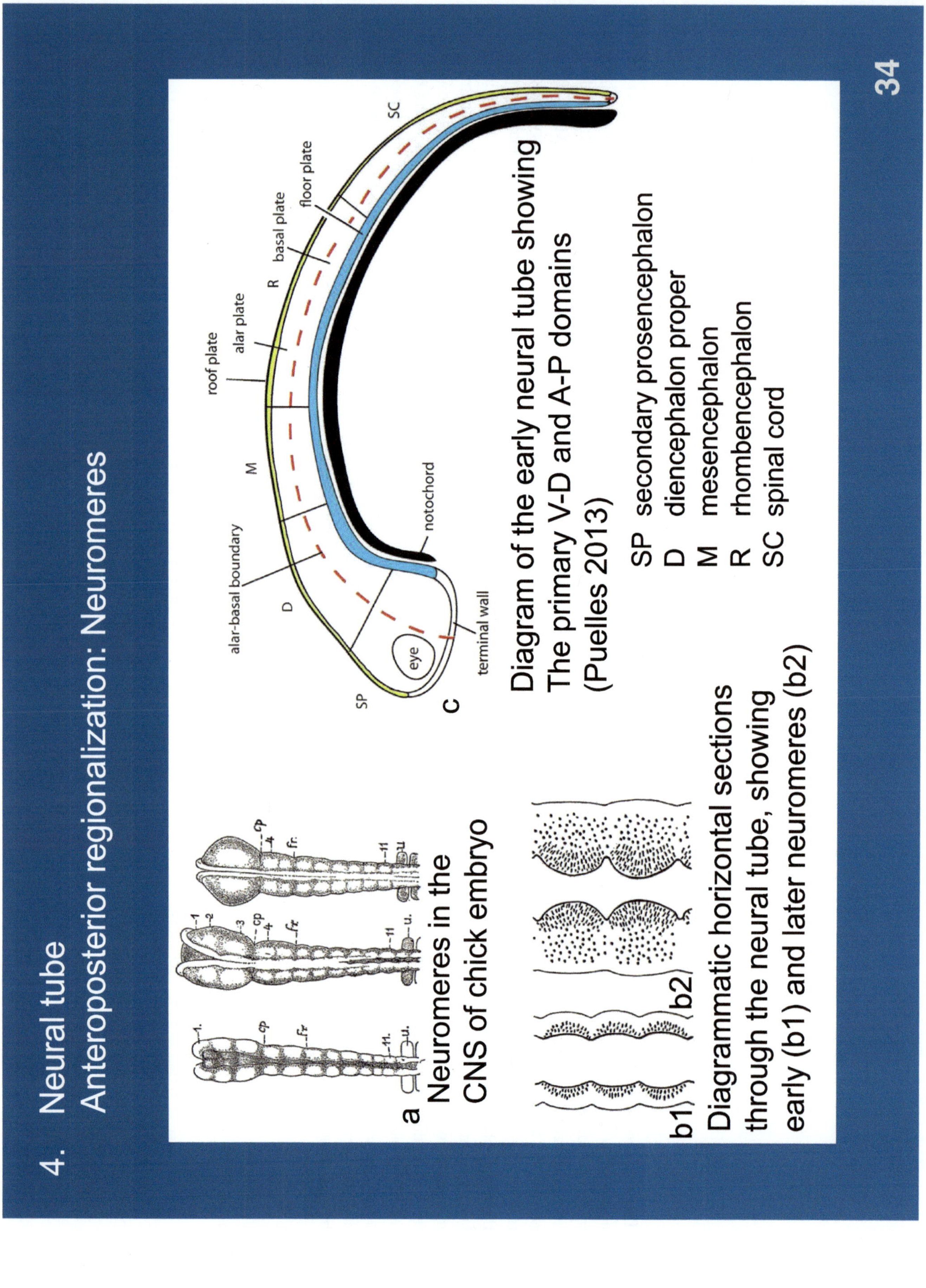
4. Neural tube
Anteroposterior regionalization: Neuromeres
a
Neuromeres in the CNS of chick embryo
b1
b2
Diagrammatic horizontal sections through the neural tube, showing early (b1) and later neuromeres (b2)
c
SP
D
M
R
SC
eye
terminal wall
notochord
alar-basal boundary
roof plate
alar plate
basal plate
floor plate
Diagram of the early neural tube showing The primary V-D and A-P domains (Puelles 2013)
SP secondary prosencephalon
D diencephalon proper
M mesencephalon
R rhombencephalon
SC spinal cord
34

4.4 Anteroposterior Regionalization: Neuromeres

Plate 34

We have seen that the neural tube can be divided into four longitudinal zones, floor plate, basal plate, alar plate and roof plate. The neural tube is, however, also divisible into anteroposterior (AP) or transverse domains. These AP domains are complete rings of neural tube wall, their diameters oriented orthogonally to the axis of the brain, irrespective of its morphogenetic bendings.

In an early neural tube stage (Plate 34, Fig. c), the following five primary (or preneuromeric) AP subdivisions can be distinguished: spinal cord (SC), rhombencephalon (R), mesencephalon (M), diencephalon proper (D) and secondary prosencephalon (SP). It is important to note that the respective boundaries of these primary AP domains are permanent and can be identified in the adult brain.

As development proceeds, a further AP regionalization will occur within these primary domains, which leads to the formation of neuromeres.

Neuromeres or neural segments may manifest themselves as externally and internally visible transverse bulging bands in the neural tube wall (Plate 34, Fig. a), due to proliferative maxima at their centres and minima at the interneuromeric boundaries (Bergquist and Kãllén 1954). They typically show transiently a differential molecular profile (combination of active versus repressed genes). Curiously, some transverse domains with dimensions similar to neuromeres are differentially specified molecularly, but do not show proliferative maxima and minima, and therefore do not bulge out (Marin et al. 2008; Puelles, 2013). These morphologically indistinct transverse entities are known as *cryptoneuromeres* [initially, they were infelicitously designated by Cambronero and Puelles (2000) as 'pseudorhombomeres']. The number of neuromeres seems to be uniformly conserved across vertebrates (Nieuwenhuys 1998a).

Plate 34, Fig. b1 shows that in early stages the external boundaries of the overt neuromeres correspond with external constrictions and intraventricular ridges. Due to local intrasegmental proliferation and differentiation processes, the overt neuromeres may manifest themselves during further development as intraventricular bulges, separated by ventricular sulci (Plate 34, Fig. b2).

Note that all of the transversely oriented domains, the primary ones as well as the neuromeres, include 'their own' sectors in the floor-, basal-, alar- and roof plates (Fig. c).

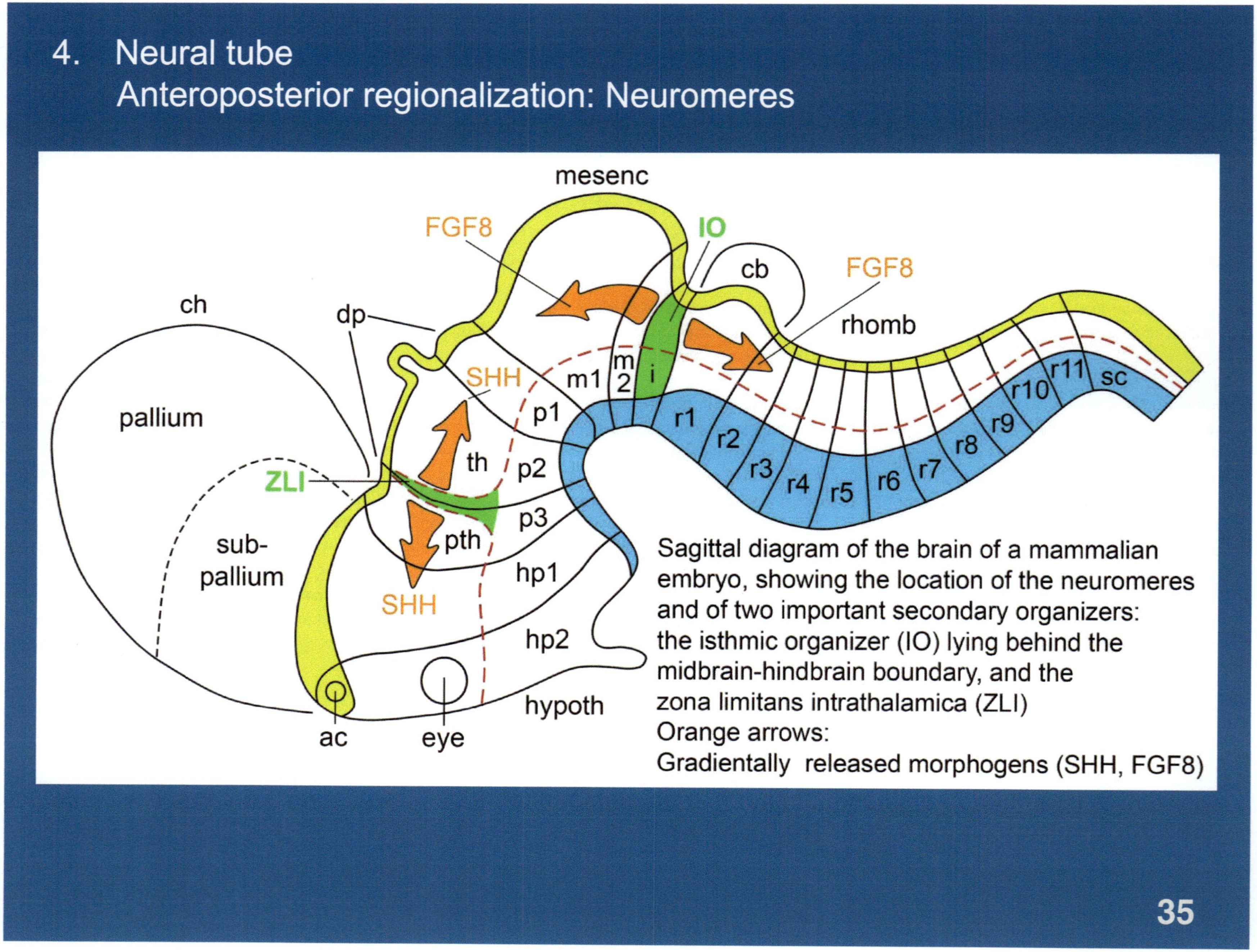
4. Neural tube
Anteroposterior regionalization: Neuromeres
mesenc
FGF8
IO
cb
FGF8
rhomb
ch
dp
pallium
SHH
m1
m 2
i
p1
th
p2
ZLI
p3
sub-pallium
pth
hp1
SHH
hp2
hypoth
ac
eye
r1
r2
r3
r4
r5
r6
r7
r8
r9
r10
r11
sc
Sagittal diagram of the brain of a mammalian embryo, showing the location of the neuromeres and of two important secondary organizers: the isthmic organizer (IO) lying behind the midbrain-hindbrain boundary, and the zona limitans intrathalamica (ZLI)
Orange arrows:
Gradientally released morphogens (SHH, FGF8)
35

Plate 35

Plate 35 shows a recently updated map of neuromeric subdivisions of the brain of a mammalian embryo (extrapolatable in principle to other vertebrates). The spinal cord encompasses 31 neuromeres (termed myelomeres) in the mouse, but their number varies considerably among vertebrates. The rhombencephalon consists of 11 rhombomeres (r1-r11). It has already been mentioned that His considered the isthmus, i.e. the narrowed sector intercalated between the rhombencephalon and mesencephalon (Plate 27, left picture: III) as a separate division of the brain. This region consists of a single isthmic neuromere (i), which is sometimes denoted as r0 (making a total of 12 rhombomeres). The mesencephalon is composed of two neuromeres (mesomeres m1, m2). The prosencephalon consists of five prosomeres, the caudal three of which (p1–p3) form the diencephalon proper (dp). The two most rostral prosomeres are denoted as hypothalamo-telencephalic prosomeres 1 and 2 (hp1, hp2). As their names indicate, these two prosomeres participate in the formation of the telencephalon as well as the hypothalamus. Three of the brain neuromeres, r1, m1 and hp1, form dorsal expansions which develop, respectively, into the cerebellum (cb), the superior colliculus (sc; mammals) or mesencephalic tectum (nonmammals), and the cerebral hemispheres (ch).

The various regional neuromeres start to become patterned within the primary neural tube domains by differential molecular signalling from several discrete secondary AP organizers, and by mutual interactions between differently specified neighbouring fields. The positions of two of these secondary AP organizers, the isthmic organizer (IO), lying just behind the midbrain-hindbrain boundary (MHB) and the zona limitans intrathalamica (ZLI), are indicated in Plate 35.

The position of the MHB is initially defined by the early gradient of Wnt in the neural plate (Plate 21) and the expression of antiWnt acting genes, such as *Six3* (Plate 23). Later on, the mutually antagonistic homeodomain transcripton factors *Otx2* and *Gbx2* play a crucial role in MHB positioning. *Gbx2* is expressed in the anterior hindbrain, and *Otx2* in the forebrain and midbrain. *Otx2* and *Gbx2* repress each other and the interface between their expression domains stabilizes the position of the MHB (see also caudal limit of *Six3* in Plate 23).

Two signalling molecules, FGF8 and Wnt1 are expressed on the posterior and anterior sides of the MHB, respectively. Mutations in the *Wnt1* and *Fgf8* genes have shown that their coded morphogens play important roles in the patterning of the midbrain and anterior hindbrain (Wnt1 acts mainly as a mitogenetic factor, conditioning protracted growth of the caudal midbrain and the rostral hindbrain; the doubleFGF8 gradient, which originated at the IO is relevant for molecular regionalization of both regions).

The ZLI (or mid-diencephalic organizer) is formed by a narrow stripe of *Shh*-expressing cells, situated between the presumptive thalamic (th) and prethalamic (pth) territories. As shown in Plate 35, its position is marked by the peculiar, dorsally directed 'spur' of the zona limitans. The ZLI acts as a local organizer of diencephalic development by secreting the SHH signalling protein. The position of the ZLI is, just like that of the MHB, determined by early Wnt signalling and singular antagonistic effects (Puelles and Martinez 2013).

4. Neural tube
A-P regionalization
expression patterns of some genes

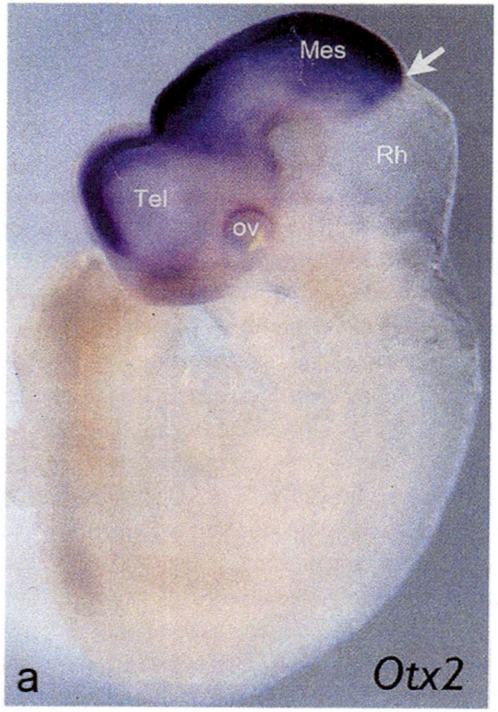

Otx2 - expression in midbrain and forebrain of a mouse embryo of 10,5 days' gestation

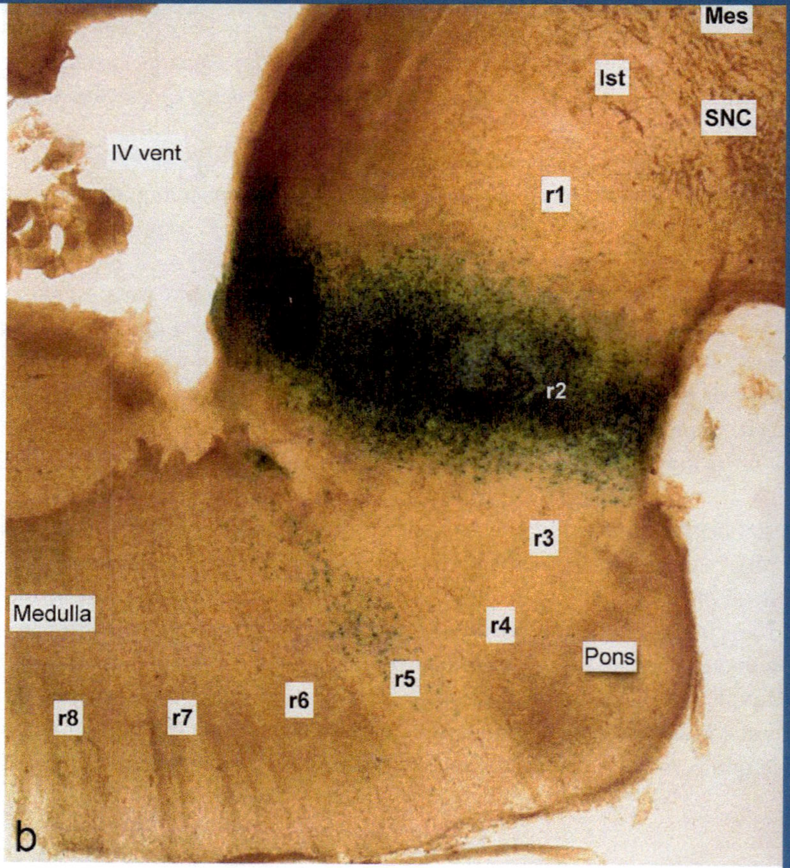

Hoxa2 - expression in rhombomere 2 of a mouse embryo 18,5 days' gestation

36

Plate 36

Plate 36 illustrates the molecular delimiting potential of some genes in the neural tube of mouse embryos.

Figure a shows a mouse embryo of 10.5 days' gestation in which the expression of *Otx2* is strictly confined to the midbrain and forebrain, with a sharp isthmic border (white arrow). This boundary persists in the adult brain, revealing that contrary to classic descriptions, the midbrain does not border directly on the pons (which forms in r3–r4; compare Plate 35).

Figure b shows the well-delimited LacZ-labelled progeny in the second rhombomere of a *Hoxa2-Lacz* transgenic E 18.5 mouse embryo.

4. Neural tube
A-P regionalization

Diagrammatic representation of the rhombencephalon of the chick.
The segmental nested expression of hox genes is indicated (left) .
The segmental origin of the branchiomotor nuclei of V, VII, IX and X and their exit points are indicated to the right.
The trigeminal (mV), facial (mVII) and glossopharyngeal cranial nerves (mIX) project into the first (b1), second (b2) and third (b3) branchial arches, respectively.

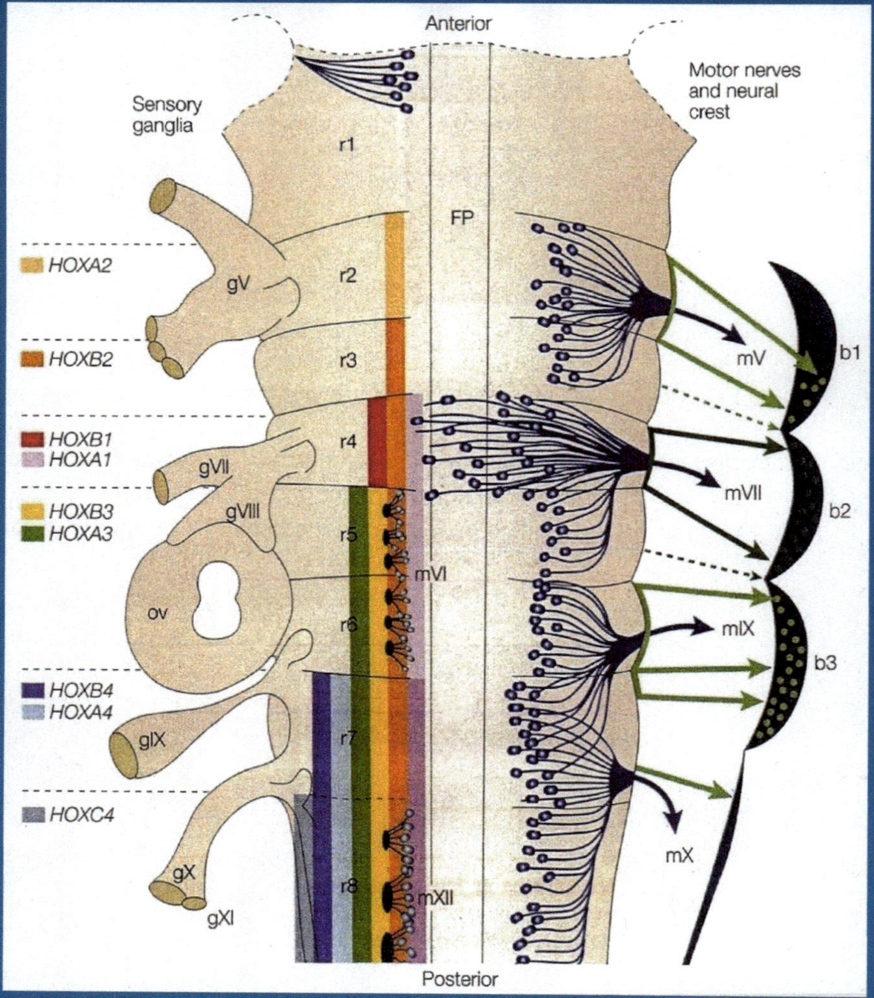

Kiecker and Lumsden (2005)

37

Plate 37

The segmentation of the hindbrain into rhombomeres is controlled in part by *Hox* genes. Most of these genes are expressed in a domain, which extends from the caudal end of the neural tube to a rostral limit in the spinal cord or rhombencephalon (different each one), but some members of this gene family, such as for instance *HOXA2* and *HOXB1*, have strictly unisegmental expression domains (Plate 37). The boundary zones between the rhombomeres act, at least during a certain developmental period, as cell lineage restrictors. This means strictly that the neuroepithelial cells from adjacent neuromeres do not intermix across those limits. As a consequence, most neurons produced within a given segment remain restricted to its mantle zone (radial migration only), though some populations migrate tangentially into other segments (see Sect. 10).

Note that the apparent alar origins of the mixed cranial nerves (sensory roots at left side; motor roots at right side) are associated to the paired even rhombomeres (r2, r4, r6, r8). This is because the primordia of neighbouring cranial sensory ganglia migrate from the neural crest along these neuromeres, and their sensory afferents induce early on the 'doors' that both motor and sensory axons use to cross the brain surface. Branchiomotor neurons situated in uneven rhombomeres have to navigate into the closest door in the adjacent segment (note these cells migrate tangentially into alar positions after sending out their axons). Somatic motor nerves produce their own doors in a more ventral position. Note also that the trochlear motor nucleus is restricted to the isthmic rhombomere, rostral to r1 and its root singularly exits dorsally through the roof (its axons are repelled by the local floor plate cells).

Molecular evidence of neuromeric origins of motor nuclei shows that they often have a double neuromeric origin (particularly the branchiomotor columns depicted on the right). Some other motor populations have plurisegmental origins across medullary cryptorhombomeres (e.g., hypoglossal, ambiguous and dorsal motor vagal nuclei). (The large rhombomere 8 in Plate 37 includes in fact the cryptorhombomeres r8–r11.)

Fate mapping studies indicate that most if not all transverse cytoarchitectonic boundaries present in the hindbrain (as well as all fibre decussations) relate to the overt or cryptic interneuromeric limits. In this sense, there is a suggestion that all long sensory columns (trigeminal, solitary, vestibular, cochlear) probably consist of as many segmental units as the respective plurisegmental topography suggests. Even the reticular formation seems to have segmental compartments.

4.5 Summary of the Neural Tube

1. The early neural tube consists histologically of a proliferative, pseudostratified, high columnar epithelium, which is molecularly patterned into longitudinal and transverse domains.
2. The ventrodorsal patterning of the lateral walls of the neural tube is induced by the floor plate (FP) and the roof plate (RP), which act as opposed signalling centres.
3. The morphogens secreted by FP and RP form concentration gradients that specify at various distances different progenitor cell fates in a dose-dependent fashion, via transcriptional activators and repressors acting on different classes of genes (Kiecker and Lumsden 2012).
4. The neuromeres formed within the primary antero-posterior divisions are patterned by differential molecular signalling from local secondary organizers (including the isthmic organizer (IO), and the zona limitans intrathalamica (ZLI), plus added mutual interactions between differentially specified local fields.
5. The brain (of all vertebrates?) is composed of 19 neuromeres: twelve rhombomeres (isthmus or r0 plus r1–r11), two mesomeres (m1, 2), three diencephalic (p1–p3), and two hypothalamic or secondary prosencephalic (hp1, 2).

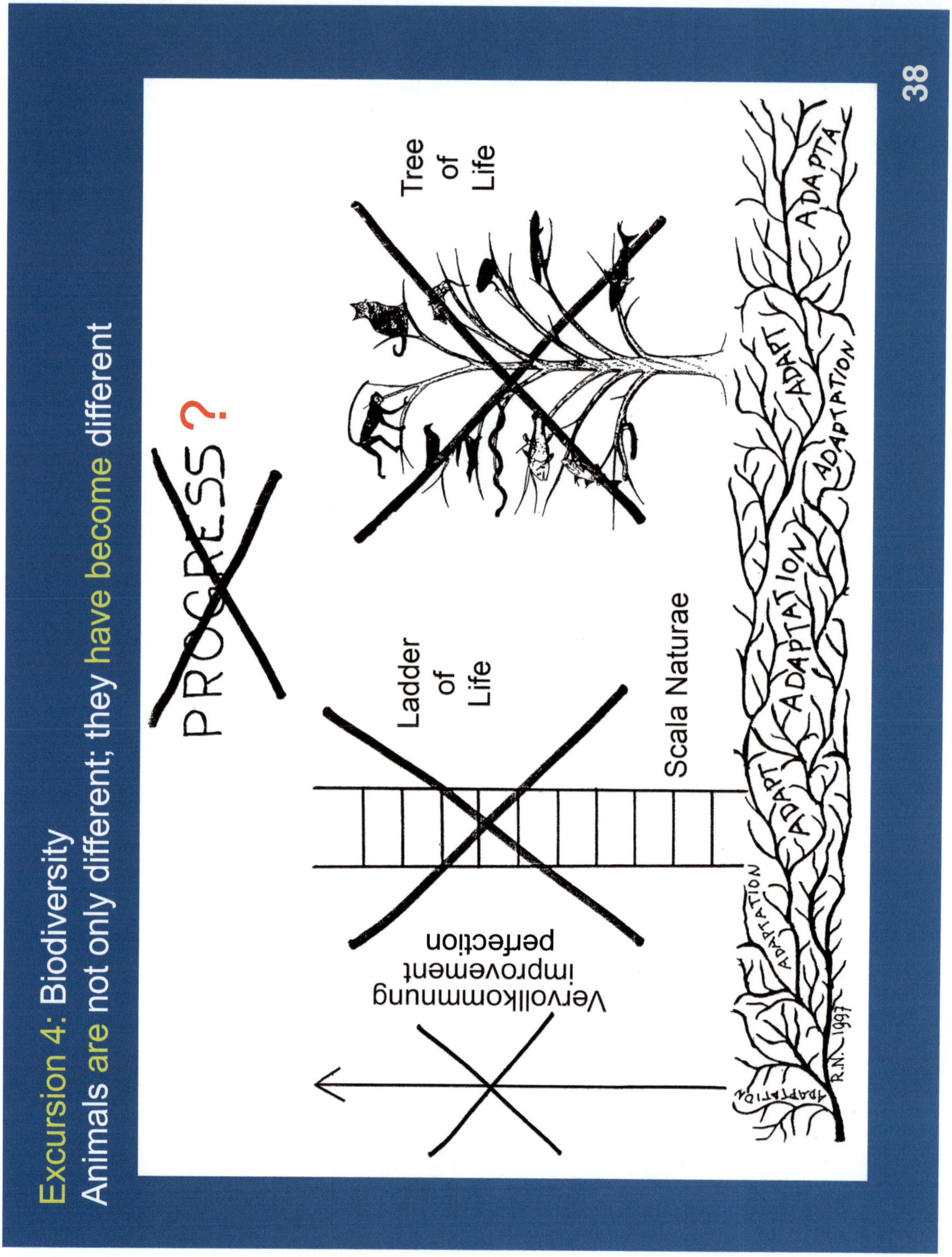
Excursion 4: Biodiversity
Animals are not only different; they have become different
38
PROGRESS ?
Vervollkommnung
improvement
perfection
Ladder
of
Life
Scala Naturae
Tree
of
Life
ADAPTATION
ADAPT
ADAPTATION
ADAPTATION
ADAPTA
R.N. 1997

Excursion 4: Biodiversity

Plate 38

When we look at the animal kingdom, the first thing that strikes the eye and the mind is its astonishing diversity. Animals are extremely diverse and so are their brains. The now generally accepted theory of evolution tells us that the members of the animal kingdom not only *are* different; they *have become* different. The multiple changes leading to diversification are to be interpreted as adaptations to new environmental situations, and the mechanism involved is the tandem dualism of genetic variability and natural selection (Mayr 1969).

Plate 38 illustrates that nowadays most evolutionists reject the idea of *evolutionary progress*. They claim that all extant species are basically equal, because they all have successfully tackled the problems of survival. So, in their biological vocabulary there is no room for terms such as 'Vervollkommnung', 'improvement', or 'perfection'. They claim that the idea of progress stems from the pre-evolutionary view that all living organisms can be arranged in a linear and hierarchical 'ladder of life' or '*scala naturae*'; moreover, evolutionary progress would imply the presence of an intrinsic driving force, apart from the external drive of selective pressure (e.g. Dawkins 1992; Maynard Smith and Szathmáry 1995). The referred tree of life, with cats and monkeys sitting on the higher branches, and cyclostomes and fish on the lower, presents, according to most evolutionists, a false picture of the history of life. In fact, there is no such thing as a tree of life, but rather just scrub.

We do not see why an increase in complexity, which manifests itself so obviously in the brains of so many taxa, could not be interpreted as 'improvement' or 'progress'. Because there is no space for a detailed argument within the frame of this course, we will restrict ourselves here to two notes.

1. According to the German zoologist Remane (1956, 1966), organs and organ systems may show two different evolutionary trends: Anpassungsreihen or trends of adaptation, and Vervollkommnungsreihen, or trends of perfection. Remane distinguished many different trends of perfection, two of which, i.e. differentiation and syn-organization, play a dominant role in the CNS. Differentiation requires little comment. It refers to the phenomenon whereby organs tend to to become more complex in structure and function during evolution. The concept of syn-organization transcends differentiation. It refers to the phenomenon whereby structures, which have developed and differentiated quite independently, may unite and become integrated into new apparatuses or functional systems. Remane suggested that during evolution of the CNS, new functional entities are formed by the increasing interlinking of groups of neurons.
2. The comparative neurophysiologist Bullock (1984, p. 473) considered that: "the brain has diversified and advanced in evolution more than any other organ", and: "This evolution was not in one line or ladder but in many radiations of descendant stocks. Nevertheless, it is a mistake to deny lower and higher forms, inasmuch as most members of later appearing classes are more advanced than earlier classes, have more distinguished parts and processes, and larger repertoires of behavior", and finally (Bullock 1993, p. 89): "Careful use of words such as 'higher', 'advanced' and 'better' is obligatory in discussions of evolution, but they need not be avoided".

Conclusion: Evolutionary progress does exist. The challenge is to develop a plausible scenario for this intriguing phenomenon, equivalent to an evolutionary exploration of potential possibiliries in morphospace.

Excursion 4: Biodiversity

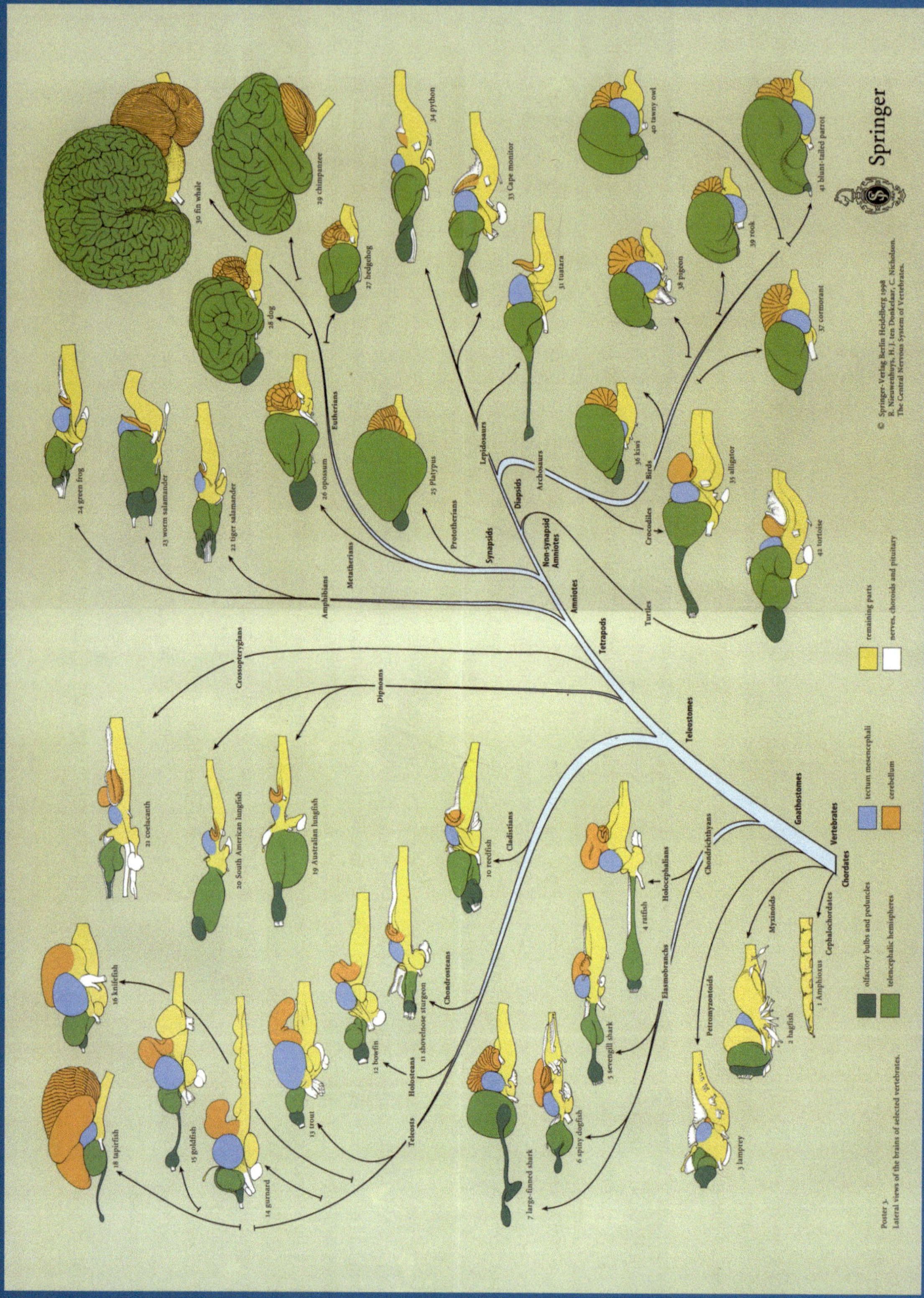

Plate 39

Plate 39 presents an overview of the evolutionary development of the brain of vertebrates. The major parts of the brains are indicated by different colours.

dark green: olfactory bulbs and peduncles
light green: telencephalic hemispheres
blue: roof of the midbrain (tectum mesencephali)
orange: cerebellum
yellow: remaining parts

It gives an impression of the astonishing diversity of the brains of vertebrates. Look, for instance, at the brain of the electrosensory tapirfish with its gigantic cerebellum (upper left), or at the brain of the fin whale, with its large and strongly folded cerebral hemispheres (upper right). We should realize, however, that the vertebrates, representing some 50,000 extant species, form only a very small segment of the animal kingdom. Margulis and Schwartz (1998) divided this kingdom into 37 phyla or stems; one of these is formed by the vertebrates or craniates; the remaining 36 phyla, encompassing together some 2,000,000 known extant species, are commonly denoted together as ‘the invertebrates’.

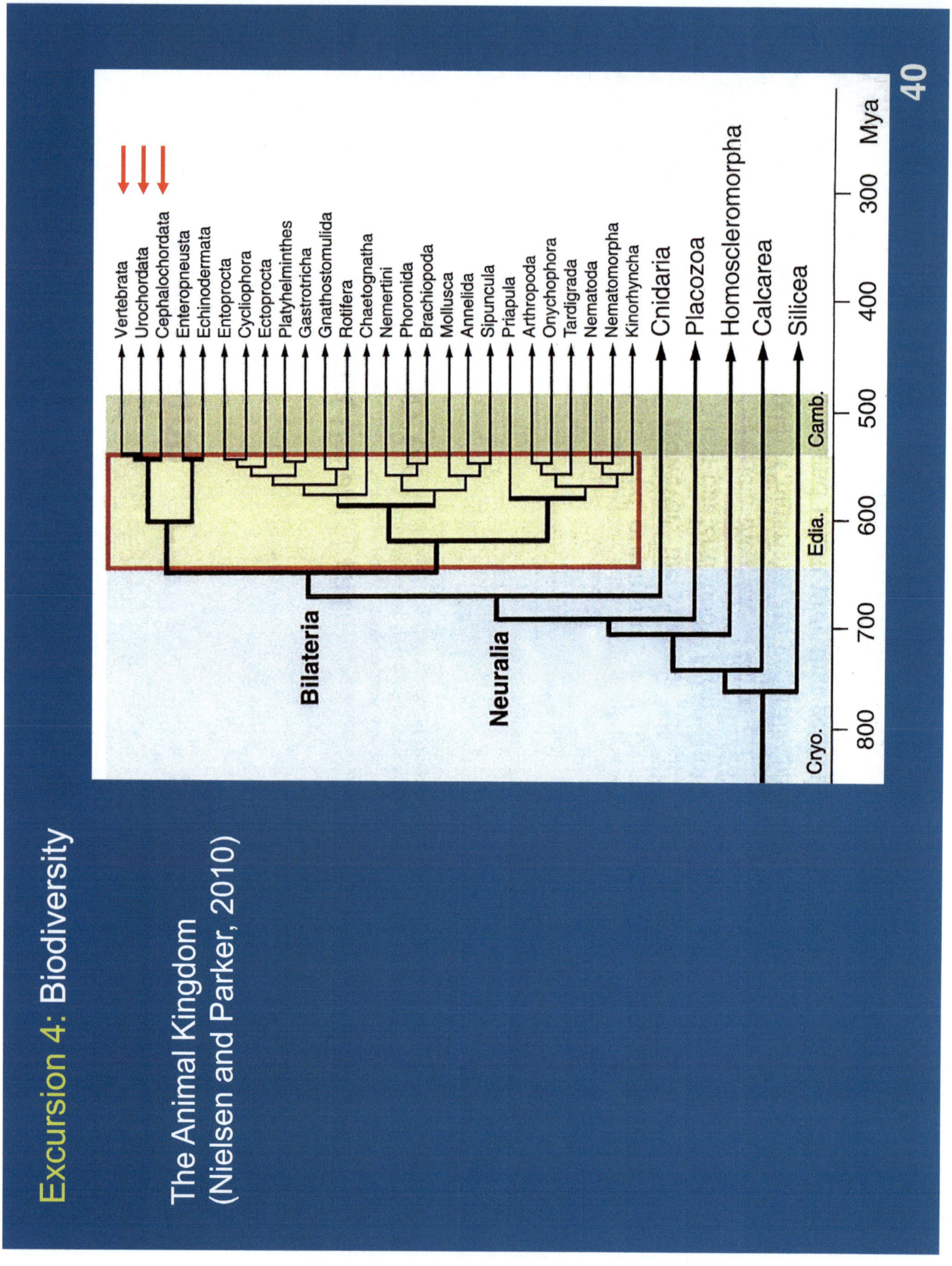
Excursion 4: Biodiversity
The Animal Kingdom
(Nielsen and Parker, 2010)
Vertebrata
Urochordata
Cephalochordata
Enteropneusta
Echinodermata
Entoprocta
Cycliophora
Ectoprocta
Platyhelminthes
Gastrotricha
Gnathostomulida
Rotifera
Chaetognatha
Nemertini
Phoronida
Brachiopoda
Mollusca
Annelida
Sipuncula
Priapula
Arthropoda
Onychophora
Tardigrada
Nematoda
Nematomorpha
Kinorhyncha
Cnidaria
Placozoa
Homoscleromorpha
Calcarea
Silicea
Bilateria
Neuralia
Cryo.
Edia.
Camb.
800
700
600
500
400
300
Mya
40

Plate 40

Plate 40, derived from Nielsen and Parker (2010), reflects our current views on the phylogenetic relationships of the various animal phyla. It is based on morphological as well as molecular evidence. Although the number of phyla is somewhat smaller than that suggested by Margulis and Schwarz (30 vs. 37), it emphasizes once again the quantitatively modest position of the vertebrates in the vast animal kingdom. The plate is of particular interest to us, because it shows an entirely newly created taxonomic category, the *Neuralia*, i.e. the animals endowed with a *nervous system*. It will be seen that this new clade encompasses, in addition to the cnidarians or coelenterates, all bilaterian animals. It would be a serious mistake to neglect all of these wonderful creatures, and confine ourselves entirely to 'our own' phylum. But what to do? Our space is limited! Earlier in this course, we paid already some attention to the CNS of amphioxus (Excursion 3, plates 28–30). This was done because the phylum Cephalochordata, to which amphioxus belongs, was until recently generally considered to represent the closest living relatives of the vertebrates. However, our evolutionary history has been thoroughly revised in the meantime, as indicated by the red arrows in Plate 40. Analysis of new genomic data has indicated that the tunicates (or urochordates) *and not* the cephalochordates are the closest living relatives of the vertebrates (Delsuc et al. 2006). Let us take another excursion and pay some attention to at least this group of evertebrates, and therewith to our 'new' relatives. To this end, we will have to travel to Helgoland, quite a journey from Leipzig but, as we shall see, very rewarding indeed.

Excursion 5: Helgoland: Evertebrates
41
AWI
Alfred-Wegener-Institut für
Polar- und Meeresforschung
in der Helmholtz-Gemeinschaft
Institute for Polar and Marine Research

Excursion 5: Helgoland: Evertebrates

Plate 41

The small island of Helgoland represents 'die einzige Hochseeinsel Deutschlands' (the only German deep-sea island). It is situated in the North Sea at a distance of some 40 miles from the German coast. The connections with the continent are excellent, with no less than four different daily ferries. The peculiar mode of debarking ('Ausboten'), shown in the—early twentieth century—upper left-hand picture, is still in practice albeit in a slightly different mode. Helgoland is famous for its fresh air, its rich bird life, and the Alfred-Wegener-Institute for Polar and Marine Research. This Institute has excellent facilities for the study of tunicates (and of marine invertebrates in general).

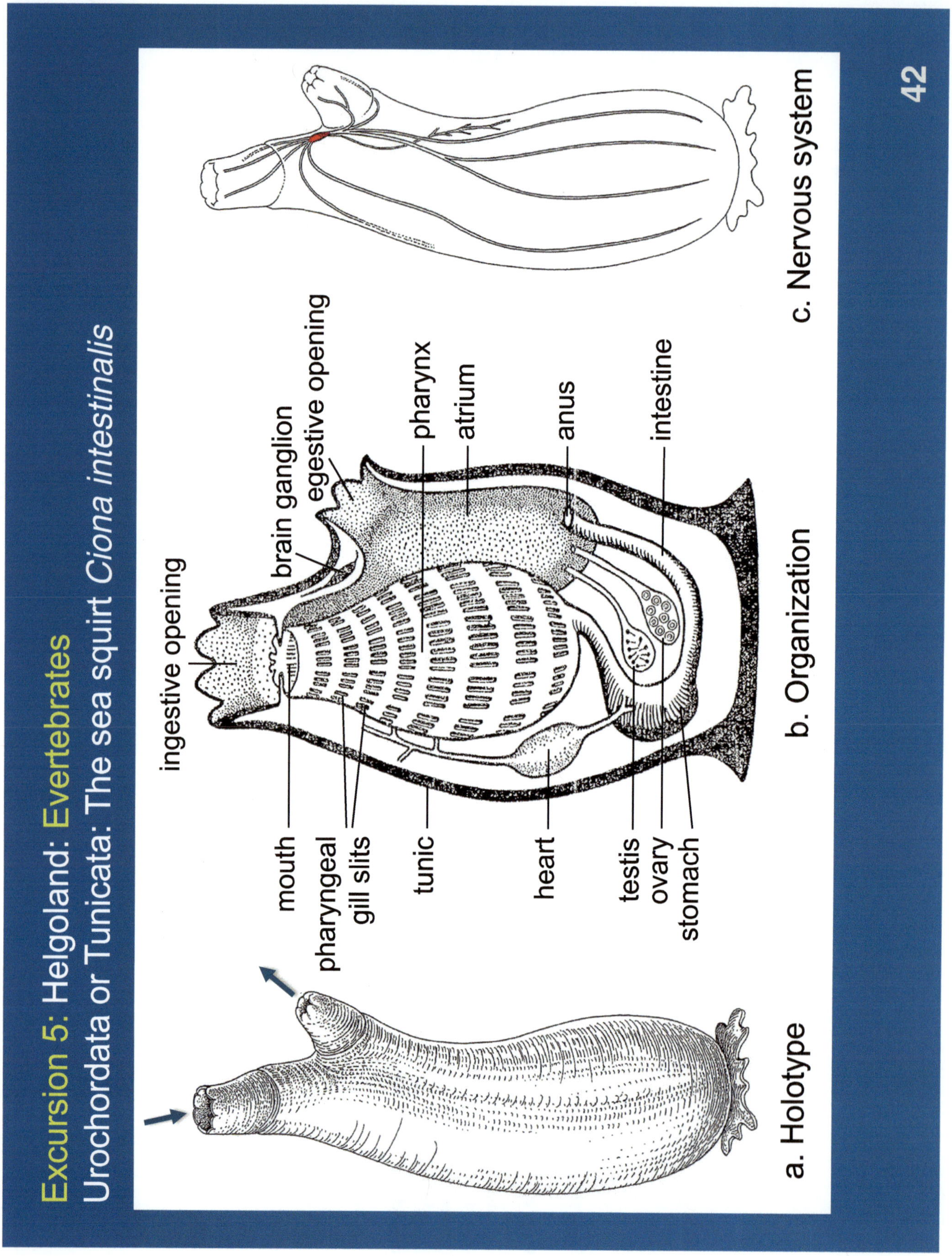
Excursion 5: Helgoland: Evertebrates
Urochordata or Tunicata: The sea squirt Ciona intestinalis
a. Holotype
ingestive opening
brain ganglion
egestive opening
pharynx
atrium
anus
intestine
mouth
pharyngeal
gill slits
tunic
heart
testis
ovary
stomach
b. Organization
c. Nervous system
42

Plate 42

Tunicates are soft-bodied marine animals, encased in a gelatinous or leathery tunic; this tunic, which gives the phylum its name, is a secretion product of the epidermis. Within the tunic is the body that contains a large perforated pharynx or branchial sac. This organ filters phytoplankton from the incoming sea water and exchanges oxygen and carbon dioxide with the sea water.

The phylum Tunicata is divided into three classes, the Ascidia, the Appendicularia or Larvacea and the Thaliacea. In what follows, we will confine ourselves to the ascidians or sea squirts, which are by far the most familiar and best-studied tunicates.

Adult ascidians, such as *Ciona intestinalis* (Plate 42, Fig. a), are bag-like sessile animals attached to rocks or other underwater hard surfaces. They have two funnel-like processes, the incurrent siphon and the excurrent siphon (arrows in Fig. a). The large pharynx is lined with cilia, which, beating inward, produce a steady current of water that enters the mouth via the incurrent siphon and passes out through the numerous pharyngeal slits (Plate 42, Fig. b). Food particles are trapped by the mucus in the pharynx and transported toward the opening of the digestive tract. The filtered water enters the atrium, a large chamber, which surrounds the pharynx, and flows outward via the excurrent siphon. The atrium also serves as an exit for faeces, because the alimentary canal discharges into it. So, the ascidians are typical filter feeders, just like the cephalochordates and many other invertebrate groups.

The CNS of ascidians is represented by a small ganglion (shown in red in Plate 42, Fig. c), which is strategically situated between the two siphons. Some nerves, all with mixed sensory and motor components, radiate out from the ganglion. Histologically, the ascidian cerebral ganglion is a typical invertebrate ganglion. It is a solid structure with a central core of fibres and neuropil and an outer shell of nerve cell bodies.

It should be stated that these creatures show very little resemblance to vertebrates. However, their kinship to typical vertebrates becomes evident if their developmental history is taken into consideration.

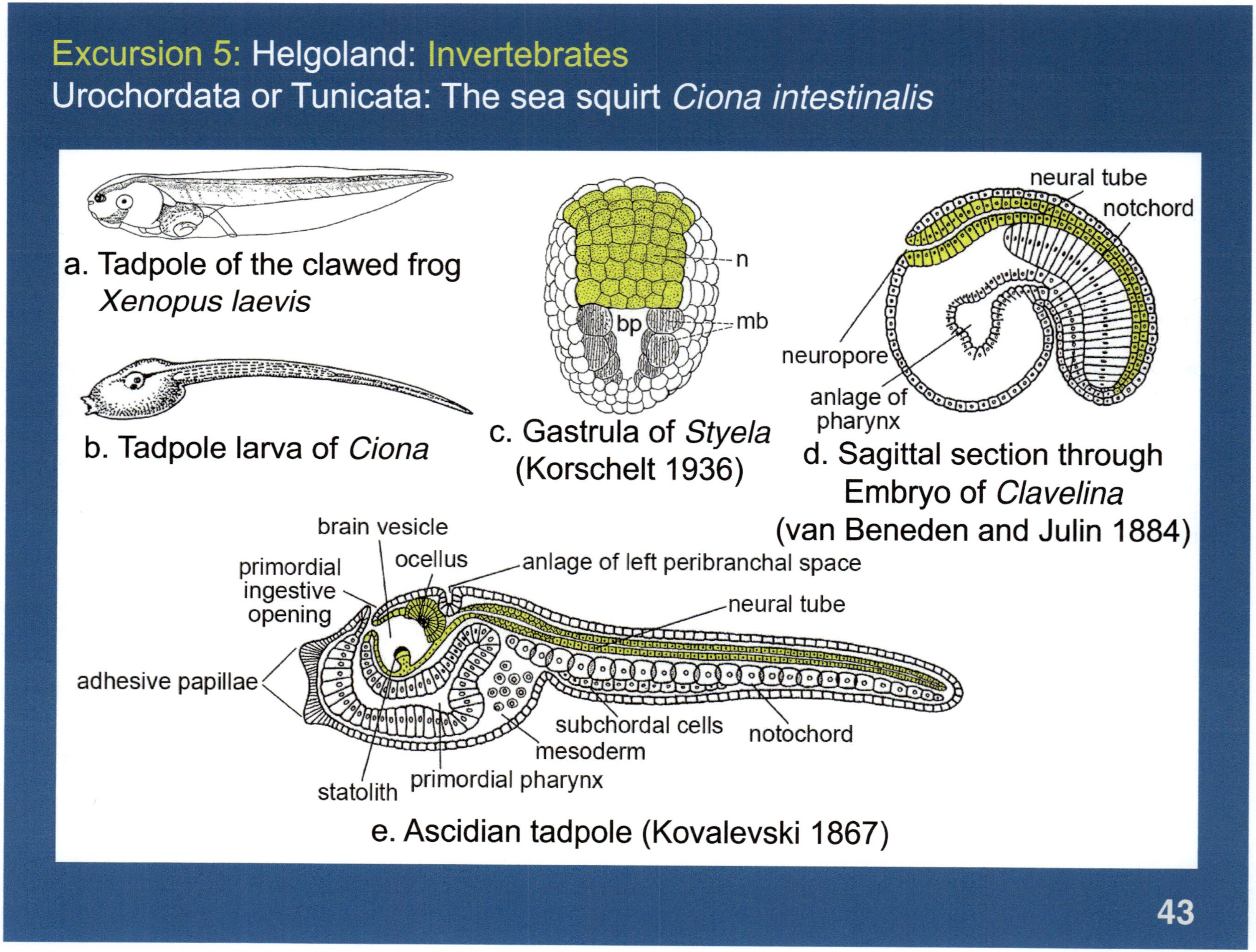
Excursion 5: Helgoland: Invertebrates
Urochordata or Tunicata: The sea squirt Ciona intestinalis
a. Tadpole of the clawed frog Xenopus laevis
b. Tadpole larva of Ciona
n
bp
mb
c. Gastrula of Styela (Korschelt 1936)
neural tube
notchord
neuropore
anlage of pharynx
d. Sagittal section through Embryo of Clavelina (van Beneden and Julin 1884)
brain vesicle
ocellus
primordial ingestive opening
anlage of left peribranchal space
neural tube
adhesive papillae
subchordal cells
notochord
mesoderm
statolith
primordial pharynx
e. Ascidian tadpole (Kovalevski 1867)
43

Plate 43

Ascidians show, just as most other marine invertebrates, an indirect development. They pass through a larval stage that differs in almost every respect from the adult. These larvae are free swimming. They look like tiny amphibian tadpoles (Plate 43, Fig. a, b), and, most remarkably, they have an axial notochord flanked by muscular tissue, and a dorsally situated CNS (Fig. e). As documented by Fig. c, this CNS develops from a typical neural plate (n), situated in front of a blastopore (bp). During further development, the neural plate is transformed into a hollow neural tube (Fig. d). This mode of development of the CNS of ascidians closely resembles that in vertebrates and differs radically from that in non-chordate invertebrates. In all non-chordates, the nervous system develops from multiple zones or patches of proliferative ectoderm. The presumptive nerve cells detach from these anlagen and migrate inwardly to participate in the formation of ganglia. Hence, all invertebrate ganglia and ganglionic complexes are solid structures that cannot be equated with a neural tube (Korschelt, 1936); in that sense, non-chordates have a ganglionic nervous system (GNS) and no central nervous system (CNS), whereas chordates have both CNS and GNS.

The CNS of larval ascidians consists of an expanded, thin-walled rostral part, a spindle-shaped intermediate part, and a thin, elongated caudal part (Kovalevsky 1866; Meinertzhagen and Okamura (2001; Fig. e). The rostral part is commonly designated as 'sensory vesicle', because parts of its otherwise mainly ependymal wall are differentiated into three sense organs: a ventral otolith or statolith formed by a single cell, (2) a group of fusiform cells in the left ventrocaudal wall of the ventricle, which is variously interpreted as a pressure or motion detector, and (3) a large left-sided photoreceptor organ. This so-called ocellus has separate lens-, pigment-, and receptor cells.

The intermediate part, known as the visceral ganglion, contains a cluster of large neurons. A short transitional zone between the visceral ganglion and the sensory vesicle is designated as the neck region.

The caudal part forms a nerve cord, which is situated directly dorsal to the notochord throughout its extent. The nerve cord does not contain neuronal cell bodies (but see below); it consists only of ciliated ependymoglial cells. In cross sections generally four of such cells are visible.

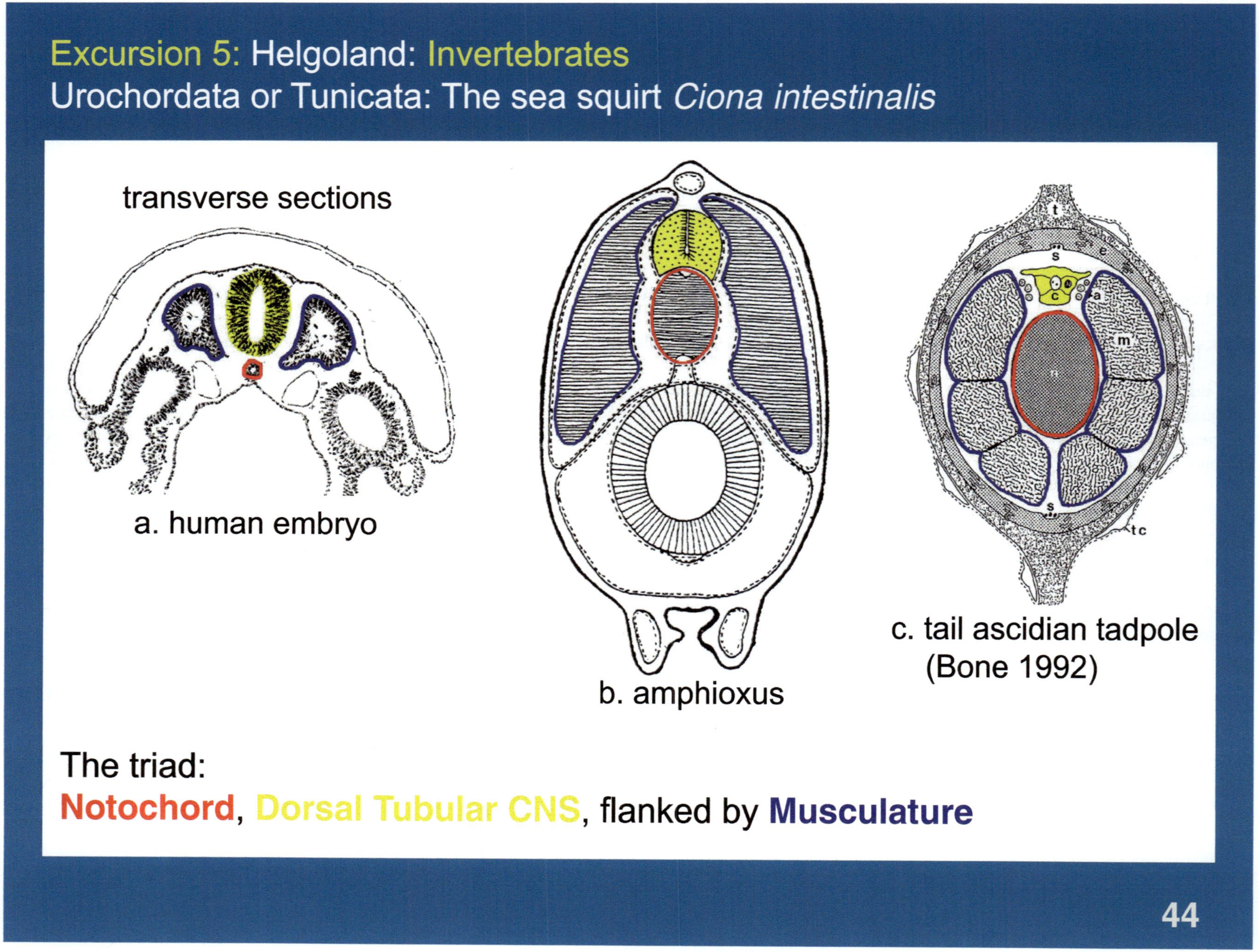
Excursion 5: Helgoland: Invertebrates
Urochordata or Tunicata: The sea squirt Ciona intestinalis
transverse sections
a. human embryo
b. amphioxus
c. tail ascidian tadpole
(Bone 1992)
The triad:
Notochord, Dorsal Tubular CNS, flanked by Musculature
44

Plate 44

This Plate shows transverse sections through the trunk region of a human embryo (**a**) and an adult amphioxus (b), as well as a similar section through the tail of an ascidian tadpole (c). It serves to illustrate the characteristic chordate structural triad, including a central notochord, a dorsally situated tubular CNS, and a bilateral musculature, flanking the other two structures.

Excursion 5: Helgoland: Invertebrates

Urochordata or Tunicata: The sea squirt *Ciona intestinalis*

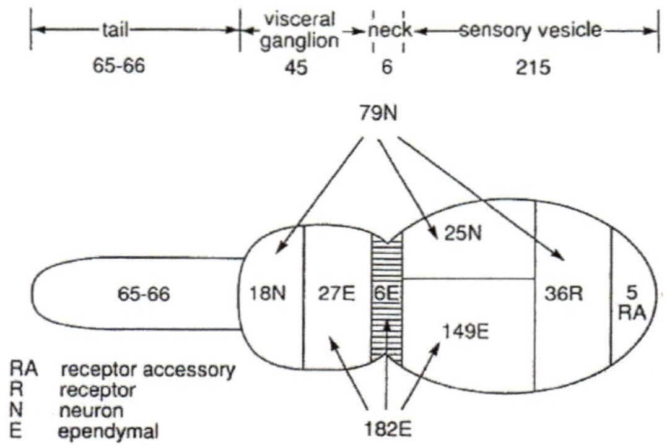

1. Cell counts in larval CNS of *Ciona* (Nicol and Meinertzhagen 1991)

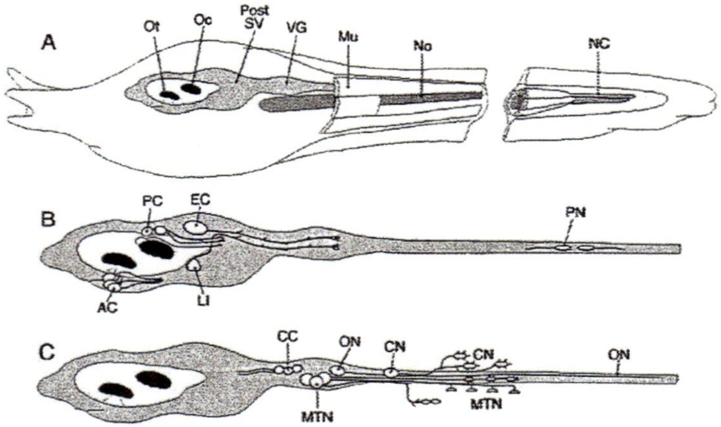

2. Identified neurons in larval CNS of *Ciona* (Imai and Meinertzhagen 2007a)

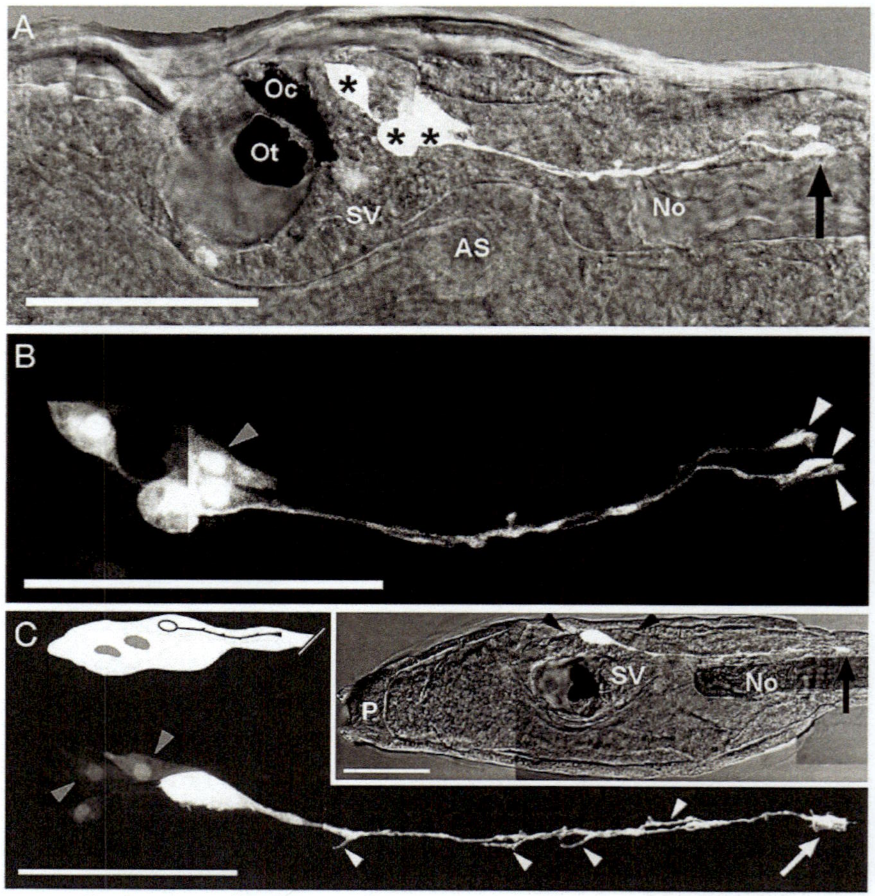

3. Lateral view of neurons in the sensory vesicle of a *Ciona* tadpole (Imai and Meinertzhagen 2007a)

45

Plate 45

Plate 45 shows the results of cell counts in the larval CNS of the ascidian *Ciona intestinalis* and some features of its neuronal organization.

A detailed study of Nicol and Meinertzhagen (1991) has shown that the total number of cells in the larval CNS of *Ciona* is about 330. The distribution of these cells over the various parts of the CNS is shown in the upper part of Fig. 1. The lower part of the same figure specifies four different cell types and indicates how many of the cells of each of these types occur in the various regions. The cell types distinguished include: receptor accessory cells (RA), receptor cells (R), neurons (N), and ependymal cells (E). Most of the elements (182, about 70 %) are classified as ependymal cells. Five cells are accessory cells of the sensory structures (RA). The remaining 79 cells are interpreted as neurons; 36 of these, which are directly associated with one of the sense organs, are designated as receptor cells (R). The visceral ganglion contains 18 neurons, and the remaining 25 neuronal elements form two clusters in the posterior region of the sensory vesicle. The nerve cord in the tail consists of 65–66 cells.

Plate 45, Figs. 2 and 3 are derived from Imai and Meinertzhagen's (2007a) study of neurons in the larval CNS of *Ciona*.

Figure 2A presents an outline of a larva, showing the location of the CNS and its major compartments, posterior sensory vesicle (PostSV), visceral ganglion (VG), and nerve cord (NC), relative to the muscle bands (Mu) and notochord (No). The cavity of the sensory vesicle is marked by the pigment spots of the ocellus (Oc) and otolith (Ot).

In Fig. 2B and C, the results of the reconstruction of a number of individual neurons are diagrammatically represented. They show, among other things, that nerve cells associated with the otolith (AC) and the ocellus (Pc) pass their axons caudally, and that the branched axon of a large, so-called eminens cell (EC), reaches the caudal part of the visceral ganglion. Neurons in the visceral ganglion include motor neurons (MTN), which directly innervate the tail musculature via typical end plates; there are other apparently motor neurons (CN, ON), the axons of which also pass caudally, and small cells (CC) with ascending axons. The axons which descend from the visceral ganglion to the tail region do not pass caudally within the nerve cord, but rather travel in bilateral nerve bundles situated directly lateral to (outside) the cord (see Plate 45, Fig. C). It is of note that Imai and Meinertzhagen (2007a) discovered for the first time the somata of some neurons in the nerve cord. They designated these elements, because of their flattened appearance, as planate neurons (PN).

Plate 45, Fig. 3 presents a detailed picture of some neurons (asterisks) in the dorsoposterior part of the sensory vesicle. Figures 3A and B reveal the projection of three neurons terminating in the caudal visceral ganglion (black arrow and white arrowheads). Figure 3C shows a single prominent eminens neuron in the sensory vesicle. Short filopodia-like processes protrude from the axon throughout its length (white arrowheads) and from the terminal (white arrow). The **inset** shows that the axon projects backward to the base of the tail (black arrow).

We considered it important to inform you about the present state of knowledge concerning the structure of the CNS of the ascidian tadpole larva. It may be concluded that, although the investigations of Imai and Meinertzhagen have revealed numerous highly relevant details, we are still far removed from a complete picture of the wiring of this highly interesting larval brain.

It should be added that Imai and Meinertzhagen (2007b) have also studied the peripheral nervous system of *Ciona* larvae. This study revealed, among many other interesting things, that the apical adhesive papillae (see Plate 43, Fig. e) contain numerous putatively chemosensory neurons, the axons of which pass posteriorly to enter the caudal part of the sensory vesicle.

We have seen that the brain of the ascidian tadpole has three parts: an anterior sensory vesicle, a short intermediate neck region, and a caudal visceral ganglion (Plate 45, Fig. 1). Gene expression studies have led to the suggestion that these three brain parts correspond to the forebrain + midbrain, isthmus region and hindbrain, respectively, of the vertebrate brain (Imai et al. 2009). Thus, an ascidian *Otx* homologue comes to expression in the sensory vesicle, expression of *Pax 2/5/8* and *Fgfrk 8/17/18* has been found in the neck region, whereas the visceral ganglion harbours the expression domains of several *Hox* genes (Imai et al. 2009). However, a detailed analysis of all known data concerning gene expressions in and around the neck region of ascidian larvae, has brought Holland (2009) to the conclusion that it is premature to equate this region with the vertebrate isthmic organizer (IO; see Plate 35), or, in other words, that the expected organizer properties are not fully evident, irrespective that a midbrain-hindbrain boundary is present.

The visceral ganglion (actually a misnomer because the known functions of its neurons are somatic rather than visceral in nature) steer and coordinate the locomotory movements of the tail musculature and therewith the motility of the larva. It is generally agreed that this motility serves only a single purpose, viz. the dispersal of the progeny of the sessile adults. The sensory vesicle plays a crucial role in the selection of a suitable place to settle.

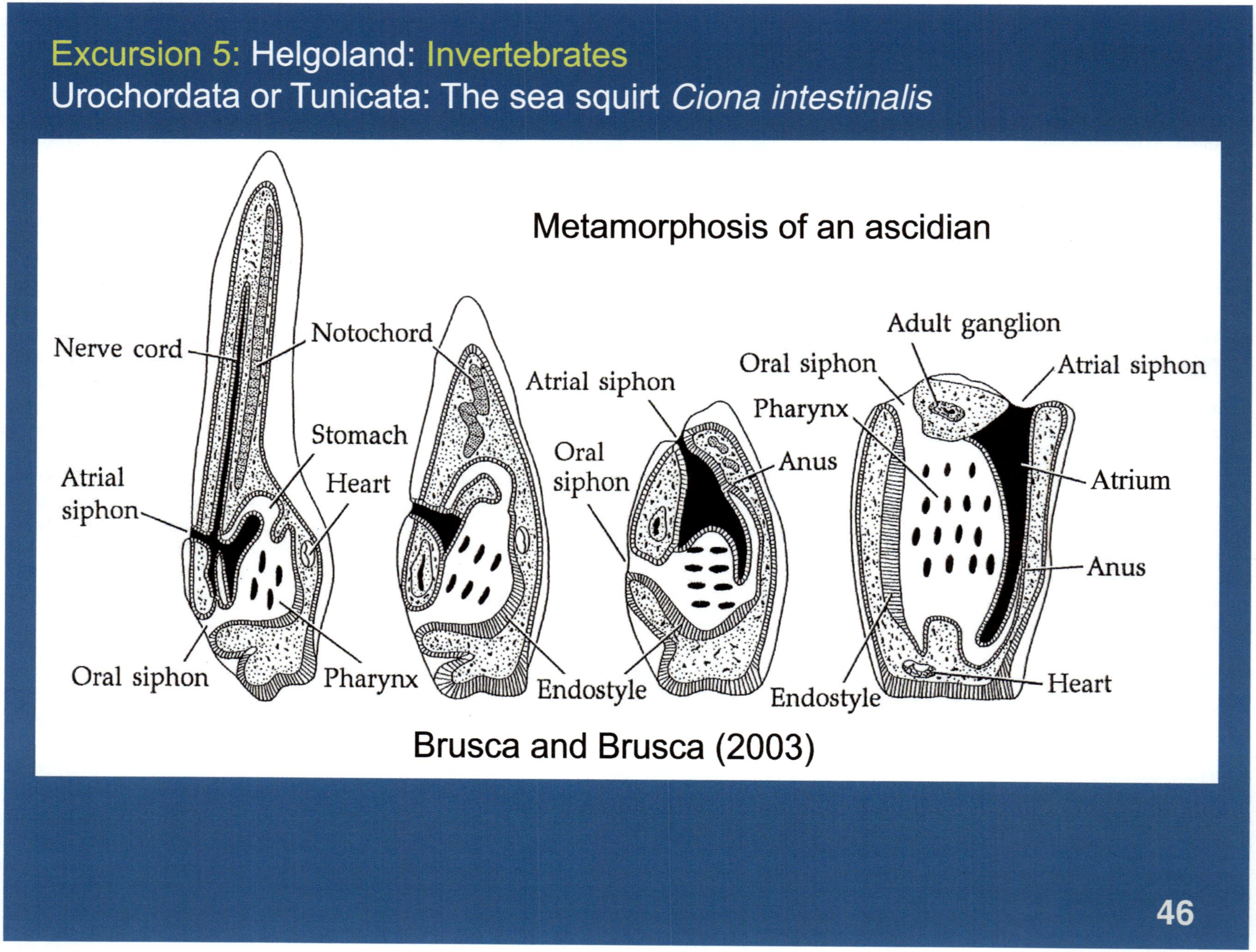
Excursion 5: Helgoland: Invertebrates
Urochordata or Tunicata: The sea squirt Ciona intestinalis
Metamorphosis of an ascidian
Nerve cord
Notochord
Stomach
Atrial siphon
Heart
Oral siphon
Pharynx
Atrial siphon
Oral siphon
Endostyle
Adult ganglion
Oral siphon
Pharynx
Anus
Endostyle
Atrial siphon
Atrium
Anus
Heart
Brusca and Brusca (2003)
46

Plate 46

The life of an ascidian as a free-swimming larva lasts only one to two days. Once a suitable 'landing place' is selected, the settling larva contacts the substratum with its anterior end, and attaches itself by its adhesive papillae. The metamorphosis from larva to adult, which begins immediately after attachment, is accompanied by dramatic changes in the body plan (see Plate 46). Tail, notochord, and nerve cord are resorbed; the viscera and the primordial siphons undergo a remarkable rotation that brings these organs to their adult position; the atrium and pharynx expand, and the filter mechanism becomes functional. The sense organs and most larval neurons disappear. The neurons, which form the adult nervous ganglion, are largely supplied from larval ependymal cells of the dwindling sensory vesicle (Horie et al. 2011).

It has already been mentioned that the life cycle of ascidians is by no means exceptional. Many marine invertebrates show an indirect development and there can be no doubt that most of the larvae produced by all of these creatures are swallowed by predators before attaining metamorphosis. The very early metamorphosis of ascidians and the consequent acquisition of a protective leathery jacket presumably evolved as an emergent predator avoidance mechanism (Lacalli 2013: personal communication).

The remarkable fact that tunicate larvae share a number of key characters with cephalochordates and vertebrates, viz. a dorsal tubular CNS, and a central notochord, flanked by muscles (Plate 44), characters which are entirely lacking in adult tunicates, have led to the theory that the vertebrates have evolved by paedomorphosis (retention of an early developmental form into adult life) from a tunicate tadpole-like organism (Garstang 1928; Berrill 1955; Romer 1972, for short: GBR). It should be emphasized that, even if this theory is correct, the last common ancestor of tunicates and vertebrates must have differed considerably from present-day ascidian tadpoles. It may be remembered that in these larvae, neurons situated in the brain innervate the tail-musculature directly, without intervention of spinal motor neurons, and that the axons of these neurons do not descend within, but rather laterally to the nerve cord; relations which all deviate markedly from those seen in vertebrate spinal premotor pathways (Bone 1992). The observed neuronal pattern rather recalls hindbrain somato- and visceromotor neurons with axons exiting via cranial nerves; note that we commented above that the molecular profile of the visceral ganglion is comparable to the vertebrate hindbrain. Nevertheless, the popularized version of the GBR theory runs smoothly:

"Once upon a time, long, long ago, there were some brave ascidian tadpoles, who refused to attach themselves to a rock, to throw away their brains, and to become dull and dumb filter-feeders; they remained free swimming creatures who defied their hostile and dangerous environment, eventually developed limbs, went on land, became brainier and brainier, began to put on airs, and:...... here WE are!!!".

Reality is more complex, as the comparative anatomical and embryological considerations of Lacalli (2005), and the molecular data assembled by Holland (2007, 2009, 2014) have conclusively shown.

Everything in biology becomes more complex on closer scrutiny.

Ladies and gentlemen, it must be admitted that our excursion to Helgoland and to tunicate biology, has got completely out of hand!! Back to the continent, back to Leipzig, and back to the CNS of vertebrates!!

5. The axis of the brain; its flexures and correlates

- The flexures of the brain axis change during development

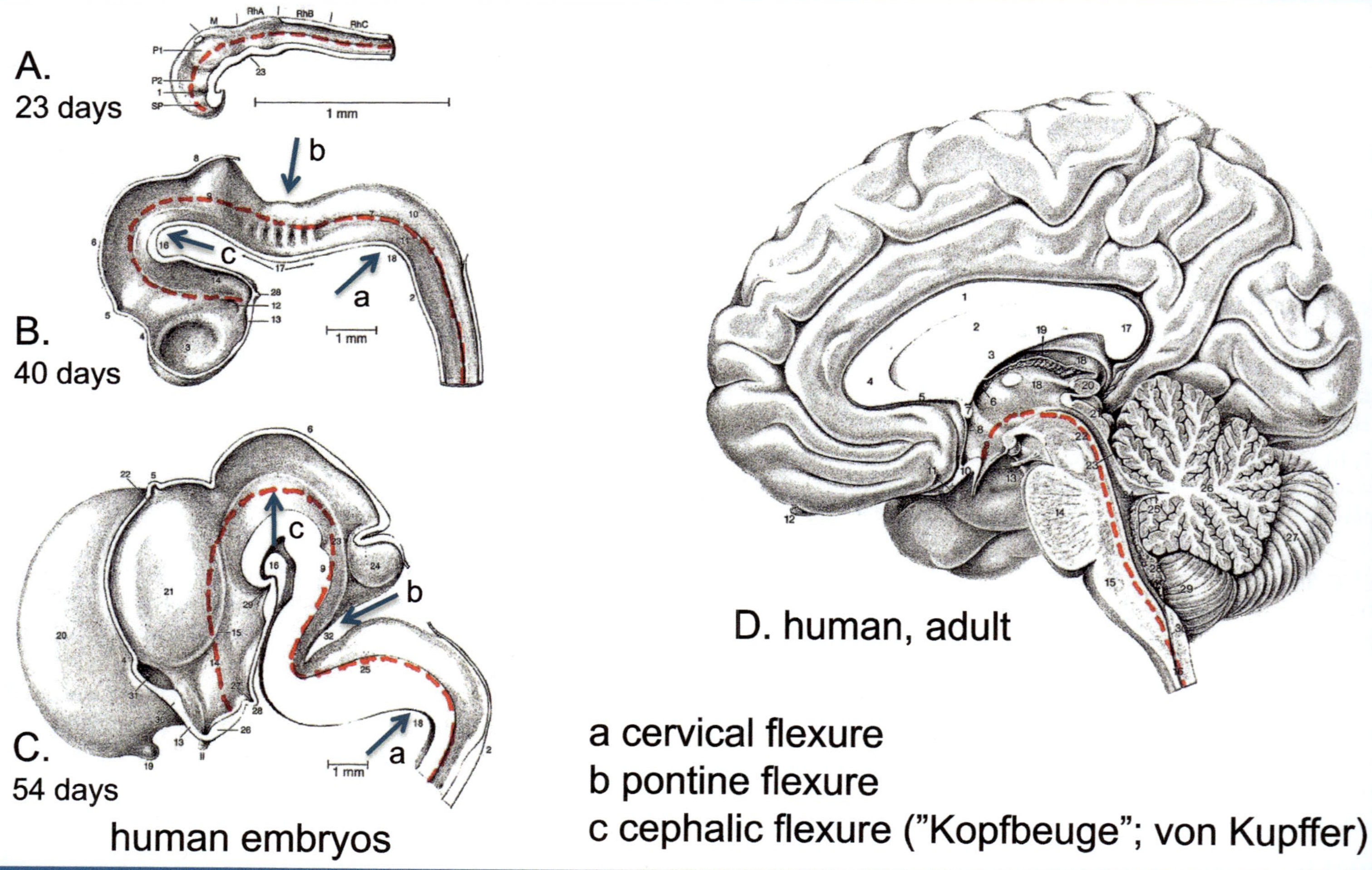

5 The Axis of the Brain, Its Flexures and Correlates

5.1 Flexures of the Axis: Changes During Development

Plate 47

It has already been discussed that the brain is essentially a tube to which a long axis can be attributed (Plate 13). In amphioxus, the neural tube and its axis are perfectly straight, even in the adult stage (Plate 28), but in all vertebrates the corresponding structures are more or less curved in the median plane.

This bending of the brain starts very early in development. In general, three different flexures, the cephalic flexure, the cervical flexure, and the pontine flexure, can be distinguished.

The ventrally concave *cephalic flexure*, or 'Kopfbeuge of von Kupffer', becomes manifest even before the closure of the neural tube and is of high morphological significance (Plate 47, Fig. A). The less marked *cervical flexure*, which, just as the cephalic flexure, is concave ventrally, appears roughly at the juncture of the rhombencephalon and the spinal cord (Figs, B, C). The *pontine flexure*, which differs from the other two in that it is ventrally convex, manifests itself in the middle of the hindbrain (Figs. B, C). It is highly variable across vertebrates. In most mammals it eventually attains such a depth that the morphologically dorsal sides of the rostral and caudal hindbrain parts approach each other (Fig. C). During later development, the cervical and pontine flexures gradually decrease in depth, but the cephalic flexure remains manifest throughout development and can be clearly observed in the adult brain (Fig. D).

R. Nieuwenhuys and L. Puelles, *Towards a New Neuromorphology*,
DOI 10.1007/978-3-319-25693-1_5

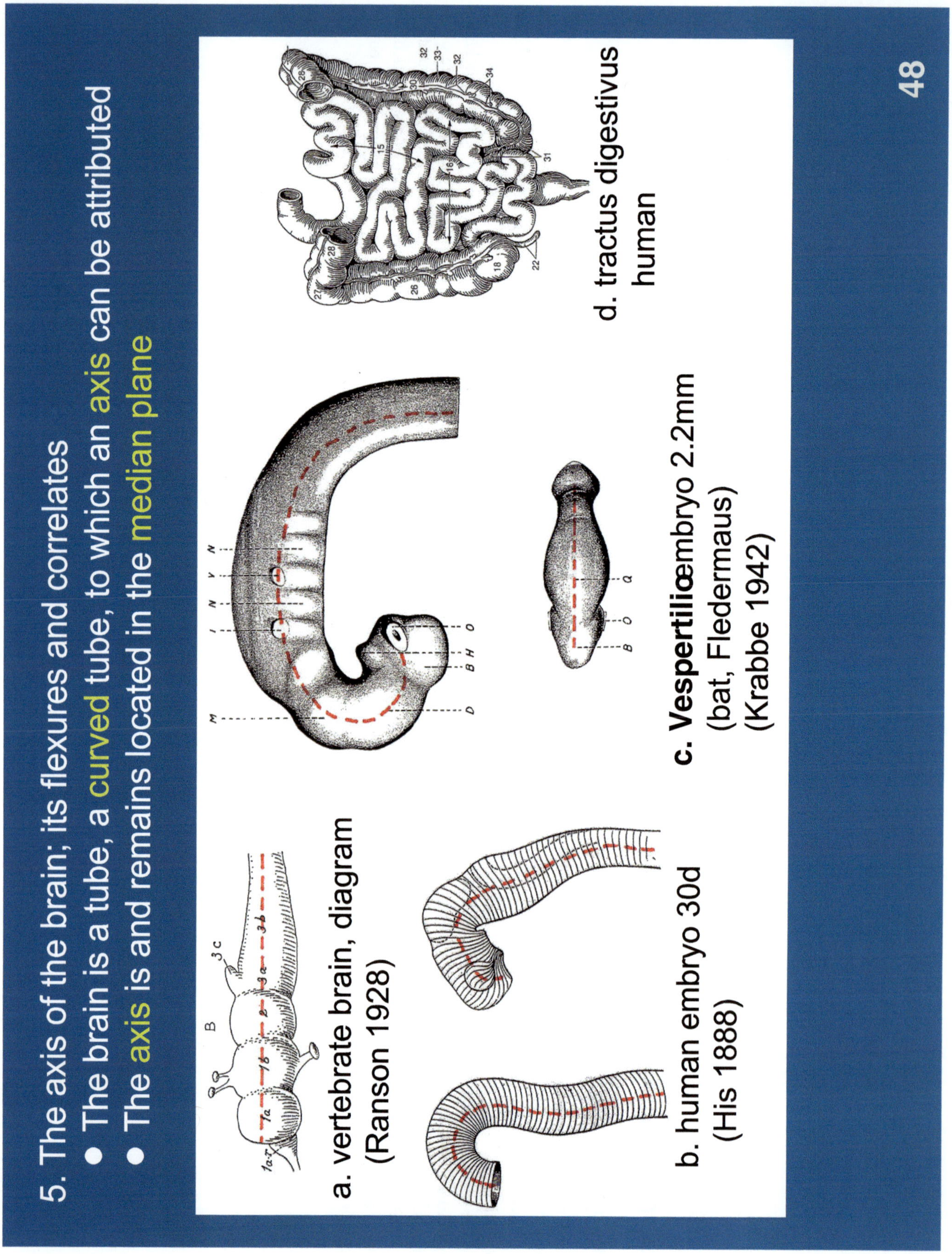
5. The axis of the brain; its flexures and correlates
● The brain is a tube, a curved tube, to which an axis can be attributed
● The axis is and remains located in the median plane
a. vertebrate brain, diagram
(Ranson 1928)
b. human embryo 30d
(His 1888)
c. Vespertilio embryo 2.2mm
(bat, Fledermaus)
(Krabbe 1942)
d. tractus digestivus
human
48

Plate 48

Plate 48 presents some additional pictures related to the flexures of the brain.

Fig. a shows once again how Ranson (1928) placed the so-called five secondary brain vesicles, telencephalon (1a), diencephalon (1b), mesencephalon (2), metencephalon (3a), and myelencephalon (3b), *for the sake of simplicity*, in strict rostrocaudal order along a straight axis, a configuration which in reality never occurs.

Fig. b represents a classic picture by the noted Leipzigian neuroembryologist Wilhelm His (1831–1904), in which he shows in a very terse way that the brain of an early human embryo can be envisioned as a curved tube.

Fig. c shows lateral and dorsal views of the brain of an early bat embryo, derived from Krabbe (1942). The cephalic and cervical flexures are clearly visible in the lateral view. The dorsal view has been added to emphasize that the curvatures of the developing brain are confined to a single plane, and that hence, the horizontal graphic projection of the brain's axis presents itself as a straight line (losing crucial morphological information).

Fig. d shows that in contrast, the convolutions of the human digestive tract, another typical tubular organ, are by no means confined to a single plane.

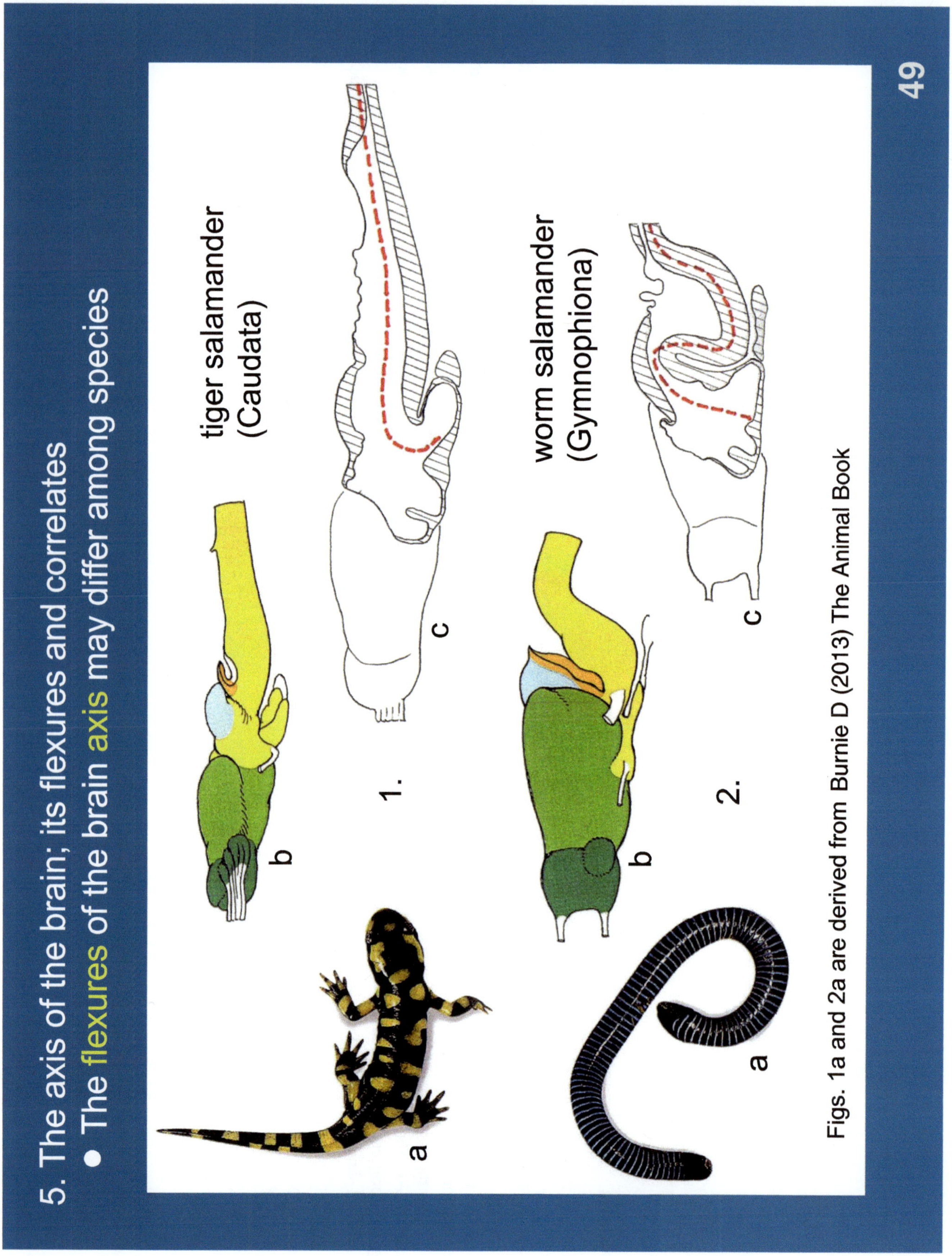
5. The axis of the brain; its flexures and correlates
• The flexures of the brain axis may differ among species
tiger salamander
(Caudata)
worm salamander
(Gymnophiona)
a
b
c
1.
a
b
c
2.
Figs. 1a and 2a are derived from Burnie D (2013) The Animal Book
49

5.2 Flexures of the Axis: Differences Among Species

Plate 49

The curvature of the brain does not only vary during development; it may also vary considerably between different adult species. For the demonstration of such an interspecies difference, we selected two amphibians, the caudatan (Urodele) *Ambystoma tigrinum* (tiger salamander), and a gymnophionan worm salamander. All known species of this group resemble each other closely with regard to both overall appearance and general structure of the CNS. It may be added that worm salamanders are, as their name indicates, worm-like, burrowing animals with rudimentary eyes ('Blindwühler'), and without any indication of limbs. They live in the tropics and are very common on the island of Sri Lanka.

Plate 49 shows holotypes (Figs. 1a and 2a), lateral views of the brains (Figs. 1b and 2b), and schematic median sections through the brains (Figs. 1c and 2c) of the two animals selected. (The bulbous head of the worm salamander, shown in Fig. 2a, is on its curved, right end.)

Comparison of Figs. 1b and 2b shows that the brain of the worm salamander seems to be rostrocaudally compressed. The olfactory system is strongly developed in these animals; consequently the olfactory bulbs (dark green) and the—largely olfactory—cerebral hemispheres (light green) are strongly developed, pressing the whole forebrain backwards. The cephalic flexure is massive and the rhombencephalon is resultantly strongly bent upon itself in the 'compressed' gymnophionan brain (Fig. 2c), whereas such a curvature is entirely lacking in the brain of the tiger salamander (Fig. 1c).

5. The axis of the brain; its flexures and correlates

- Flexures of the axis: a historical error and its correction

1. Two introductory figures

A anterior
P posterior
R rostral
C caudal
D dorsal
V ventral

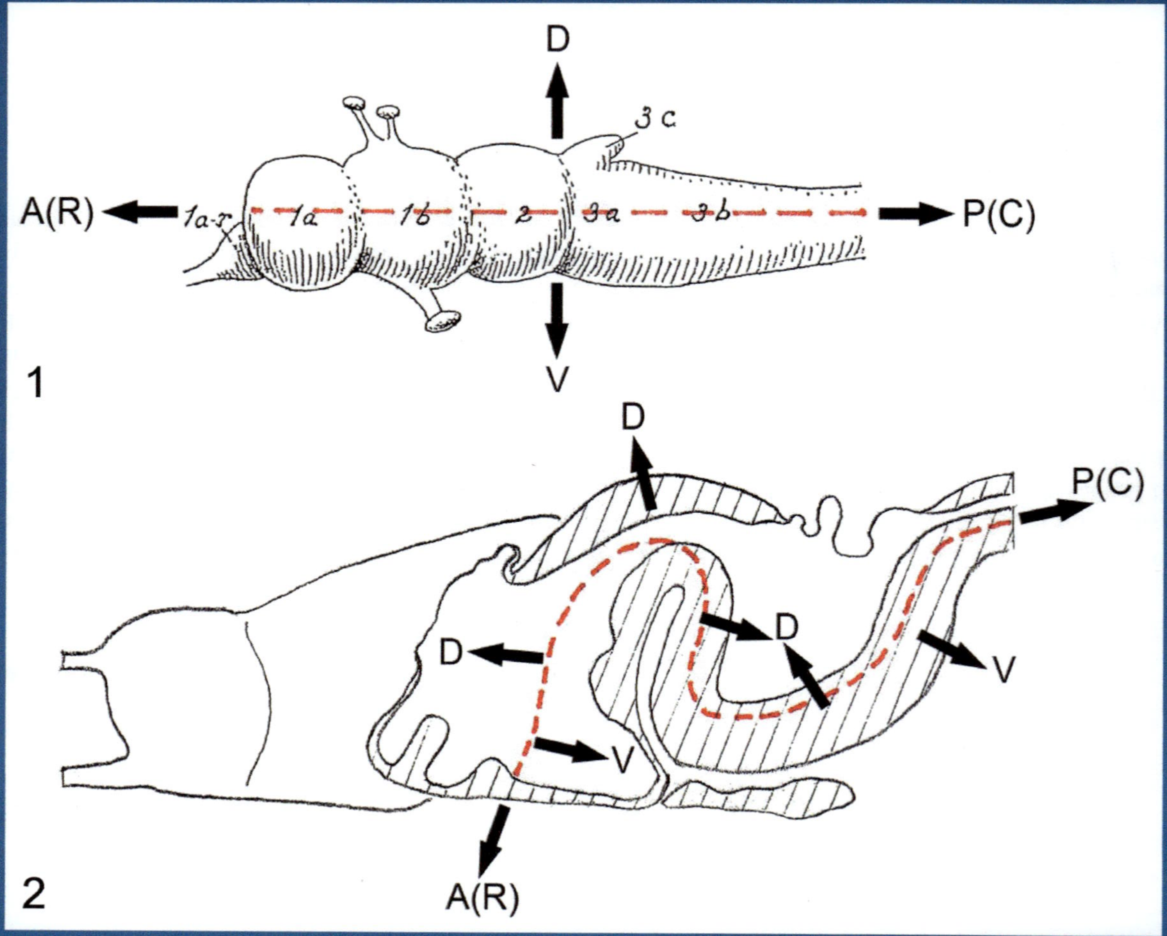

50

5.3 Flexures of the Axis: A Historical Error and Its Correction

Plate 50

The historical error designated in the title of this section, has to do with the course of the anterior part of the axis of the brain, and will be detailed below.

By way of introduction to the problem, we show here once again, the artificially stretched embryonic 'Ranson-brain' (Plate 50, Fig. 1), and the strongly curved gymnophionan brain just discussed (Plate 50, Fig. 2). In the 'Ranson-brain', the axis is represented by a straight line and hence, the constant cephalic flexure is erroneously lacking. In the gymnophionan brain, on the other hand, the course of the axis is correctly indicated, including the pronounced cephalic flexure. Note that the axis ends just behind the optic chiasm, and is strictly longitudinal topologically, although its rostral part seems vertically oriented topographically (relative to the surface on which the brain rests).

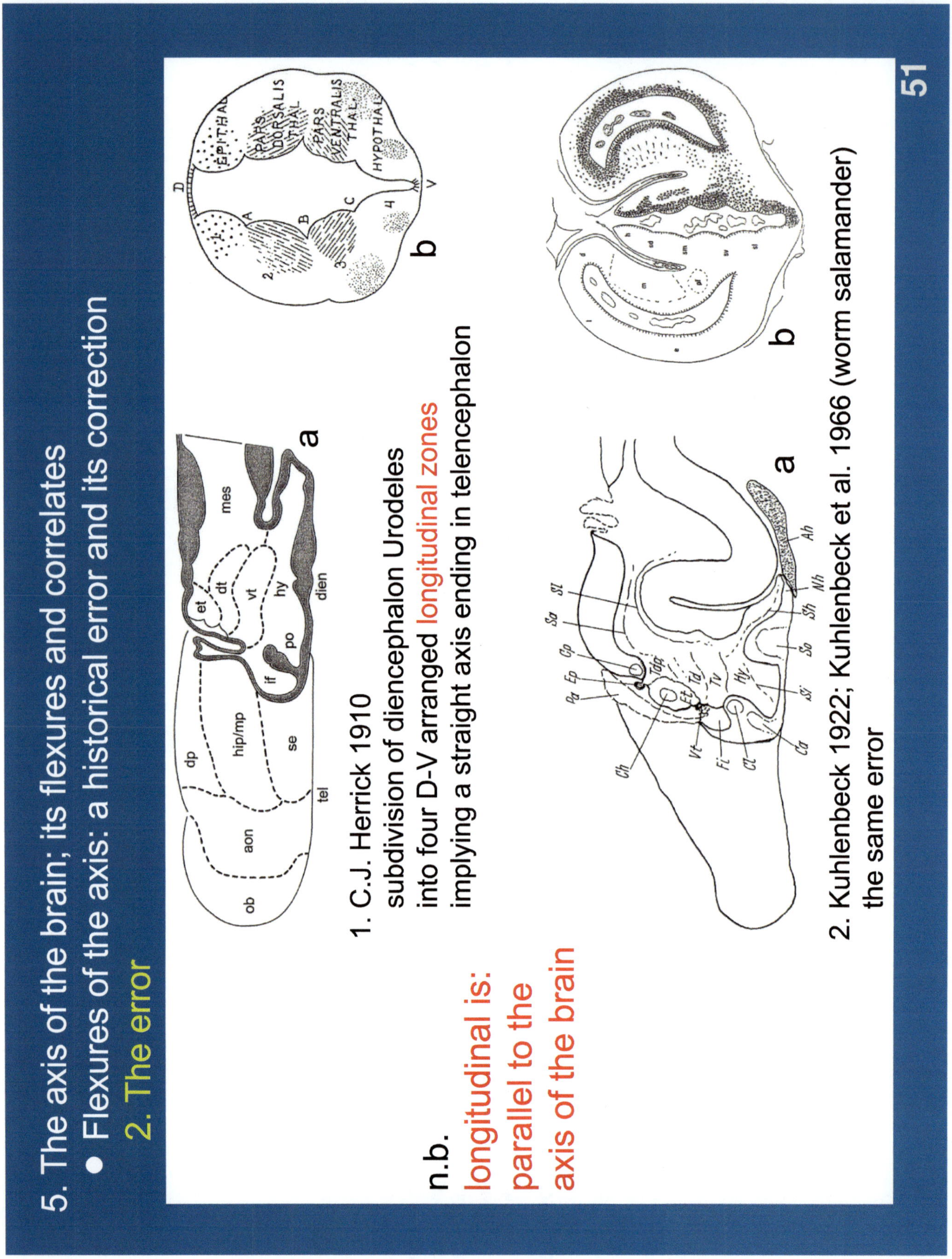
5. The axis of the brain; its flexures and correlates
• Flexures of the axis: a historical error and its correction
2. The error
n.b.
longitudinal is:
parallel to the
axis of the brain
ob
aon
dp
hip/mp
se
tel
lf
po
et
dt
vt
hy
dien
mes
a
EPITHAL
PARS DORSALIS THAL
PARS VENTRALIS THAL
HYPOTHAL
b
1. C.J. Herrick 1910
subdivision of diencephalon Urodeles
into four D-V arranged longitudinal zones
implying a straight axis ending in telencephalon
a
b
2. Kuhlenbeck 1922; Kuhlenbeck et al. 1966 (worm salamander)
the same error
51

Plate 51

Now then, the announced error. At the beginning of the 20th century, the concept prevailed that the vertebrate brain stem is essentially composed of four bilateral longitudinal zones (two alar and two basal, each pair attending to somatic versus visceral sensory and motor components of the cranial nerves). In 1910, the noted American comparative neuroanatomist Charles Judson Herrick made an attempt at extending this at the time exciting functional concept into the rostral part of the brain. Studying adult caudate amphibians, he proposed that the diencephalon of these animals is, just like the brainstem, composed of four longitudinal cellular zones or ‘columns’: epithalamus, dorsal thalamus, ventral thalamus, and hypothalmus. His evidence was that the boundaries of these zones seemed marked by distinct ventricular sulci, just as occurred with the brainstem columns (Plate 51, Fig. 1a, b). However, in characterizing the diencephalic zones as ‘longitudinal’ Herrick (1910) arbitrarily used a straight forebrain axis ending in the telencephalon (also supposed to be divisible into four columns by him) that disregarded the cephalic flexure and the standard forebrain axis previously defined by His (1893a, 1904). Herrick’s ‘longitudinal diencephalic sulci’ are ostensibly orthogonal relative to the sulcus limitans of His and the cephalic flexure.

The also noted German comparative neuroanatomist Hartwig Kuhlenbeck, who studied the brain of gymnophionans (Kuhlenbeck 1922; Kuhlenbeck et al. 1966), fully confirmed the correctness of the findings of Herrick for this amphibian group (Plate 51, Fig. 2a, b). In a series of later studies, summarized in Kuhlenbeck (1973), he reported to have found corresponding morphological relations in the diencephalon of representatives of all classes of vertebrates.

The ‘four-storey-concept’ of Herrick and Kuhlenbeck, just outlined, holds that the four diencephalic zones are longitudinally oriented, running parallel to an arbitrarily drawn straight axis (see Plate 50, Fig. 1). This model has been widely accepted subsequently as applicable to all vertebrate groups. Nevertheless, it is fundamentally wrong, due to the initial error of disregarding the systematically visible cephalic flexure in interpreting what is longitudinal in the forebrain. It can be added that recent work with gene markers has not identified any forebrain axial pattern ending in the telencephalon, whereas the axial ending in the hypothalamus has been supported conclusively (Puelles et al. 2012).

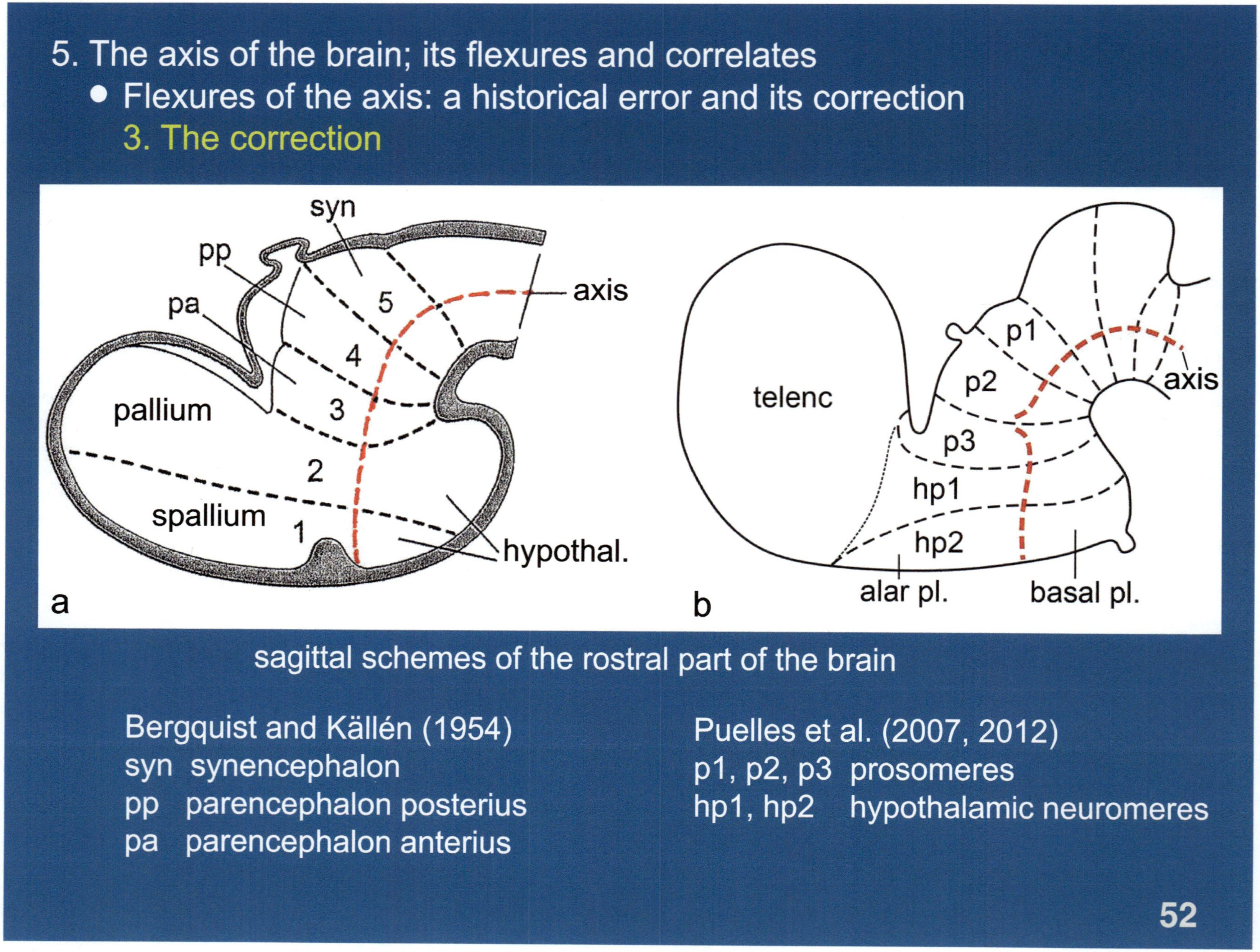
5. The axis of the brain; its flexures and correlates
Flexures of the axis: a historical error and its correction
3. The correction
syn
pp
pa
pallium
spallium
1
2
3
4
5
axis
hypothal.
a
telenc
p1
p2
p3
hp1
hp2
axis
alar pl.
basal pl.
b
sagittal schemes of the rostral part of the brain
Bergquist and Källén (1954)
syn synencephalon
pp parencephalon posterius
pa parencephalon anterius
Puelles et al. (2007, 2012)
p1, p2, p3 prosomeres
hp1, hp2 hypothalamic neuromeres
52

Plate 52

In contrast, the embryological studies of Bergquist and Källén (1954), Keyser (1972), Bulfone et al. (1993), Hauptmann et al. 2002), and many others (review in Puelles et al. 2012) have clearly shown that the Herrick-Kuhlenbeck forebrain zones only partly belong to the diencephalon (the hypothalamus lies rostral to it, according to the true axis) and represent in fact transverse neuromeric derivatives which are hence oriented perpendicular to the portion of the longitudinal axis of the brain that lies in front of the cephalic flexure. These spatial relations become immediately clear if we take into account the curvature of the neural tube at the juncture of the midbrain and the forebrain, and the concomitant ventral bending of the brain axis (Plate 50, Fig. 2). These salient features are emphasized in the general brain models of Bergquist and Källén (1954) and Puelles et al. (2007), which have been shown before (Plates 8, 10).

In Plate 52, the rostral portions of these two brain models are shown once again.

The brain model of Bergquist and Källén (1954) was based on extensive embryological studies in representatives of all major groups of vertebrates.

Focusing on patterns of cellular proliferation and migration, these authors found that in embryonic brains there are transversely oriented bands of high mitotic activity that intersect with longitudinal zones showing similar neurogenetic features. By interference, a network made up of squares with a high proliferative activity could be distinguished. During further development these central proliferative areas also manifested themselves as centres of radial migration into the developing mantle layer, and were therefore designated as migration areas (Puelles et al. 1987 later termed them radial histogenetic areas). It was found that these migration areas show a remarkable consistency in both number and pattern throughout the vertebrate kingdom; hence they were considered to be fundamental morphological units of the brain wall, providing a sound basis for the eventual homologization of the neural grisea they produce. In the prosencephalon five transverse bands intersecting with two longitudinal zones were observed (Plate 52, Fig. a). Transverse band 5 represents the synencephalon or pretectal region; transverse band 4 represents the parencephalon posterius, which gives rise to the epithalamic habenular nuclei and the dorsal thalamus, whereas transverse band 3, which represents the parencephalon anterius, includes the ventral thalamus. Transverse bands 2 and 1 extend from the hypothalamus into the telencephalon, in which the former would give rise to the pallium and the latter to the subpallium, according to these authors. The boundary separating the prosencephalic longitudinal zones follows the brain axis and terminates rostrally directly behind (under) the optic chiasm.

The model of Puelles et al. (2007; updated *in idem* 2012), which is principally based on studies of the expression patterns of a large number of developmental regulatory genes in mouse and chick embryos, resembles the Bergquist-Källén model. An important difference is however, that in the model of Puelles et al., the first hypothalamic neuromere (hp1), which corresponds with the second transverse band of Bergquist and Källén, participates in the formation of the entire telencephalic hemispheres, rather than only in the formation of the pallium (Plate 52, Fig, b).

5. The axis of the brain; its correlates

(a) the sulcus limitans of His

(b) the, molecularly defined, alar/basal boundary

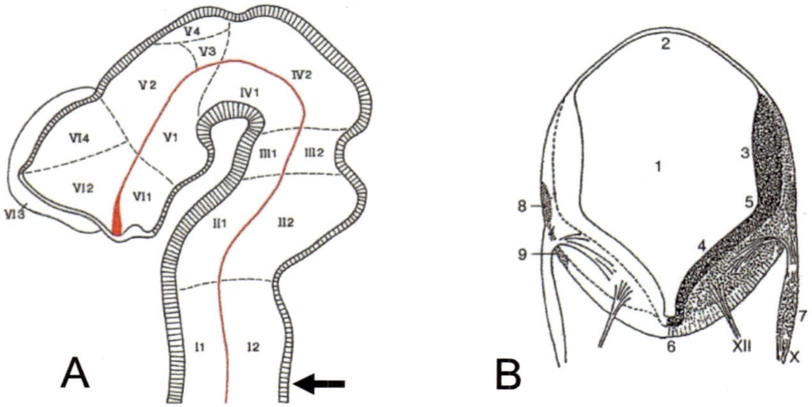

"Zwei seitliche Längsfurchen bezeichnen frühzeitig die Grenze der ventralen und der dorsalen Hälfte des Gehirnrohres (der Grundplatten und der Flügelplatten meiner früheren Aufsätze). Für diese schlage ich die Namen der Grenzfurchen oder Sulci limitantes vor. Alle motorischen Kerne liegen ventral- bezw. medialwärts von den Grenzfurchen, die sogen. sensibeln Kerne dagegen dorsal- bezw. lateralwärts davon."

"Two lateral longitudinal sulci indentify early on the boundary between the dorsal and ventral halves of the neural tube (the basal and alar plates of my earlier reports). For these I propose the names boundary slulci or sulci limitantes.
All motor nuclei lie ventral, i.e. medial to the limiting sulci, whereas the so-called sensory nuclei lie dorsal, i.e. lateral to them."

53

5.4 Correlates of the Axis: Sulcus Limitans and Alar/Basal Boundary

Plate 53

The axis of the brain is of course a construct and does not exist as an anatomic entity since it actually refers to the imagined central axis of the 'brain cylinder' (Plate 48, Fig. b). However, it has two possible structural correlates, the *sulcus limitans* and the *alar/basal boundary*.

The *sulcus limitans* was first observed in human embryos by Wilhelm His (1893b). He mentions that this longitudinal groove marks bilaterally the boundary between the dorsal and ventral halves of the neural tube, and he adds that all motor nuclei in the brain are situated ventral to this sulcus, whereas all sensory nuclei are situated dorsal to it.

In Plate 53, Fig. A, which has been shown before (Plate 27), the course of the sulcus limitans is shown in red in a human embryo of about one month.

Plate 53, Fig. B shows the sulcus limitans (red arrow) in a transverse section through the myelencephalon, at the level of the black arrow in Plate 53, Fig. A.

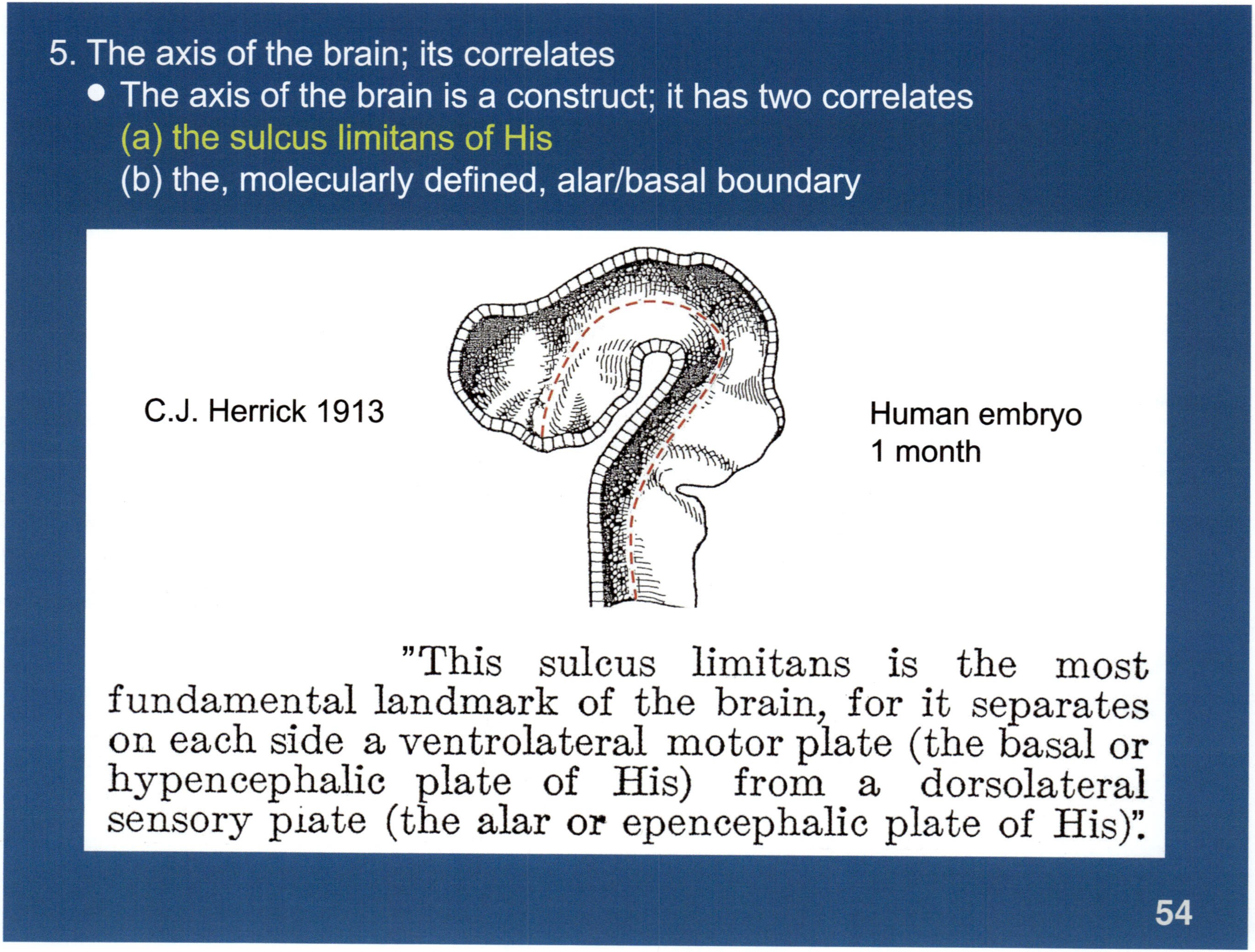
5. The axis of the brain; its correlates
The axis of the brain is a construct; it has two correlates
(a) the sulcus limitans of His
(b) the, molecularly defined, alar/basal boundary
C.J. Herrick 1913
Human embryo
1 month
"This sulcus limitans is the most fundamental landmark of the brain, for it separates on each side a ventrolateral motor plate (the basal or hypencephalic plate of His) from a dorsolateral sensory piate (the alar or epencephalic plate of His)".
54

Plate 54

As shown by Plate 54, the sulcus limitans of His was proclaimed by Charles Judson Herrick (1913, p. 285) to be *the most fundamental landmark of the brain*, because it separates a motor zone from a sensory zone (though this did not impede him defying it straightforwardly in the columnar concept of the diencephalon). This sulcus plays a prominent role in the concept enunciated by C. J. Herrick (1899, 1908, 1913), and J. B. Johnston (1902a, b, 1905), according to which the rhombencephalon can be subdivided into a number of functional longitudinal zones or columns. This concept will be illustrated in the next plate.

5. The axis of the brain; its correlates
- The axis of the brain is a contruct; it has two correlates
  (a) the sulcus limitans of His
  (b) the, molecularly defined, alar/basal boundary

Functional columns in the brainstem (Herrick and Johnston)

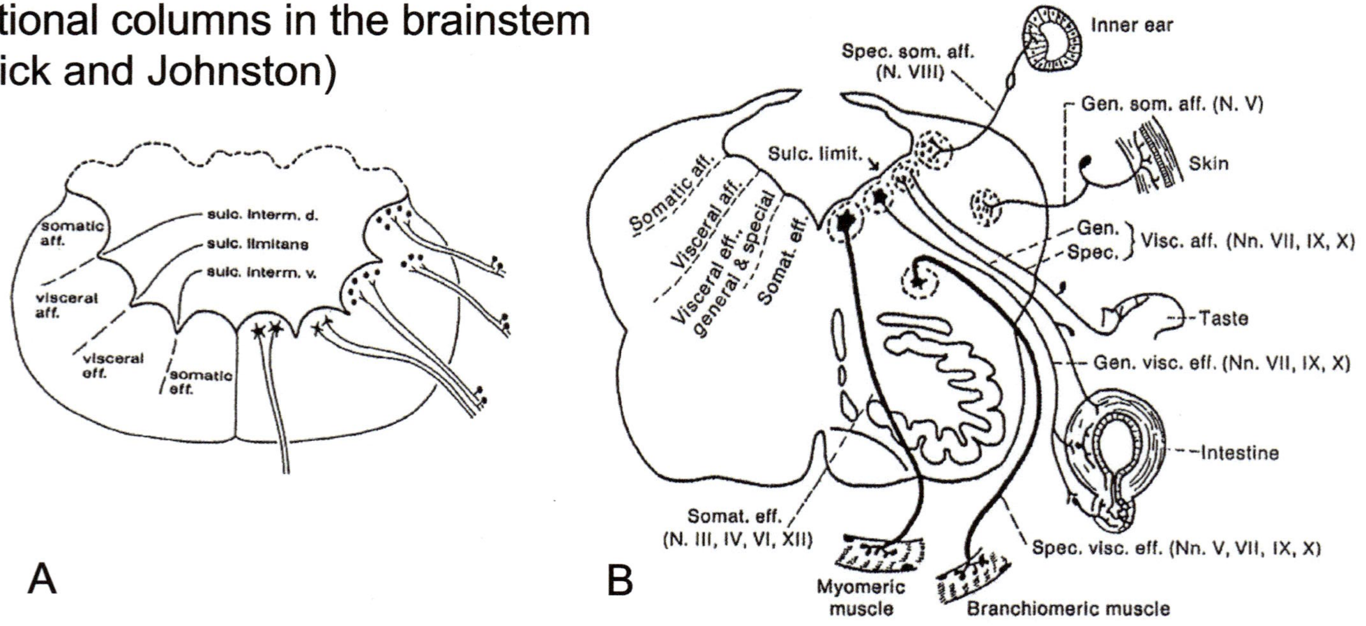

FIGURE 2 | Diagrammatic transverse sections through the rhombencephalon of an anamiote (A), and the human (B), to show the arrangement of the so-called functional columns.

55

Plate 55

We have seen that, according to His (1893b), the entire brain can be divided into a topologically dorsal sensory zone (also known as alar plate) and a topologically ventral motor zone (also known as basal plate), and that the boundary between these two zones is marked by the sulcus limitans. Studying the central relations of the cranial nerves, Herrick and Johnston found that a further functional subdivision of the rhombencephalon is possible, in that within both the sensory and the motor zone, separate somatic and visceral domains or columns can be delineated.

This 'four-functional columns-concept' is illustrated in Plate 55. It shows diagrammatic transverse sections through the rhombencephalon of an anamniote (A) and the human (B). The conceived location of the four functional columns, somatic sensory, visceral sensory, visceral motor and somatic motor, is indicated in both figures. Note these classic schemata did not contemplate the recently discovered tangential migration of visceromotor neurons into the alar plate.

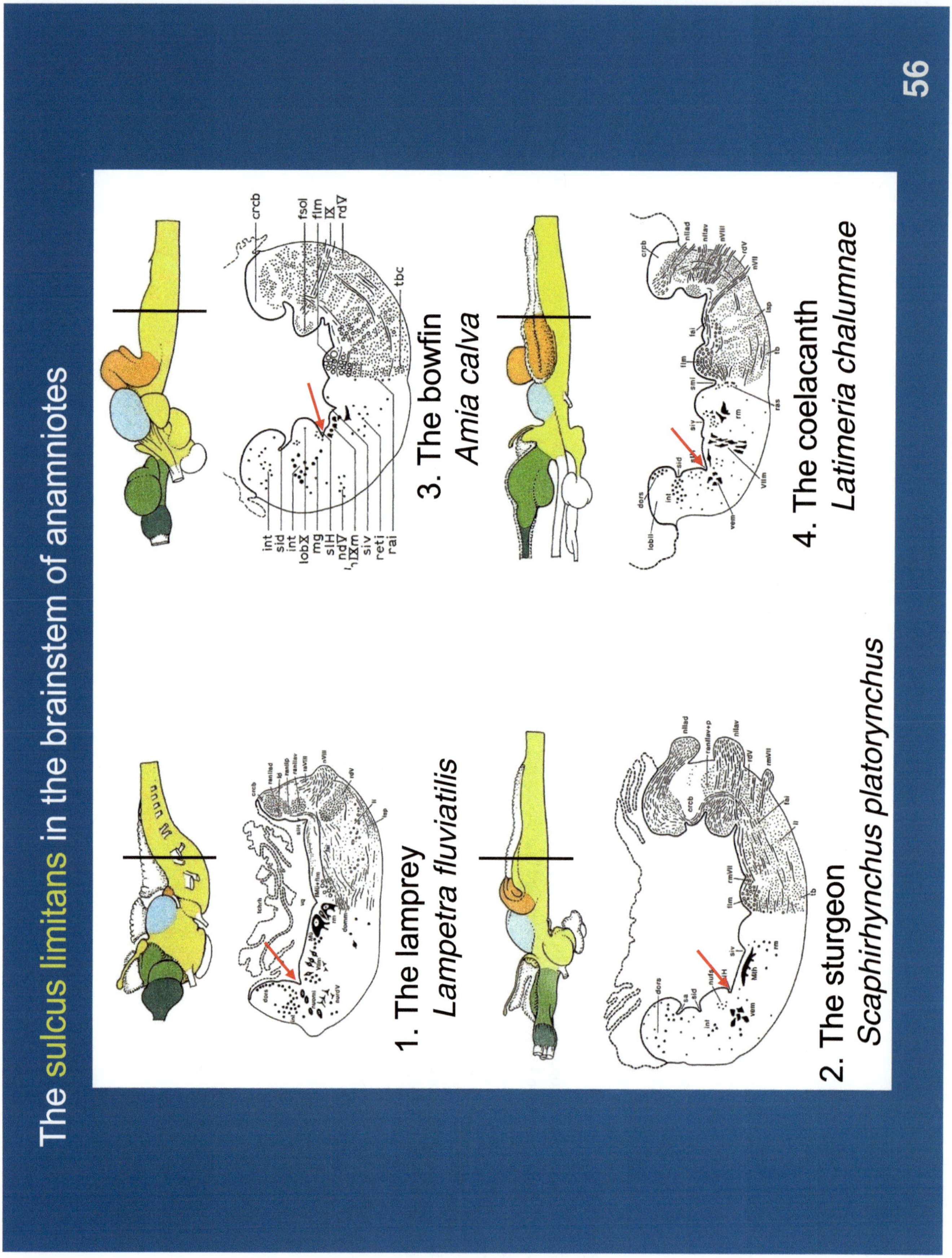
The sulcus limitans in the brainstem of anamniotes
56
1. The lamprey
Lampetra fluviatilis
2. The sturgeon
Scaphirhynchus platorynchus
3. The bowfin
Amia calva
4. The coelacanth
Latimeria chalumnae
crcb
fsol
flm
IX
rdV
tbc
int
sid
int
lobX
mg
slH
ndV
IXm
siv
reti
rai

Plate 56

Back to the sulcus limitans!

Within the frame of an extensive research program on the structural and functional relations of the brainstem, Nieuwenhuys and collaborators (for review see: Nieuwenhuys 2011b) analyzed the structure of this brain part in adults of 17 different anamniote species. It was found that a distinct sulcus limitans is present in all of them and that, again in all, the sulcus limitans ends, either in the rostral part of the rhombencephalon or in the most caudal part of the mesencephalon.

Plate 56 presents lateral views of the brains of four of the species investigated, with transverse sections through the rhombencephalic parts of these brains taken at the levels, indicated by vertical lines. It will be seen that the sulcus limitans manifests itself in all of these sections as a deep and distinct indentation of the ventricular surface (red arrows).

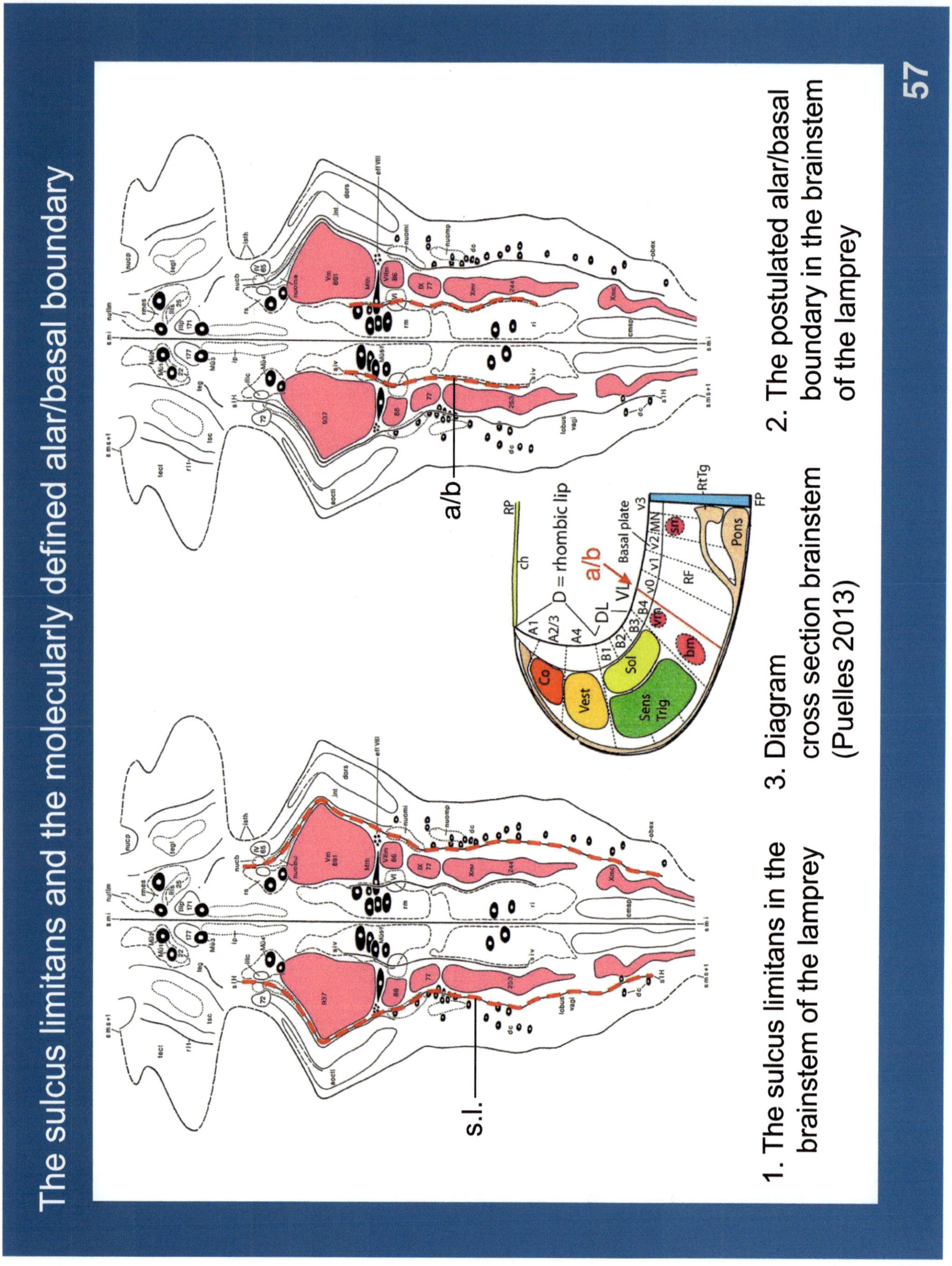
The sulcus limitans and the molecularly defined alar/basal boundary
s.l.
a/b
1. The sulcus limitans in the brainstem of the lamprey
2. The postulated alar/basal boundary in the brainstem of the lamprey
3. Diagram cross section brainstem (Puelles 2013)
D = rhombic lip
Basal plate
Pons
RF
Sens Trig
Vest
Sol
Co
57

Plate 57

The second possible correlate of the longitudinal brain axis when projected upon the lateral wall of the brain is the *alar/basal boundary*. It has already been mentioned that this boundary is defined by Puelles et al. (2012) as a linear equilibrium between alternative histologic fates, established by interplay of opposite ventralizing and dorsalizing signalling effects, originating from the floor plate and roof plate (Plate 33). The alar/basal boundary is sharply defined in the rhombencephalon by the expression domains of *Pax7* throughout the hindbrain ventricular zone, jointly with complementary labeling by *Cdh7* of the basal plate territory (Ju et al. 2004).

It should be realized that at rhombencephalic levels, the ventricular end of the molecularly defined alar/basal boundary does not coincide exactly with the sulcus limitans, but rather is situated slightly more ventrally. Due to this shift, the special visceromotor or branchiomotor and general visceromotor cell masses, which, according to the classic interpretation of His are situated in the basal plate, have now come to lie in the alar plate.

The shift just mentioned may be demonstrated in the rhombencephalon of the lamprey (*Lampetra fluviatilis*).

Plate 57 shows two maps (Figs. 1 and 2) of the brainstem of this species. In these maps first the cell masses and large individual neurons are projected upon the ventricular surface, and next the ventricular surface is flattened out (for details on this mapping procedure see Nieuwenhuys 1972, 2011b). The branchiomotor and visceromotor nuclei form a conspicuous column in the brainstem of the lamprey, which is shown in magenta in both charts. It will be seen that this column is situated medially to the sulcus limitans (Fig. 1), but laterally to the postulated alar/basal boundary of Puelles (Fig. 2), just as in the general scheme shown before (Fig. 3). We use the designation 'postulated', because to our knowledge this molecularly defined boundary has not yet been demonstrated in cyclostomes. The primary molecular evidence on the alar-basal boundary is based on avian and mammalian data, and includes labeling of Pax7-derived (alar) progeny (Puelles, unpublished data).

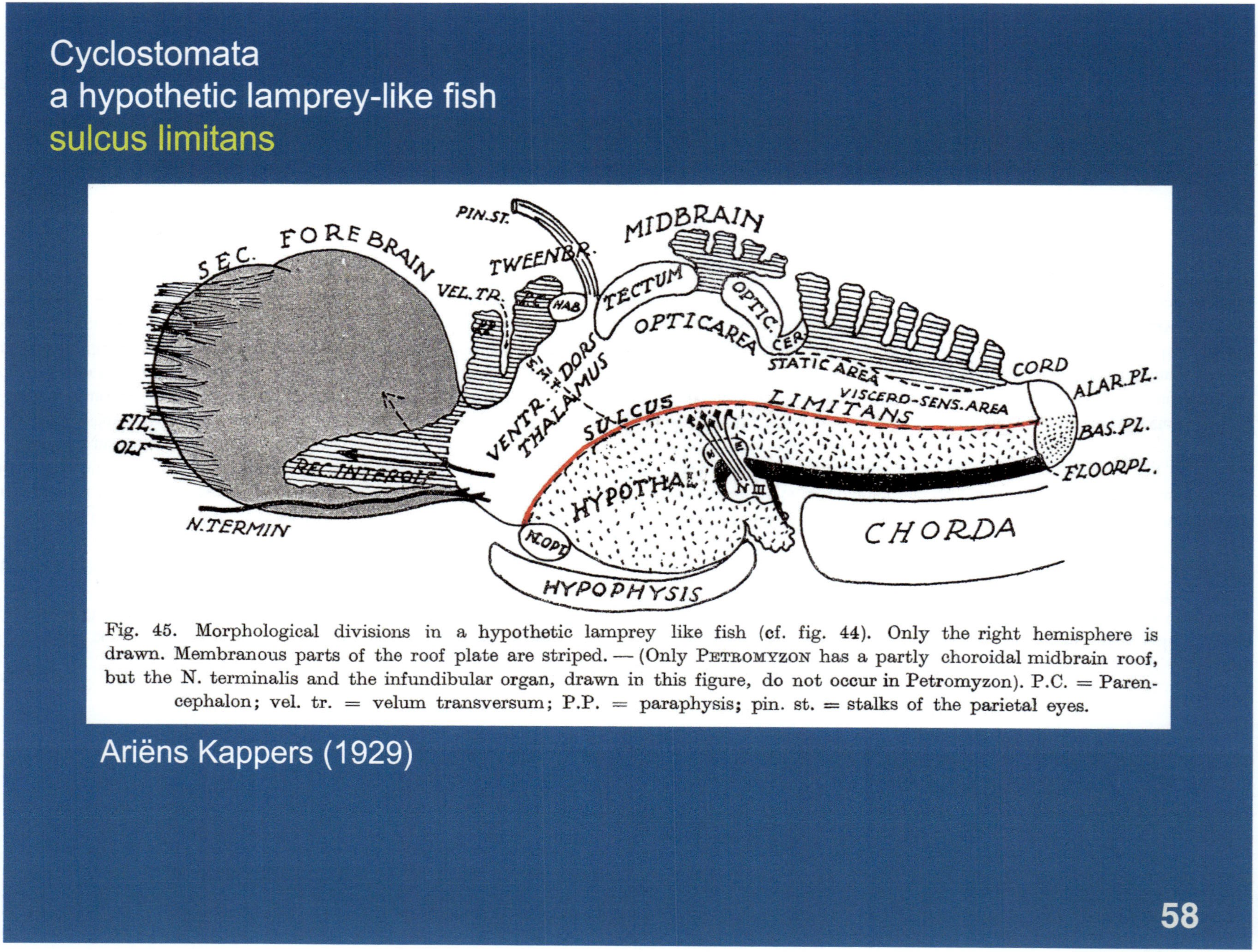

Fig. 45. Morphological divisions in a hypothetic lamprey like fish (cf. fig. 44). Only the right hemisphere is drawn. Membranous parts of the roof plate are striped. — (Only Petromyzon has a partly choroidal midbrain roof, but the N. terminalis and the infundibular organ, drawn in this figure, do not occur in Petromyzon). P.C. = Parencephalon; vel. tr. = velum transversum; P.P. = paraphysis; pin. st. = stalks of the parietal eyes.

Plate 58

It has already been mentioned that, according to our observations in a large number of anamniote species, the sulcus limitans ends either in the rostral rhombencephalon or in the most caudal part of the mesencephalon. Several authors, including His (1893a, b), who studied human embryos, Schulte and Tilney (1915), who studied cat embryos, and Kingsbury (1920, 1922), who studied shark and chick embryos, have indicated that the embryonic sulcus limitans can be traced into the rostral part of the brain. However, there is no unanimity among these authors with regard to the location of the anterior end of this sulcus. Thus, His located this anterior end in the preoptic recess (Plate 53, Fig. A), whereas Schulte and Tilney located it in the mamillary recess.

Plate 58 presents a historical picture from Ariëns Kappers (1929), showing a medial view of the bisected brain of a 'hypothetic lamprey like fish'. It will be seen that in this scheme the sulcus limitans ends just rostral to the optic nerve.

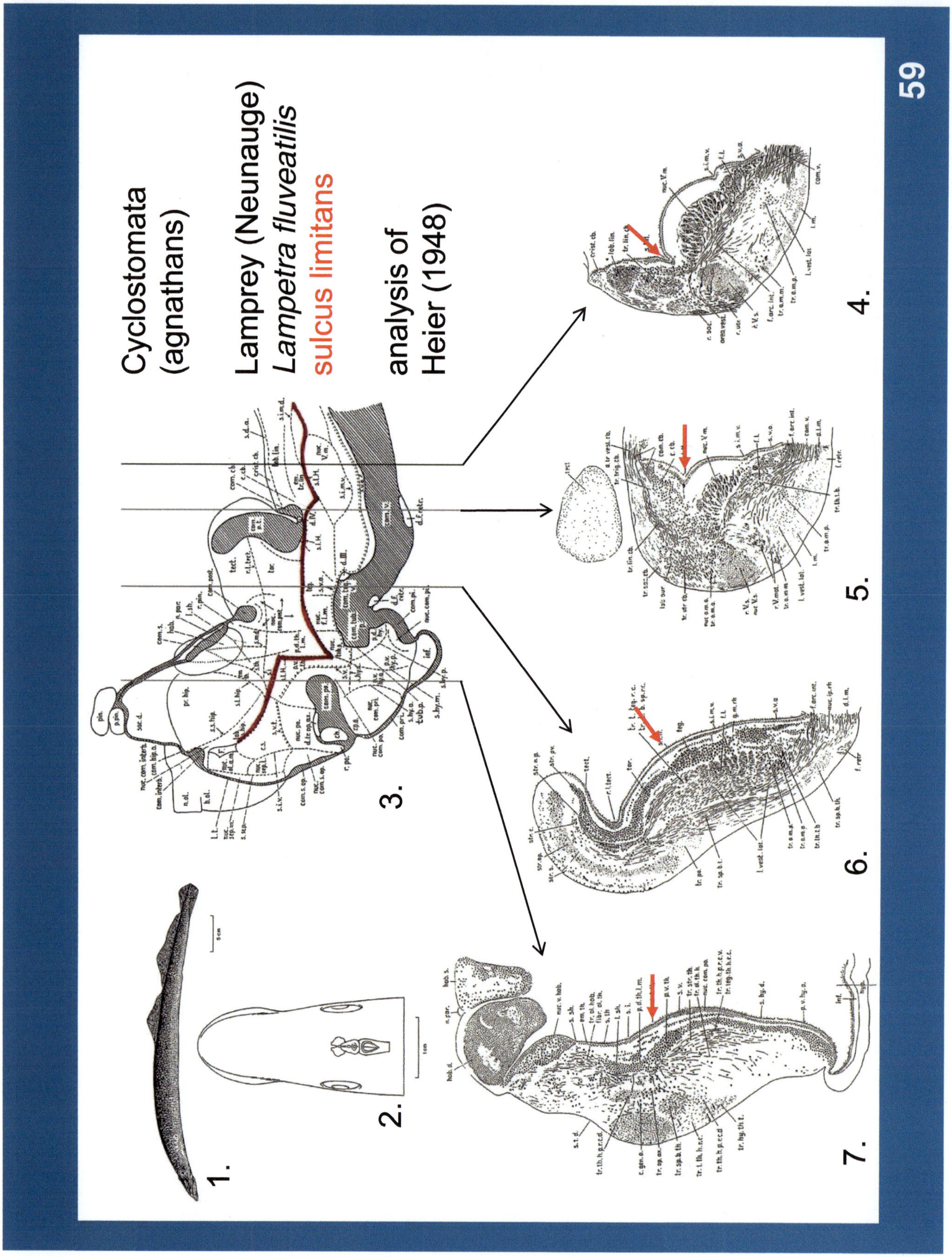
59
Cyclostomata
(agnathans)
Lamprey (Neunauge)
Lampetra fluveatilis
sulcus limitans
analysis of
Heier (1948)
1.
2.
3.
4.
5.
6.
7.

Plate 59

With regard to the course and rostral ending of the sulcus limitans, the Norwegian investigator Per Heier (1948), who made a detailed study of the structure of the CNS of the lamprey *Lampetra fluviatilis*, arrived at quite another conclusion than Ariëns Kappers.

Plate 59 shows the species in question (Fig. 1), a dorsal view of the head with the tiny brain in position (Fig. 2), and a medial view of the bisected brain (Fig. 3), in which the sulcus limitans is marked in red. Figures 4–7 show transverse hemisections through the brain at the levels indicated in Fig. 3.

If we follow the caudorostral course of the sulcus limitans in Heier's pictures the following observations can be made:

1. At rhombencephalic levels, the sulcus limitans is represented by a deep and distinct ventricular groove (Figs. 4 and 5: red arrows). This observation tallies with our own in the same species (Nieuwenhuys 1972; Plate 56, Fig. 1).
2. At mesencephalic and diencephalic levels, the sulcus limitans is labeled as marking the dorsal border of the motor tegmentum and the ventral thalamus, respectively. It is of note that, although Heier indicates the position of the sulcus limitans at these levels (Figs. 6 and 7: red arrows), the ventricular surface does not show any groove at the sites indicated.
3. The most rostral part of the sulcus limitans is held to mark the dorsal border of the corpus striatum and enters the interventricular foramen (marked with an F in Fig. 3). The intrahemispheric course of the sulcus under discussion has not been recorded by Heier.

Although Heier's observations concerning the course of the sulcus limitans deviate from those of most other investigators, they are not unique. Thus, Bok (1926) published a general plan of the vertebrate brain in which, just as in Heier's study, the sulcus limitans marks the boundary between the corpus stratum and the pallium. Bok based this particular feature of his general plan on observations in human embryos.

If we survey the pertinent literature, it must be concluded that the data on the course and termination of the more rostral parts of the sulcus limitans are controversial.

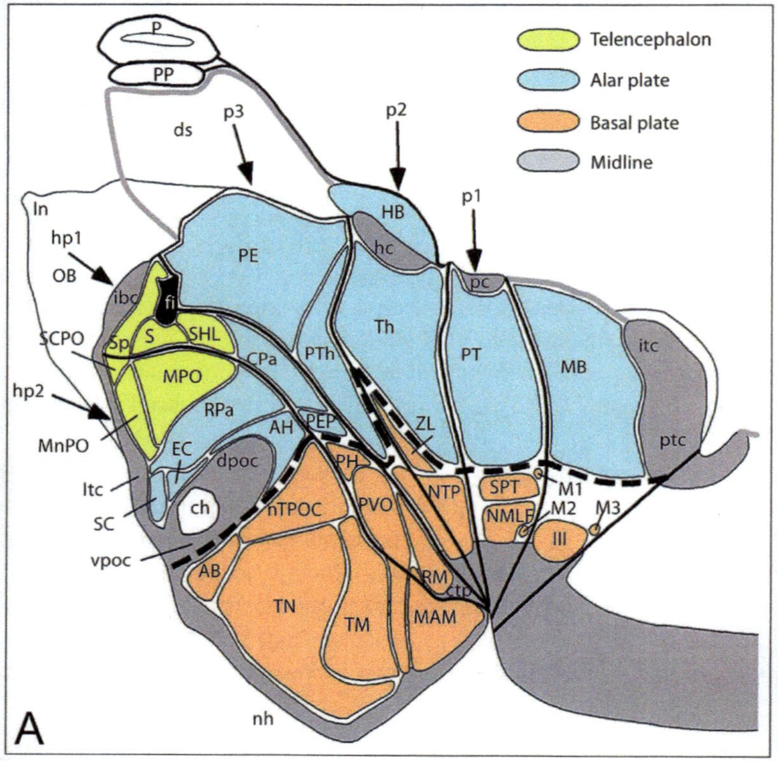

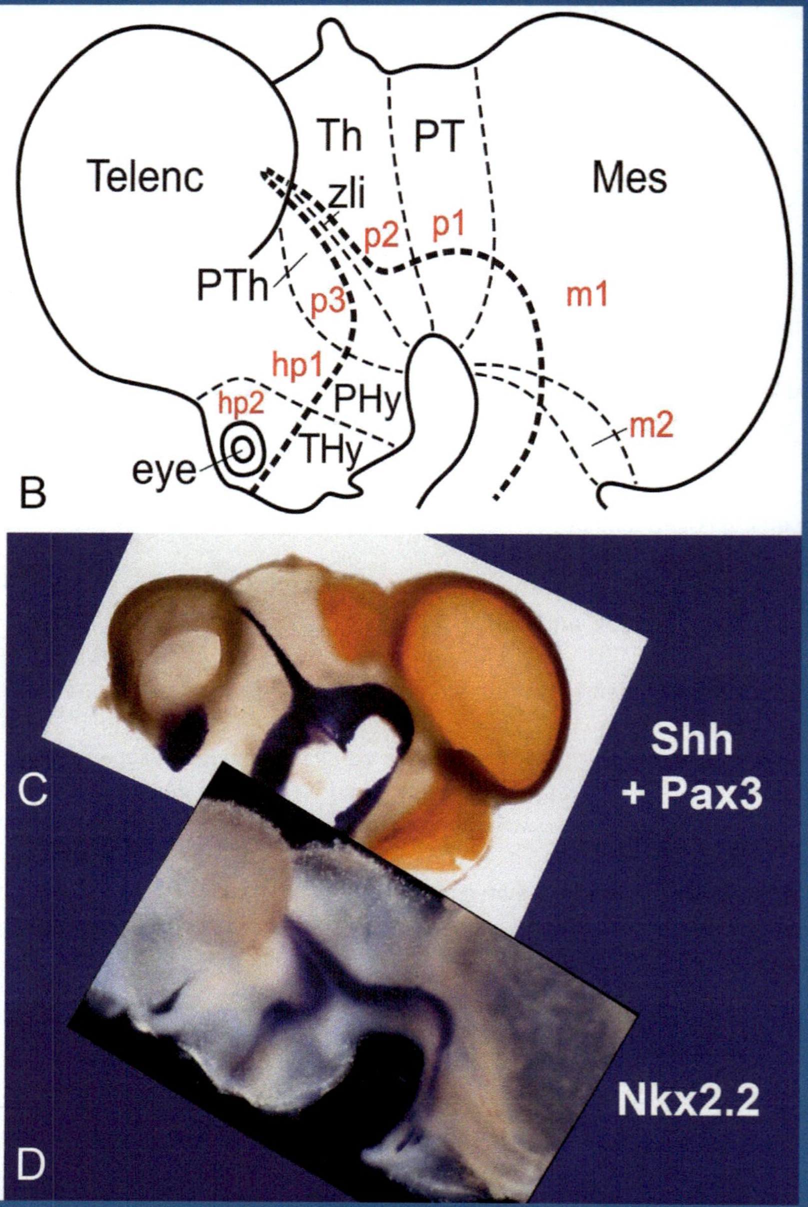

A,B: Rostral end of the a/b boundary (heavy dashed lines) in the adult lamprey (A; from Pombal et al., 2009) and in a chick embryo of 6 days of incubation (B)
C: Expression of Shh in the chicken midbrain and forebrain basal plate with a clear cut a/b boundary.
D: Expression of Nkx2.2 along the chicken a/b boundary (and around the zli Shh-positive central core)

60

Plate 60

Gene expression studies in the lamprey and in chick and mouse embryos have clearly shown that the alar/basal boundary can be traced throughout the rostral part of the brain. Leaving aside a remarkable apparent dorsal 'spur' of this boundary in the zona limitans intrathalamica (zli) at the border zone of prosomeres 2 and 3, the molecularly defined dorsal border of the basal territory continues beyond the midbrain, the diencephalon proper and the hypothalamus, and terminates in the postoptic recess (roughly as suggested by His 1893b). The midbrain and forebrain floor and basal plate territories express initially the ventralizing gene *Shh*, under inductive control of the underlying notochord (some areas in the tuberal hypothalamus later downregulate this signal).

The nature of the mysterious ZLI spur or spike, which in all vertebrates appears only secondarily to decorate the continuous course of the alar/basal boundary (a line which is already detectable at neural plate stages: Shimamura et al. 1995), has recently received some causal explanation. Molecular evidence suggests that the ZLI does not really represent a 'spike' of basal-like tissue that elongates dorsalward, but is rather a single transverse line of permanently alar neuroepithelium where a peculiar cascade of effects causes typical 'basal' genes to be secondarily expressed; i.e. it is a case of repatterning, which exemplifies that individual genes are neither "basal" or 'alar' as such (Martinez-Ferre et al. 2013).

Plate 60 illustrates these features in diagrams showing the rostral parts of the brain of the lamprey (Fig. A), and of a chick embryo at 6 days of incubation (Fig. B). Figs. C and D show some photomicrographs of the expression patterns of several relevant genes in the rostral part of the brain of chick embryos at 6 days of incubation.

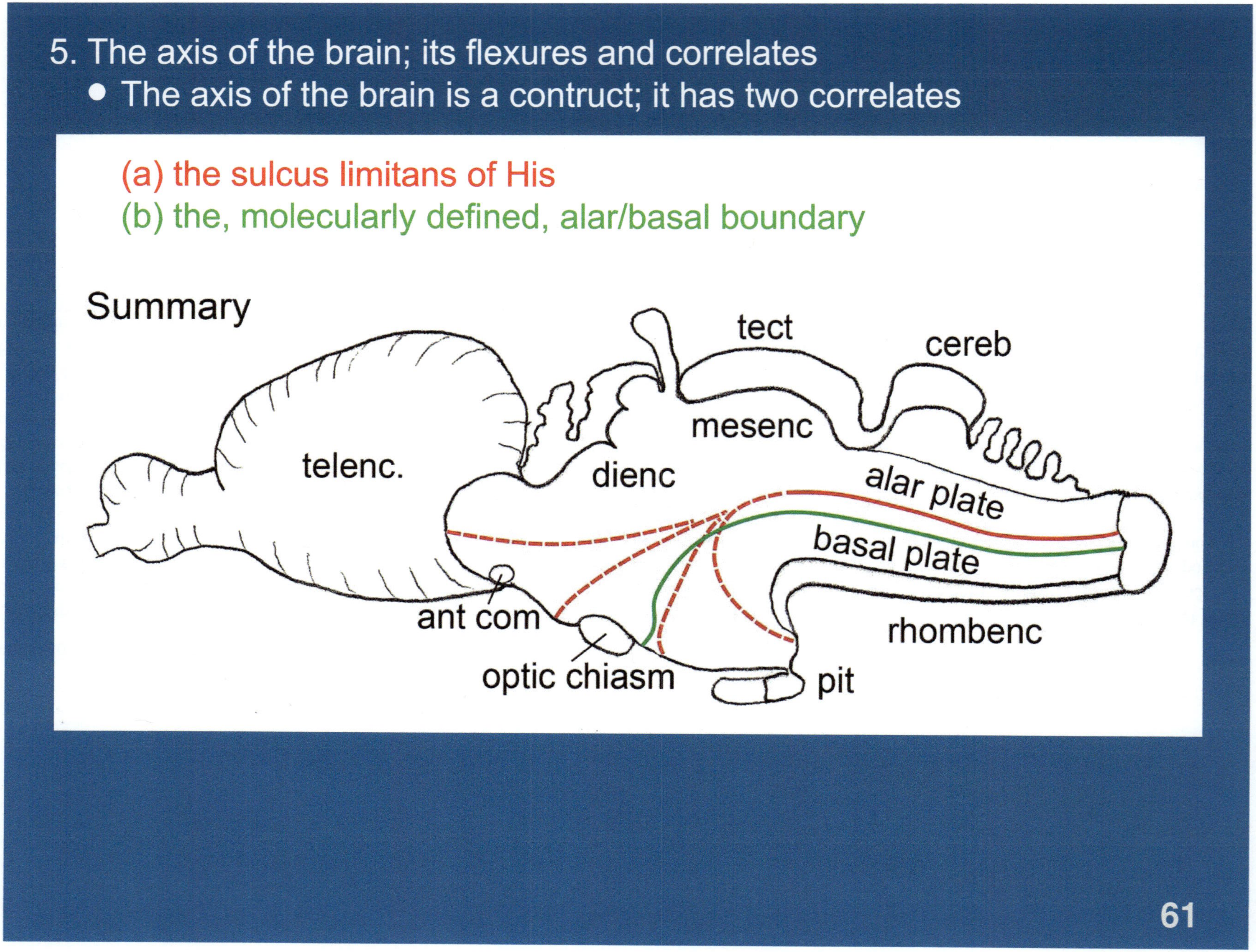
5. The axis of the brain; its flexures and correlates
• The axis of the brain is a contruct; it has two correlates
(a) the sulcus limitans of His
(b) the, molecularly defined, alar/basal boundary
Summary
telenc.
dienc
mesenc
tect
cereb
alar plate
basal plate
rhombenc
ant com
optic chiasm
pit
61

Plate 61 illustrates the following summary.

5.5 Summary of the Brain Axis

1. A longitudinal axis can be attributed to the CNS of vertebrates.
2. This axis shows curvatures, the depth of which may vary considerably among species, as well as during development of one and the same species.
3. Classically, the diencephalon was subdivided into four longitudinal, dorsoventrally arranged zones, designated as epithalamus, dorsal thalamus, ventral thalamus and hypothalamus. However, if the rostral curvature of the neuraxis (Kopfbeuge) is taken into consideration, it appears that these zones represent *topologically*, transversely oriented neuromeric derivatives forming a caudorostral series with the hypothalamus most rostral.
4. The sulcus limitans of His clearly marks the boundary between a classic motor basal plate and a classic sensory alar plate in the rhombencephalon of anamniotes ('lower vertebrates') (red solid line) (Nieuwenhuys 2011b).
5. The course and termination of the rostral part of the sulcus limitans as such is controversial in the literature (red dashed lines).
6. The molecularly defined alar/basal boundary already emerges at neural plate stages and currently represents the most reliable correlate of the axis of the brain (Puelles). In the midbrain and forebrain, this boundary correlates with a thin longitudinal band expressing the gene *Nkx2.2*, which limits the basal plate domain expressing *Shh* (Plate 60, Figs. C, D). In the hindbrain and spinal cord the equivalent boundary is marked by the ventral limit of *Pax7* expression.
7. At rhombencephalic levels the molecular alar/basal boundary does not coincide with the classic sulcus limitans.

Towards a New Neuromorphology
6. The (complex) rostral part of the brain 1

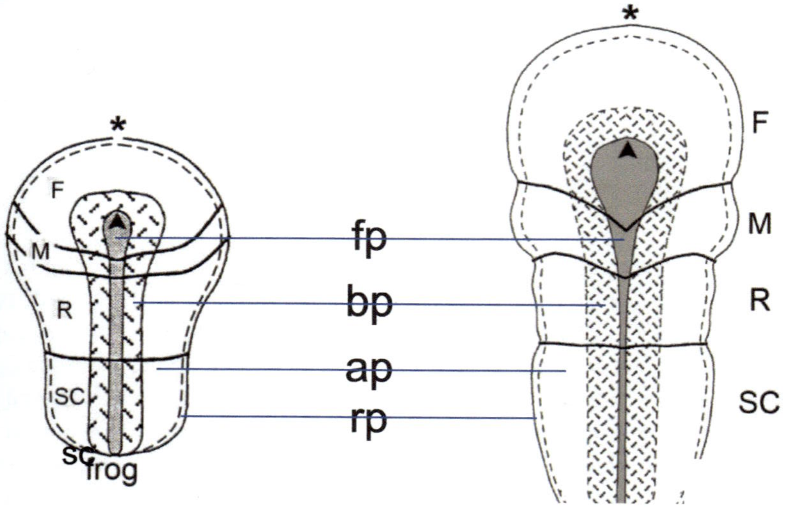

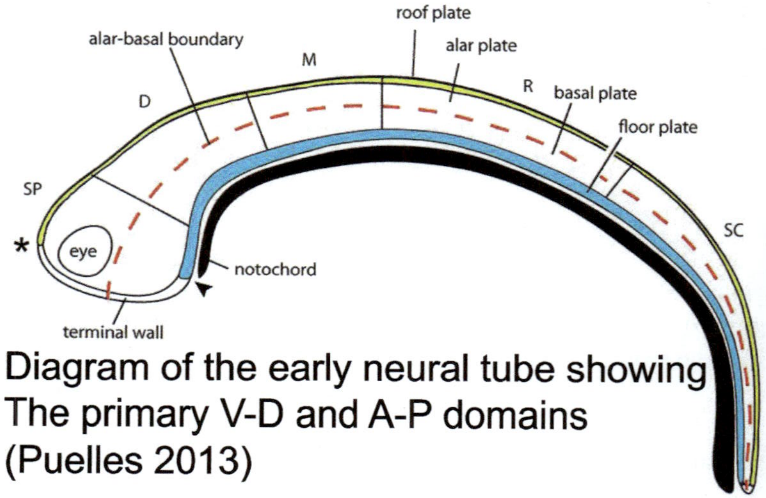

2. Diagram of the early neural tube showing The primary V-D and A-P domains (Puelles 2013)

1. The basal and alar longitudinal domains course in a semi-circle around the rostral end of the floor plate (Rubenstein et al. 1998)

F	forebrain
SP	secondary prosencephalon
D	diencephalon
M	mesencephalon
R	rhombencephalon
SC	spinal cord
ap	alar plate
bp	basal plate
fp	floor plate
rp	roof plate

6 The (Complex) Rostral Part of the Brain

Plate 62

By way of introduction, we present here two figures that have been shown before (see Plates 24 and 34).

Plate 62, Fig. 1 shows the position of the primary rostrocaudal divisions and the main longitudinal zones in fate maps of the neural plates of the frog and the mouse. Note that the rostral parts of the basal and alar plates course in a semicircle around the rostral end of the floor plate. The median zone extending from the arrowhead to the asterisk (the rostral ends of the floor and roof plates, respectively), will form the terminal wall or acroterminal area of the neural tube. This median domain falls within the tuberal and chiasmatic hypothalamus and the preoptic area (lamina terminalis).

Plate 62, Fig. 2 shows the early subdivisions in the neural tube. Note that the notochord lies under the entire length of the neural tube.

R. Nieuwenhuys and L. Puelles, *Towards a New Neuromorphology*,
DOI 10.1007/978-3-319-25693-1_6

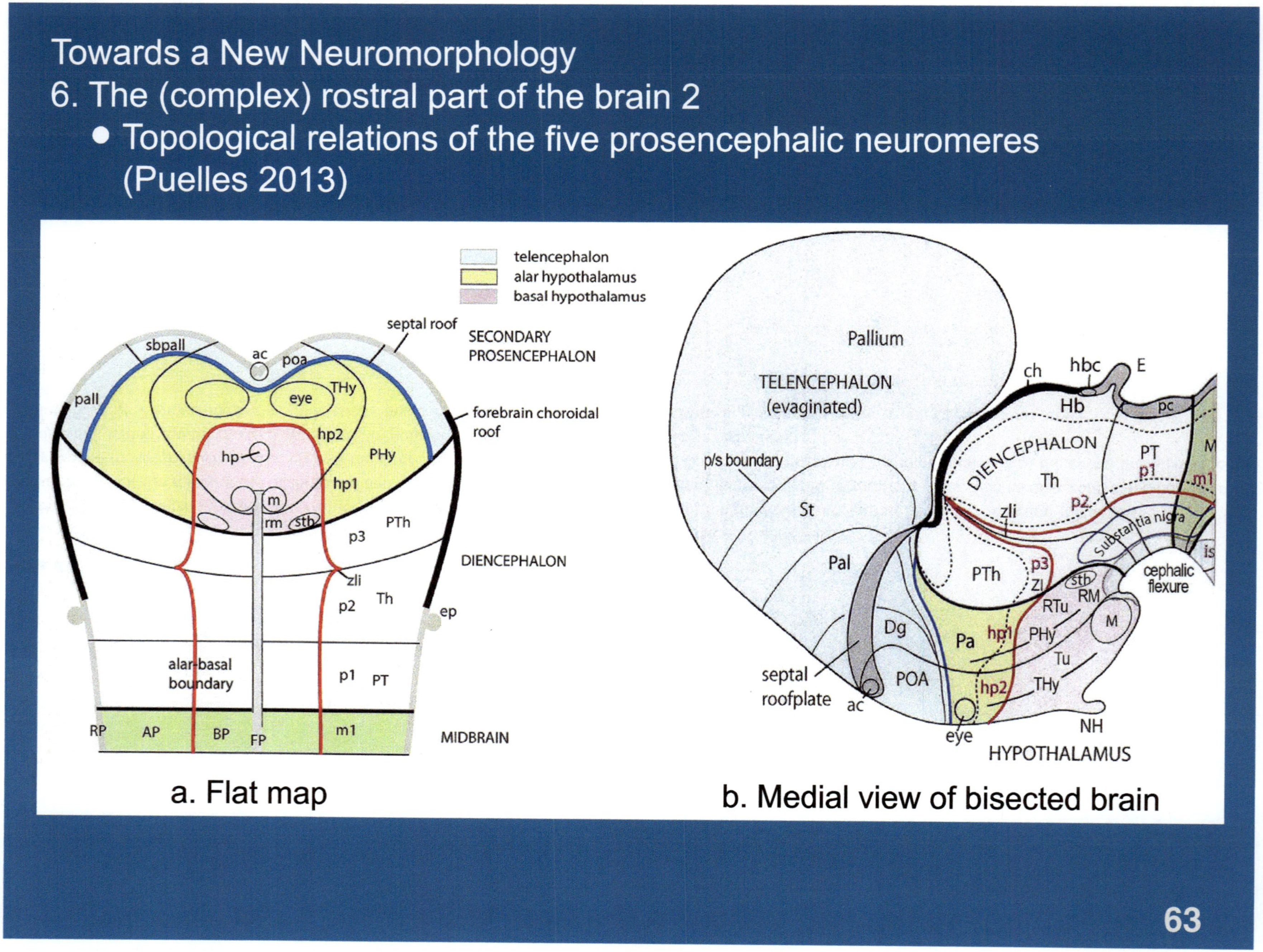
Towards a New Neuromorphology
6. The (complex) rostral part of the brain 2
• Topological relations of the five prosencephalic neuromeres (Puelles 2013)
telencephalon
alar hypothalamus
basal hypothalamus
SECONDARY PROSENCEPHALON
DIENCEPHALON
MIDBRAIN
septal roof
forebrain choroidal roof
pall
sbpall
ac
poa
eye
THy
hp2
hp
PHy
hp1
m
rm
sth
p3
PTh
zli
p2
Th
ep
alar-basal boundary
p1
PT
RP
AP
BP
FP
m1
a. Flat map
Pallium
TELENCEPHALON (evaginated)
p/s boundary
St
Pal
septal roofplate
ac
Dg
POA
ch
hbc
E
Hb
pc
DIENCEPHALON
Th
PT
p1
M
m1
zli
p2
Substantia nigra
is
cephalic flexure
PTh
p3
ZI
sth
RM
RTu
Pa
hp1
PHy
M
Tu
hp2
THy
eye
NH
HYPOTHALAMUS
b. Medial view of bisected brain
63

Plate 63

Plate 63 presents a flat map (Fig. a) and a medial view of the bisected prosencephalon (Fig. b) of a generic mammalian embryo, showing the topological relations of the fundamental anteroposterior and dorsoventral subdivisions of this part of the brain, as well as the location of some grisea (areas, nuclei) differentiating within these fundamental entities.

The following overview is principally based on the review articles of Puelles et al. (2012) and Puelles (2013), to which we refer readers for details and references.

As already discussed (Plate 52), the prosencephalon consists of five prosomeres, p1, p2 and p3, which constitute together the diencephalon proper, and the hypothalamic prosomeres hp1 and hp2, which form the secondary prosencephalon. The latter structure encompasses the telencephalon and the hypothalamus.

The neural tube ends rostrally at the terminal wall, which extends from the locus of the anterior commissure (ac) to the mamillary pouch (M, m).

The molecularly defined alar/basal (a/b) boundary is indicated in the figures by a continuous red line. Due to the cephalic flexure (Plate 63, Fig. b), this longitudinal boundary bends ventrally in the prosencephalon. In the diencephalon it runs under the zona limitans intrathalamica (zli), which marks the limit between the alar portions of p2 and p3. Rostrally the a/b boundary is continuous from left to right just under the optic stalks. The figures show clearly that the a/b boundary divides all of the five prosomeres into basal and alar domains.

The basal sectors of p1–p3 form together the diencephalic tegmentum. This region contains the rostral parts of the ventral tegmental area and the substantia nigra, both of which represent in fact continuous isthmo-meso-diencephalic entities. The alar region of p1 forms the pretectum (PT). The differentiation of this region will be discussed in some detail later on in this course. As seen in Plate 63, Fig. b, the alar region of p2 gives rise to the habenular nuclei (Hb) and the dorsal thalamus (Th). The dorsal thalamus develops into a large multinuclear complex, which entertains strong reciprocal connections with the neocortex. Due to the expansion of the dorsal thalamus the alar region of p3, which is known as the prethalamus (PTh), is gradually reduced to a thin shell covering the anterior and lateral aspects of the dorsal thalamus. Two cell masses, the zona incerta (zi) and the reticular thalamic nucleus, represent the major derivatives that develop within the confines of the prethalamic region.

The first hypothalamic prosomere (hp1) includes the morphologically caudal or peduncular hypothalamus and the bulk of the telencephalon. The term 'peduncular' hypothalamus refers to the fact that the medial and lateral forebrain bundles (forming the so-called 'telencephalic peduncle') traverse this region dorsoventrally, before bending into a longitudinal course in the rostral diencephalic tegmentum. The telencephalic part of hp1 expands enormously and evaginates into the bilateral telencephalic vesicles.

The basal peduncular hypothalamus includes the retromamillary (RM) and retrotuberal (RTu) regions, and gives rise among other entities to the subthalamic nucleus (sth), whereas the alar peduncular hypothalamus is represented by the paraventricular region (Pa). The magnocellular hypothalamic neurons of the main paraventricular nuclear complex arise from this region.

The evaginated telencephalic vesicles can be divided into the pallium and the subpallium. Differential gene expressions define the pallio/subpallial (p/s) boundary, which is already detectable in late neural plate stages. The evaginated subpallium is divisible into separate diagonal (Dg), pallidal (Pal) and striatal (St) domains. The pallium differentiates into four mediolaterally arranged zones, which are designated as medial pallium, dorsal pallium, lateral pallium and ventral pallium. In mammals, the dorsal pallium differentiates into the highly complex neocortex.

The second hypothalamic prosomere, hp2, includes the prepeduncular or terminal hypothalamus (THy), whose alar subregion includes the evaginated optic vesicle and stalk, and a small, unevaginated portion of the telencephalon occupied by the preoptic area (POA). The basal terminal hypothalamus includes the mamillary (M) and tuberal (Tu} regions, and gives rise to the neurohypophysis (NH) and the median eminence within its acroterminal area. The eye vesicles and surrounding supraoptic, suprachiasmatic and anterior hypothalamic areas develop from the alar terminal hypothalamus.

The (unevaginated) preoptic area (POA) belongs to the telencephalic subpallium as a distinct subdomain on the basis of shared and differential molecular profiles. This classification is further supported by the finding that the preoptic area contributes to a stream of tangentially migrating neuroblasts that target the telencephalic pallium, just as for the three other parts (Dg, Pal, St) of the subpallium.

Summary

1. The entire CNS lies dorsal to the notochord (i.e. is epichordal in position) (Puelles et al. 2012).
2. The prosencephalon is composed of five prosomeres, p1, p2, p3, hp1, and hp2, each of which includes its own floor plate, basal plate, alar plate and roof plate sectors.
3. The terminal wall of the neural tube extends from the rostral end of the floor plate (the mamillary region) to the rostral end of the roof plate (the anterior commissure bed). It is topologically transverse.
4. Rostrally, the right and left basal- and alar plates are directly continuous across the median terminal wall, which is thus singularly patterned along the dorsoventral dimension, irrespective of its midline topography.
5. The alar/basal boundary is continuous from left to right just under the primordium of the optic chiasm.
6. The fundamental morphological entities of the prosencephalon, i.e. the basal and alar domains of the five prosomeres can be projected to form a neural plate-like flat map.
7. The prosomeres p1–p3 form together the diencephalon proper.
8. The alar domains of p1–p3 differentiate into three nuclear complexes, the pretectum, dorsal thalamus and prethalamus, respectively.
9. The hypothalamic prosomeres hp1 and hp1 form together the secondary prosencephalon, a structure which includes the hypothalamus (with the eyes) and the telencephalon.
10. The hypothalamus is divided into peduncular (caudal) and terminal (rostral) neuromeric portions (hp1 and hp2), continuous dorsally with the telencephalic hemispheres and the preoptic area (unevaginated telencephalon), respectively.
11. The telencephalon can be divided into a rostrally situated subpallium and a caudally located pallium. Note this corrects the widely held wrong notion that the pallium lies ‘dorsal’ to the subpallium; this obsolete view was based on the assumption that the forebrain axis ends in the telencephalon.
12. Within the telencephalon eight principal cellular zones can be delineated, the subpallial preoptic, diagonal, pallidal and striatal zones, and the medial, dorsal, lateral and ventral pallial zones.

Towards a New Neuromorphology
7. The natural coordinate system of the CNS 1

The CNS of vertebrates contains its own, built-in, natural system of coordinates. This system includes two natural planes, i.e. (**a**) the ventricular and (**b**) the meningeal surfaces of the neural tube; (**c**) a set of radially oriented curves, termed radii or vectors,which connect these two surfaces; (**d**) a set of tangenital curves, which connect the floor- and roof plates of the neural tube; (**e**) the bent axis of the neural tube, and (**f**) the deformed transverse interneuromeral planes.

A. Cartesian coordinate system

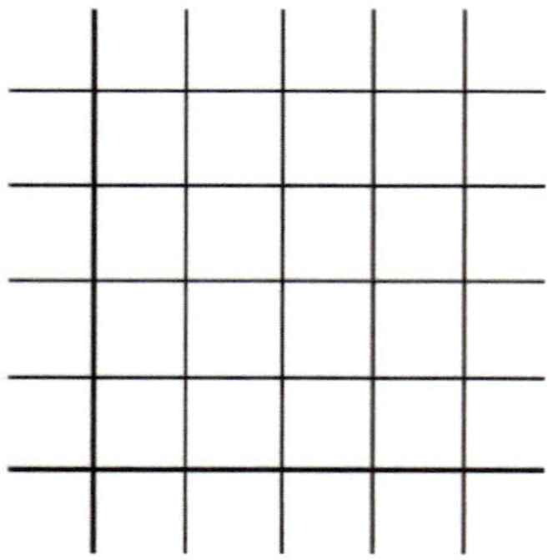

B. Natural coordinate system CNS

transverse section

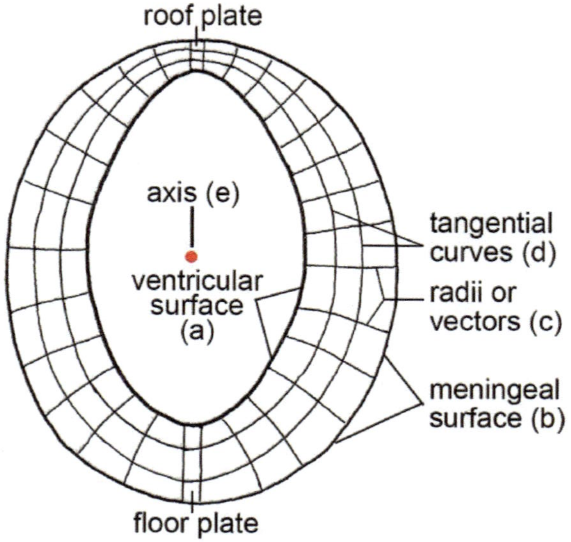

lateral view

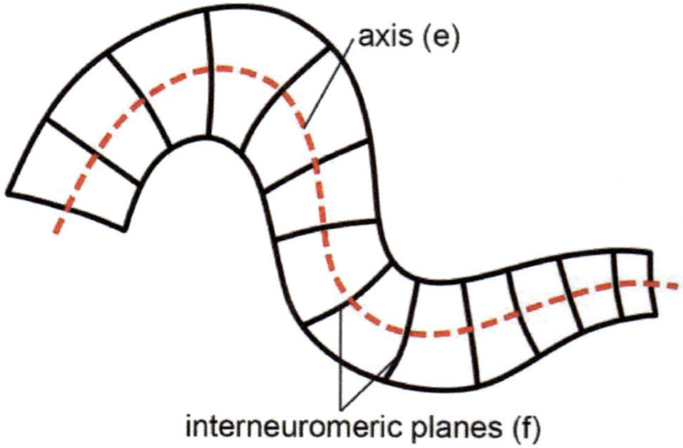

7 The Natural Coordinate System of the CNS

Plate 64

We have already pointed out that the orthogonal or Cartesian coordinate system (Plate 64, Fig. A) used in descriptive anatomy and adult brain stereotaxis, is fully inadequate to characterize the true morphological directional and positional relations within developing and adult brains, due to various accumulated morphogenetic deformations which start with the closure of the neural tube itself (Sect. 1.5; Plate 13). Fortunately, the CNS of vertebrates happens to contain its own, built-in, natural system of coordinates (Nieuwenhuys 1972, 1998b, 2009c, 2011b). Knowledge of this system is of paramount importance for understanding the fundamental morphological organization of the vertebrate neuraxis.

We have seen that during the early morphogenesis of the CNS, the neural plate is transformed into the neural tube. In this structure we distinguish the bilateral lateral plates, which are interconnected dorsally and ventrally by the roof plate and floor plate, respectively. These surround a central, fluid-filled ventricular cavity. The natural coordinate system referred to above includes (a) two natural planes, i.e. the ventricular and meningeal surfaces of the neural tube (representing the apical and basal surfaces of the neuroepithelium respectively); (b) a set of radially oriented curves, termed radii or vectors, which connect these two surfaces, (c) a set of tangential curves parallel to the ventricular and outer surfaces, which connect the floor- and roof plates and (d) a series of topologically transverse (or curved) planes representing the interneuromeric boundaries along the axial dimension. The geometric axis of the brain (e), either imagined at the centre of the ventricular cavity, or represented by the alar/basal boundary, as was discussed in Sect. 5, represents the third dimension of the natural coordinate system (Plate 64, Figs. B).

R. Nieuwenhuys and L. Puelles, *Towards a New Neuromorphology*,
DOI 10.1007/978-3-319-25693-1_7

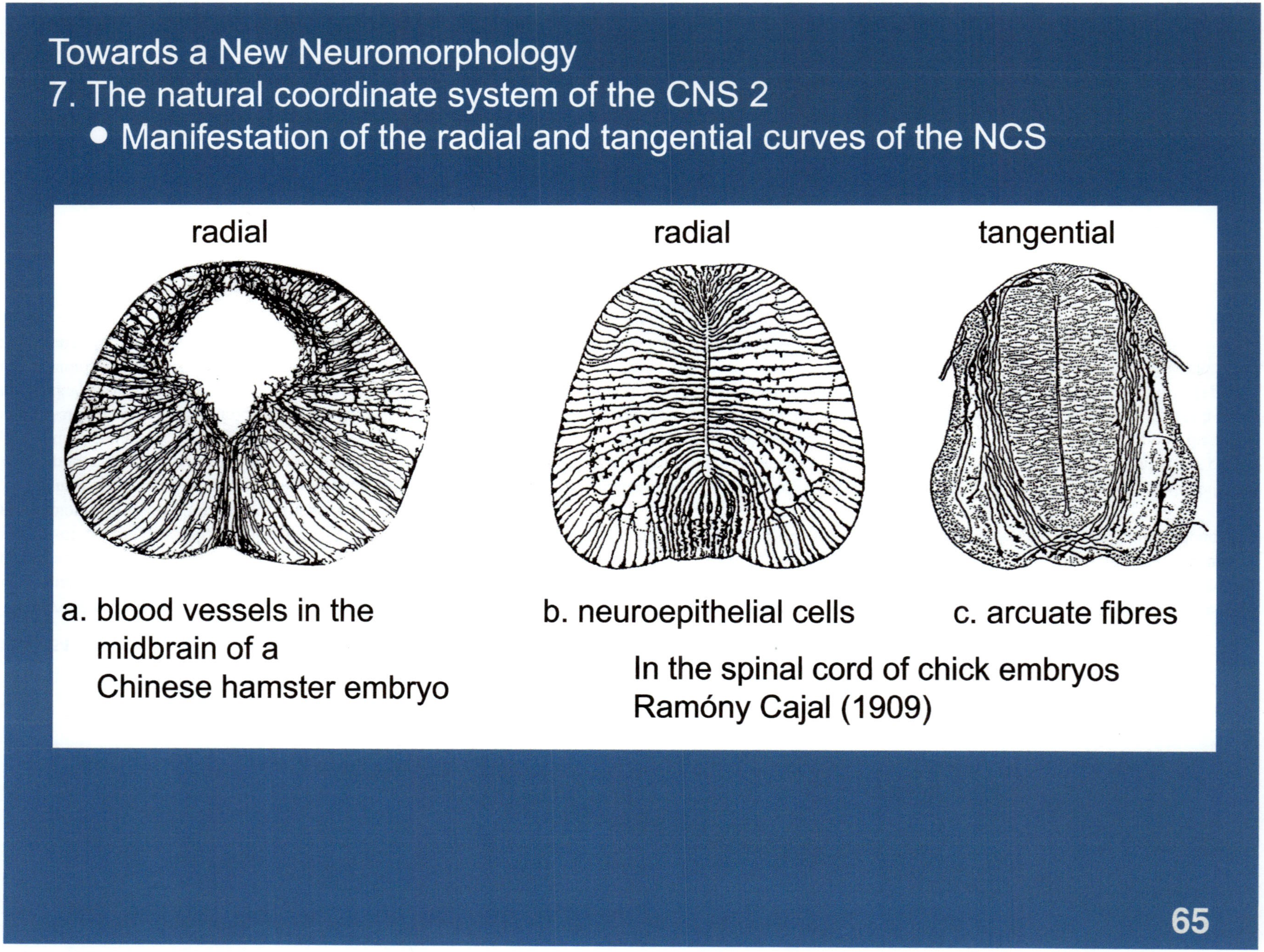
Towards a New Neuromorphology
7. The natural coordinate system of the CNS 2
• Manifestation of the radial and tangential curves of the NCS
radial
radial
tangential
a. blood vessels in the midbrain of a Chinese hamster embryo
b. neuroepithelial cells
c. arcuate fibres
In the spinal cord of chick embryos
Ramóny Cajal (1909)
65

Plate 65

Plate 65 presents some properties of the radial and tangential curves of the natural coordinate system.

The radial curves or vectors manifest themselves in the direction and orientation of:

1. the neuroepithelial cells, which form the neural plate and early neural tube (Plate 17, Fig. D);
2. the neural matrix or progenitor cells, which span the width of the lateral plates during early development (Plate 65, Fig. b);
3. the radial glial cells, which are present during early neurogenesis in all vertebrates, and throughout development in most anamniotes;
4. many penetrating blood vessels (Plate 65, Fig. a).

It is of note that the processes of the elements mentioned under 2 and 3 play an important role as guidelines during the migration of neuroblasts or their precursors (see Sect. 8). This means that many developing neurons shift their positions in a simple radial direction within the natural coordinate system. The neuroblasts and their precursors (progenitor neuroepithelial cells) are products of mitosis which predominantly occur in the immediate vicinity of the ventricular surface; consequently, this surface can be characterized as the 'starting' or 'zero' plane of neurogenesis.

The tangential component of the natural coordinate system manifests itself in the course of 'arcuate fibres', i.e. axonal processes which during early development pass ventrodorsally/dorsoventrally, directly peripheral to the matrix layer (Plate 65, Fig. c). Numerous additional arcuate fibres emerge during further development. These fibres form, just like the radial processes of the matrix cells and the radial glial elements, an important substrate for the migration of neuroblasts, though implying in this case tangentially oriented migrations. Finally, it is of note that numerous molecularly defined ('geno-architectural') boundaries of developing and adult grisea (e.g. cell strata and layers) adhere to the course of the natural radial and/or tangential curves of the NCS.

In general, the natural coordinate system manifests itself more clearly in embryonic and fetal than in adult brains. Nevertheless, according to our experience, traces of the primary organization are always detectable in the adult CNS.

In transverse sections through the brain of adult birds and mammals, stained for myelinated fibres, the natural coordinate system can generally be read off from: (1) the direction of the central fibres of cranial nerves; (2) the course of arcuate fibres; (3) the course of some blood vessels, and (4) the 'lattice' pattern formed by the interstices between the bundles of longitudinally running fibres (Plate 67). The neuromeric structural units and their topologically transverse boundaries are best visualized in sagittal sections, or in true horizontal sections with regard to each longitudinal sector of interest, due to the variable axial curvature. This scenario needs to be remembered when studying cross-sections, since often these are not properly transverse (perpendicular) to the local part of the axis.

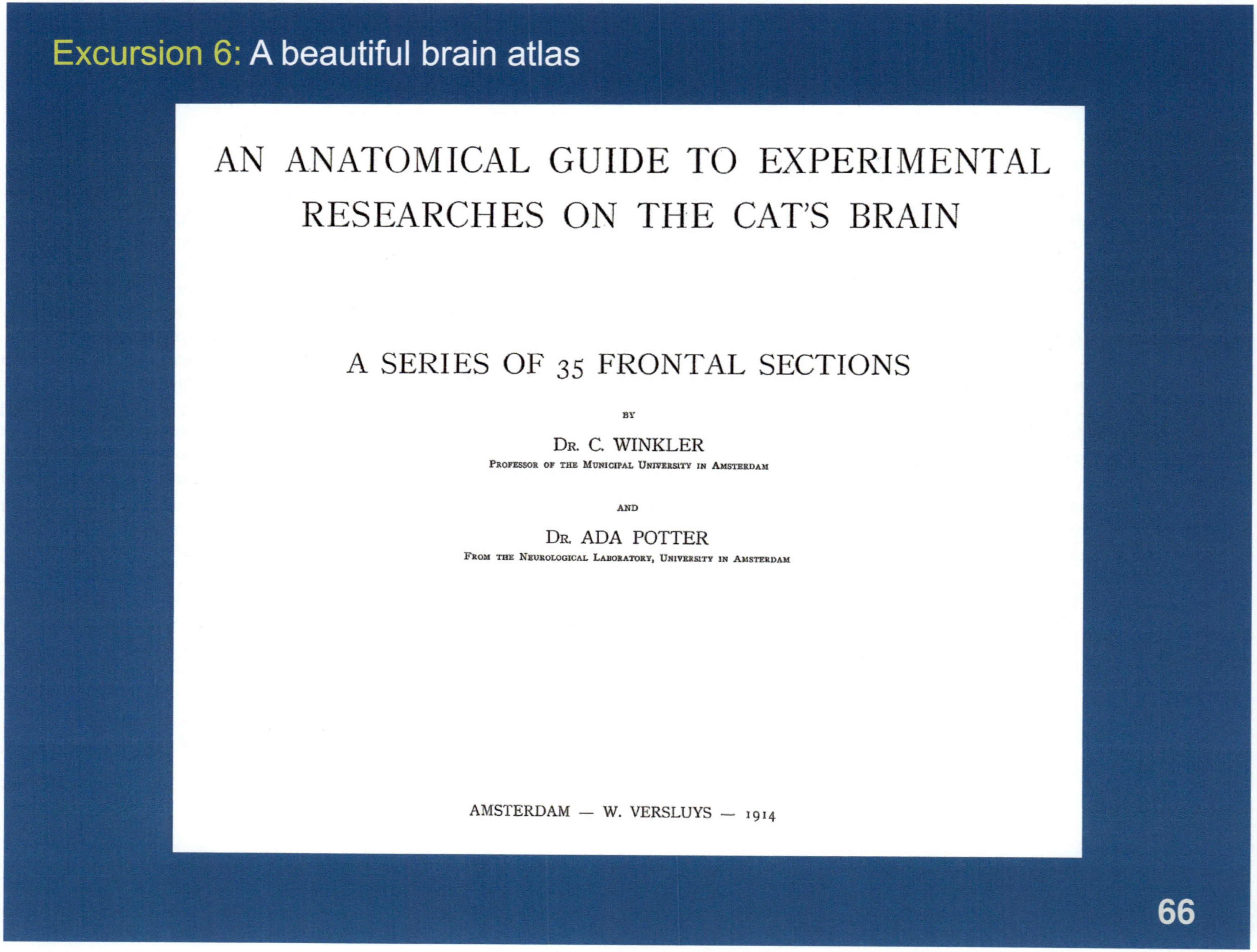
Excursion 6: A beautiful brain atlas
AN ANATOMICAL GUIDE TO EXPERIMENTAL
RESEARCHES ON THE CAT'S BRAIN
A SERIES OF 35 FRONTAL SECTIONS
BY
DR. C. WINKLER
PROFESSOR OF THE MUNICIPAL UNIVERSITY IN AMSTERDAM
AND
DR. ADA POTTER
FROM THE NEUROLOGICAL LABORATORY, UNIVERSITY IN AMSTERDAM
AMSTERDAM — W. VERSLUYS — 1914
66

Excursion 6: A beautiful brain atlas

Plate 66

This excursion has the purpose of drawing attention to the most beautiful atlas of the cat's brain, produced by Cornelis Winkler and Ada Potter (1914).

The drawings prepared with great skill and patience by the second author are based on transverse sections stained for neuronal somata with the Nissl technique and for myelinated fibres with the Weigert-Pal technique.

Excursion 6: A beautiful brain atlas

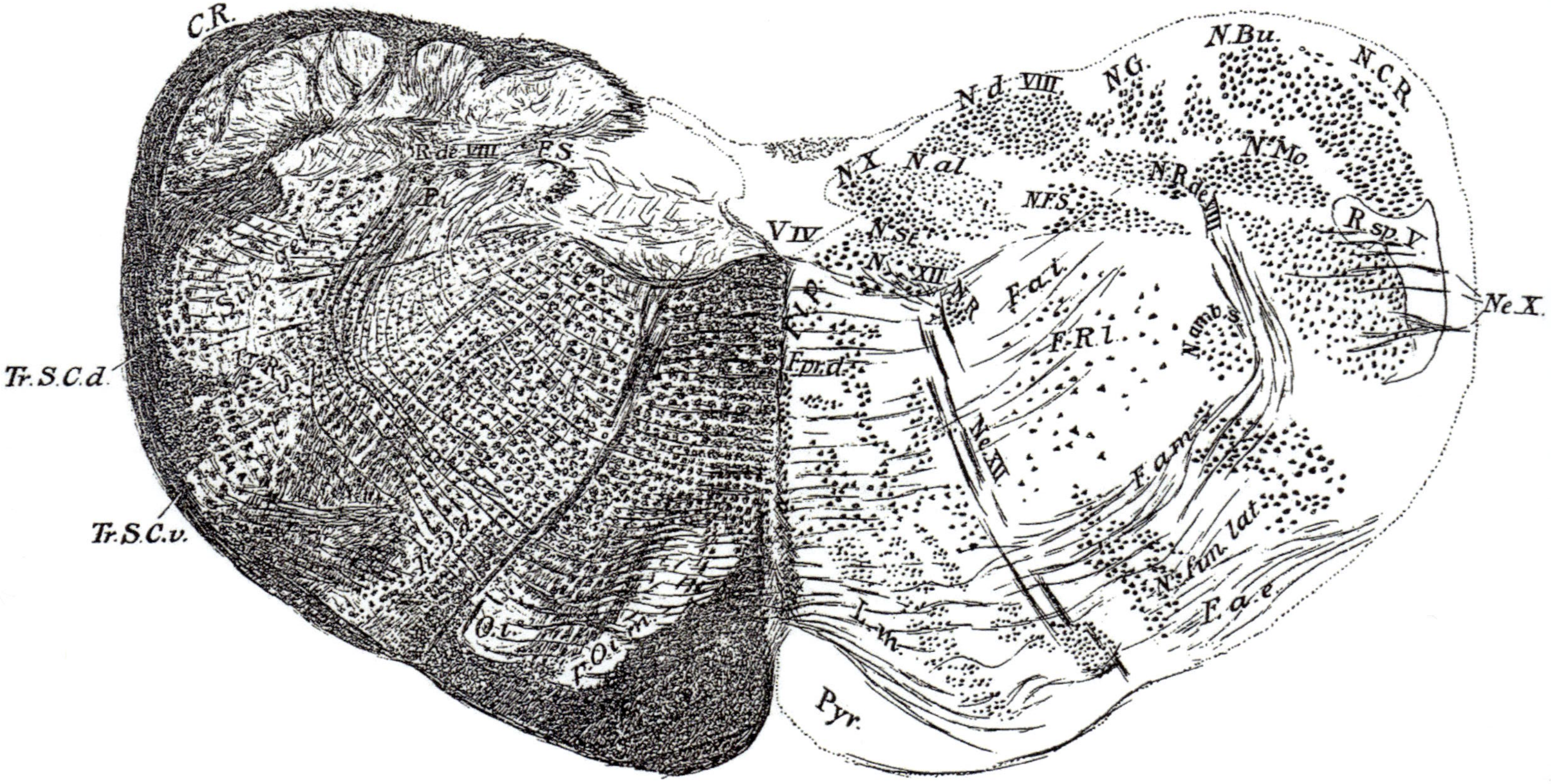

transverse section through the
rhombencephalon of the cat
(homage to Ada Potter)

67

Plate 67

Plate 67 presents one of the 35 plates of Winkler and Potter's atlas, showing a section through the caudal part of the rhombencephalon. The left half of the plate shows the fibre systems, the right half the cell masses.

Traces of the natural coordinate system can be recognized in: (1) the radial course of the fibres of the Xth and XIIth cranial nerves; (2) the tangential course of the numerous arcuate fibres passing from the alar to the basal region, and (c) the typical lattice pattern formed by the interstices between the longitudinally running fibre bundles in the reticular core of the section.

Summary

1. The CNS of vertebrates contains its own, non-cartesian, built-in, natural coordinate system.
2. The natural coordinate system includes (a) a curved long axis and two natural limiting curved planes: the ventricular and meningeal surfaces of the neural tube; (b) a set of radially oriented curves, termed radii or vectors, which connect point to point the two natural planes; (c) a set of tangentially oriented curves which connect the floor and roof plates, parallel to the limiting curved planes and (d) a set of transverse interneuromeric boundaries (visible in sagittal and horizontal sections), variously deformed from their theoretical mutual parallelism according to the axial curvature at each anteroposterior level.
3. The natural coordinate system manifests itself in (a) the orientation of neuroepithelial and neural matrix cells; (b) the course of the processes of ependymal gliocytes: (3) the course of blood vessels penetrating the brain across the meningeal surface: (4) the course of arcuate fibres, and (5) the orientation of the molecularly defined boundaries of many grisea in the developing and adult CNS, and (6) the radial and tangential migration routes of various neuronal populations.
4. A knowledge of the three-dimensional configuration of the natural coordinate system is indispensable for understanding the histogenesis and the fundamental morphological organisation of the vertebrate neuraxis.

Towards a New Neuromorphology
8. The fundamental morphological units of the CNS
8.1 Orientation

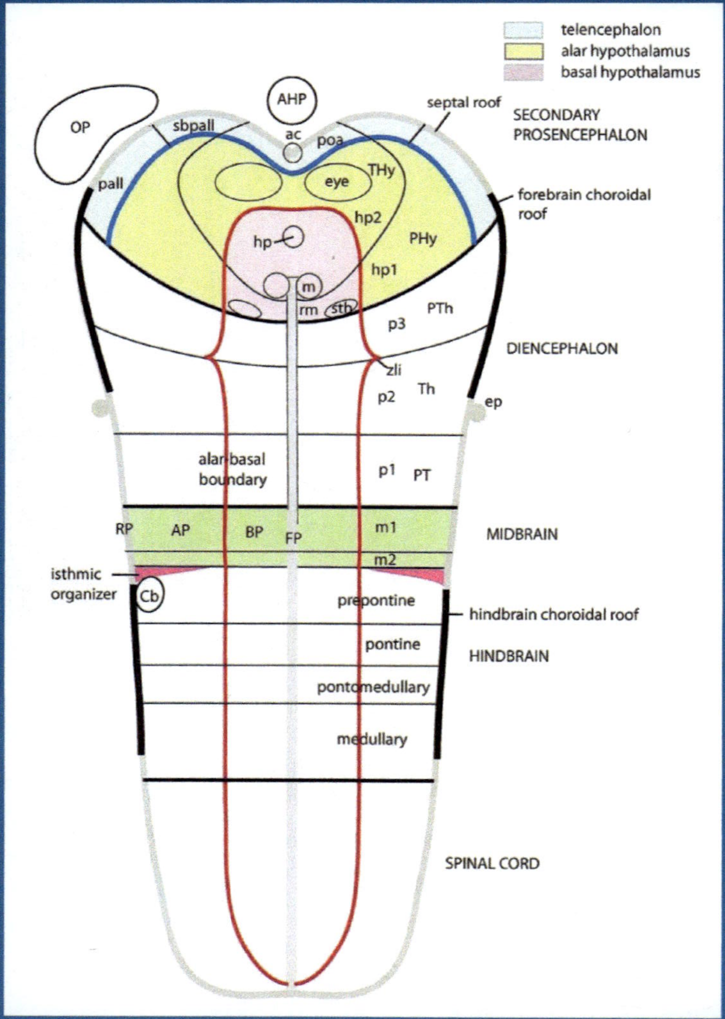

n.b. All (major) morphological domains are projected back on the ventricular surface

Flat map of the mammalian CNS (Puelles 2013)

8 The Fundamental Morphological Units (FMUs) of the CNS

8.1 Orientation

Plate 68

Plate 68 shows a model, flat map or blueprint of the (mammalian) CNS, in which the fundamental anteroposterior and ventrodorsal subdivisions are projected upon the ventricular surface (Puelles 2013). The spinal cord is shortened for simplicity. The alar/basal boundary is indicated as a thick red line. Some regions are coloured for contrast: the floor plate ends in the mamillary region and appears in grey; the midbrain with its two mesomeres is shaded in green, and the basal hypothalamus, alar hypothalamus and telencephalon are coloured pink, yellow and blue respectively.

R. Nieuwenhuys and L. Puelles, *Towards a New Neuromorphology*,
DOI 10.1007/978-3-319-25693-1_8

Towards a New Neuromorphology

8. The fundamental morphological units of the CNS

8.1 Orientation

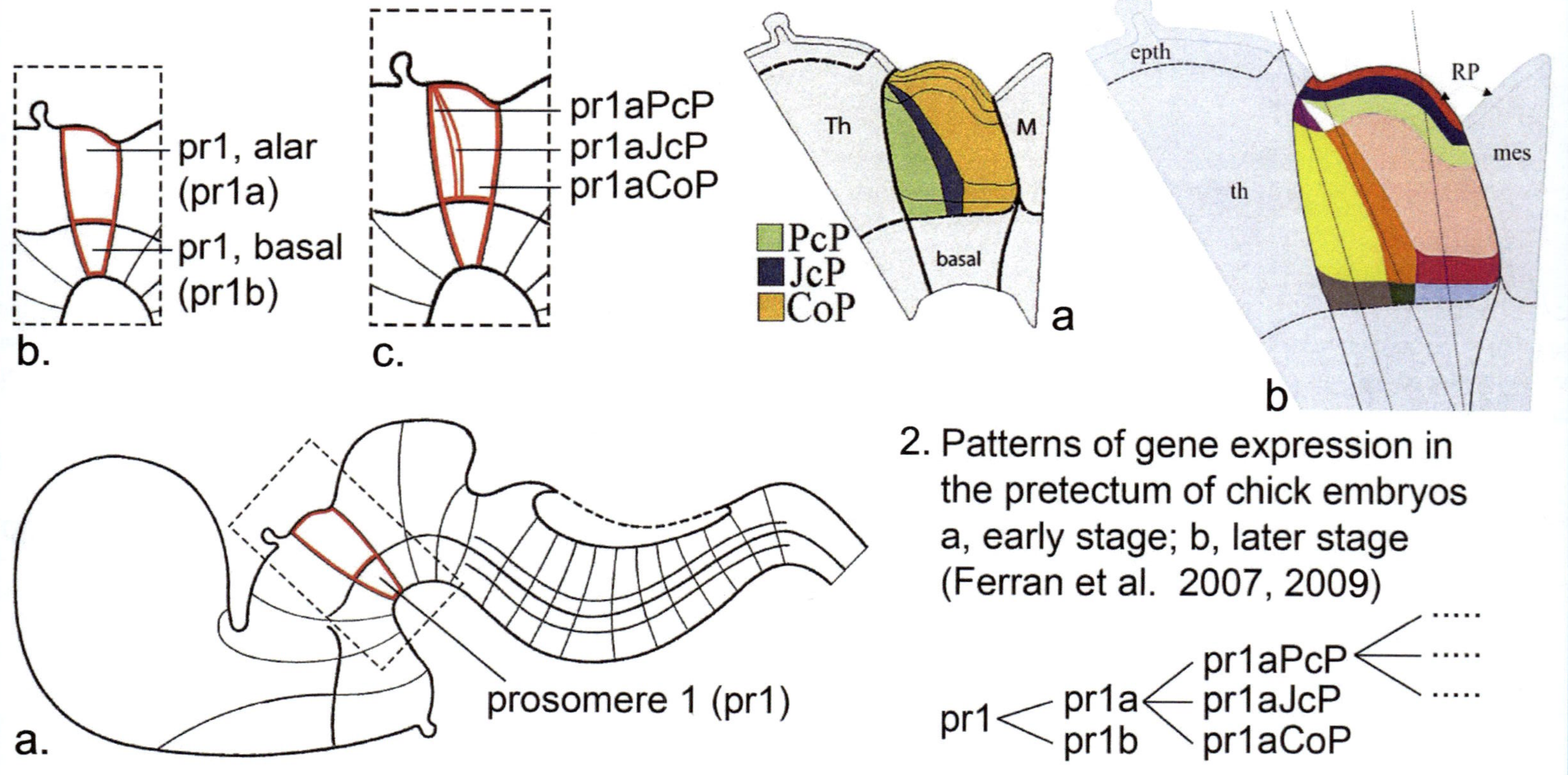

1. Schemes showing position and differentiation of the pretectum

2. Patterns of gene expression in the pretectum of chick embryos a, early stage; b, later stage (Ferran et al. 2007, 2009)

69

Plate 69

In order to elucidate the significance and the progressive differentiation of the fundamental morphological units (FMUs), we introduce here prosomere 1 as a model system.

Plate 69, Fig. 1 summarizes the early differentiation of prosomere 1 (pr1) in three schemes. Figure 1a shows the position of pr1, i.e. the most caudal of the three diencephalic neuromeres. In Fig. 1b it is seen that pr1 is subdivided from the very beginning into a basal region (pr1b) and an alar region (pr1a); the latter region is known as the *pretectum* (usually this concept also includes the local roof plate, occupied by the posterior commissure and the subcommissural organ). Figure 1c shows that the pretectum differentiates early into three rostrocaudally arranged areas, the precommisural- (pr1aPcP), juxtacommissural- (pr1aJcP), and commissural pretectum (pr1aCoP). This early tripartitioning of the pretectum is based on gene expression studies in chicken and mouse embryos (Ferran et al. 2007, 2008), as well as in *Xenopus* larvae (Morona et al. 2011), and presumably holds for all tetrapods.

Plate 69, Fig. 2 shows the molecular regionalization in the chicken pretectum at (2a) an early (2.5–3 days of incubation; HH17–18), and (2b) a later (4 days of incubation; HH23–25) developmental stage (Ferran et al. 2007). The early stage (2a) shows the tripartitioning of the pretectum, already mentioned. In (2b) the regionalization is considerably more advanced. The precommissural -, juxtacommissural -, and commissural pretectal domains are now differentiated dorsoventrally into three, three, and five subdomains, respectively. In (2c) the first steps in the progressive development (or *ontology*; cf. Puelles et al. 2013) of the pretectum of the chicken are summarized.

Because the analysis of the progressive regionalization of such areas as the pretectal region is entirely based on the localization of genes or gene products, this type of analysis is designated as *genoarchitectonics,* and the resultant parcellations as *genoarchitecture* (Ferran et al. 2009; Puelles and Ferran 2012).

In all of the figures in this plate, the developing genoarchitectonc entities are depicted as seen from the ventricular side of the neural tube. It is important to note that all of these entities represent FMUs, because they extend radially from the ventricular surface to the meningeal surface of the neural tube.

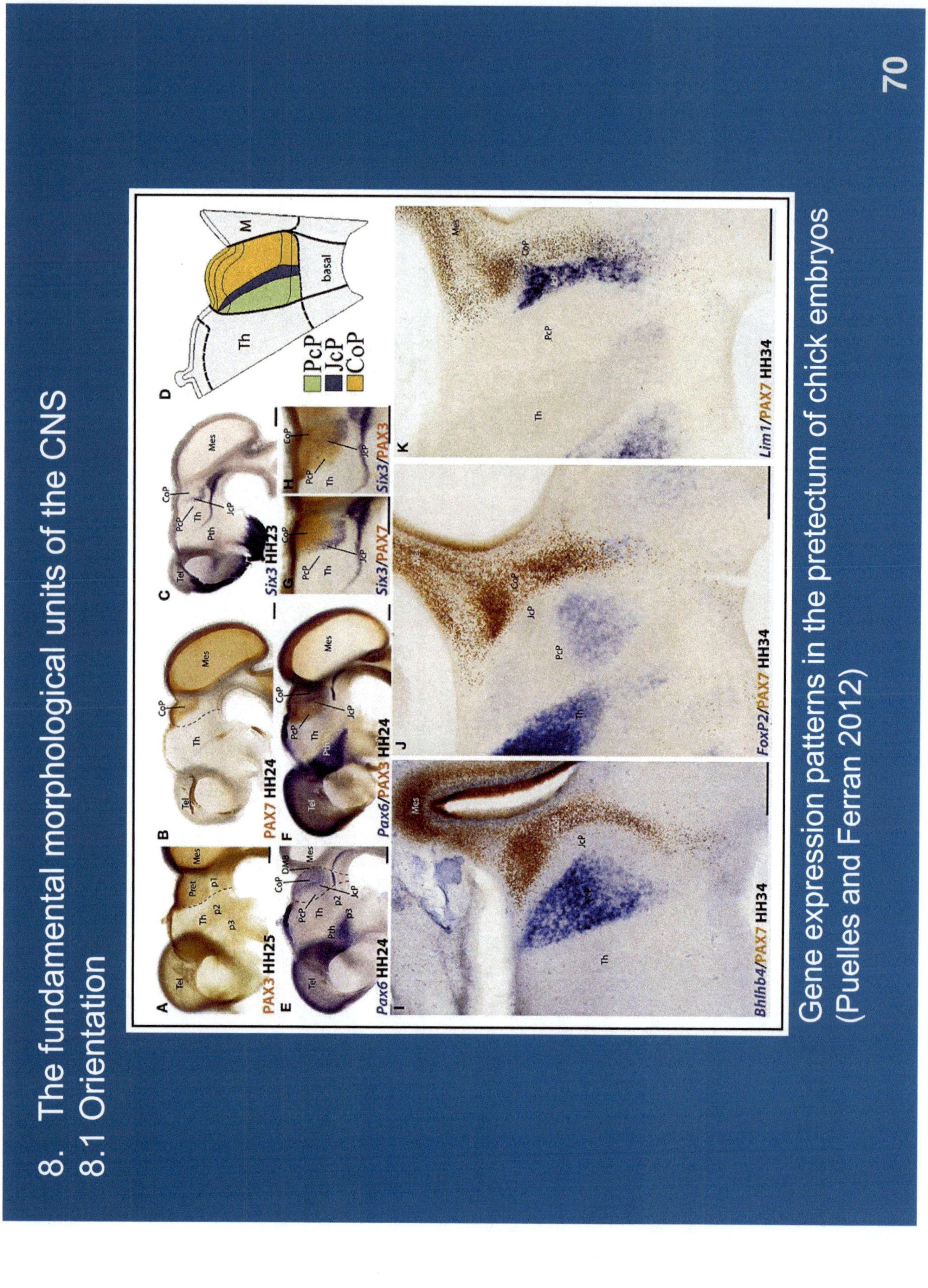
8. The fundamental morphological units of the CNS
8.1 Orientation
PAX3 HH25
PAX7 HH24
Six3 HH23
Pax6 HH24
Pax6/PAX3 HH24
Six3/PAX7
Six3/PAX3
PcP
JcP
CoP
Th
M
basal
Bhlhb4/PAX7 HH34
FoxP2/PAX7 HH34
Lim1/PAX7 HH34
Gene expression patterns in the pretectum of chick embryos
(Puelles and Ferran 2012)
70

Plate 70

Plate 70 illustrates an example of differential expression of transcription factors in the developing chick brain, in a genoarchitectonic analysis of the pretectal region made by Ferran et al. (2007, 2009). Panels (A–H) show half-brain whole-mounts of embryonic chick brains at 4–5 days of incubation (HH stages are indicated), immuno-reacted or in situ-reacted for various transcription factor genes that establish boundaries defining the pretectum and its main subdivisions. In (A), *Pax3* marks the pretecto-thalamic border (dashed line, see also (B). In (B), *Pax7* is restricted to the caudal, commissural pretectum (CoP). In (C), *Six3* labels the thin, intermediate, juxtacommissural pretectal domain (JcP). Rostral to JcP there remains the precommissural pretectal domain (PcP) adjacent to the thalamus. In (D), for orientation, the schema of the tripartitioned pretectum, intercalated between the midbrain (M) and the thalamus (Th), is shown here once again. In (E), the *Pax6* pattern in (E) is restricted to CoP (just as *Pax7* is in B), but extends ventrally all the way to the alar/basal boundary. Note that JcP and PcP are unlabeled in (E), but double labeling with *Pax6* and *Pax3* in (F) reveals their position. Figures (G) and (H) show comparison between the patterns of *Six3*, *Pax3* and *Pax7* expression in the pretectum, where in (G) the CoP domain is defined by *Pax7* whilst JcP expresses *Six3*, in(G) the whole pretectum is marked by *Pax3* (H). All these distinct areas are interpreted as differentially coded progenitor areas likely to produce different neuronal populations (i.e., as radial histogenetic units or migration areas). The panels (I–K) show three thick, mediolaterally arranged sagittal sections through the pretectum of the chick at 8 days of incubation (HH34). In all cases two markers are labeled either in blue (in situ) or in brown (see lower left corner for marker identification). The observed expressions largely pertain to postmitotic neurons born at the PcP, JcP, and CoP pretectal domains, which are now differentiating in the mantle layer, superficial to their original progenitors. Note that the respective boundaries between the three domains are well conserved (PcP, blue; JcP, unlabeled; CoP, brown in I and J), and that the JcP domain is labeled differently from the CoP (K).

8. The fundamental morphological units of the CNS
8.2 The third dimension

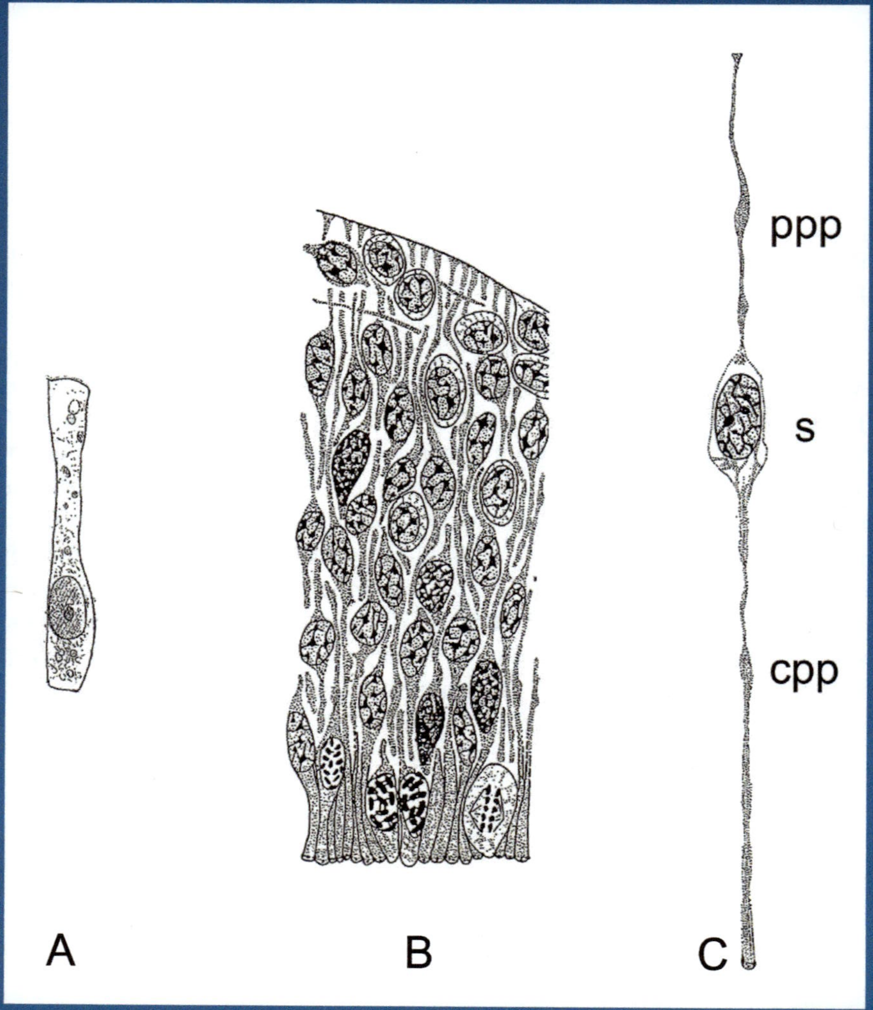

71

8.2 The Third Dimension

Plate 71

All fundamental morphological units (FMUs) in the CNS occupy their own, specific locus at the ventricular surface of the neural tube, and hence their position can be fully characterized in flat maps of the ventricular surface of the CNS (see Plate 68). However, the CNS is a three-dimensional structure and so are the FMUs. The third dimension, which is (topologically) oriented perpendicularly to the ventricular surface, is represented by the thickness of the neural wall. This thickness increases steadily during ontogeny and may attain amazing proportions (10 cm or more).

Plate 71 serves to bring the third dimension of the developing CNS into focus. It shows (Fig. A) a typical neuroepithelial cell from the neural plate of a shark embryo (His 1889); (Fig. B) a small part of a section through the neural tube of a 10-mm pig embryo (Sauer 1935b), and (Fig. C) an individual undifferentiated matrix cell from the neural tube of a pig embryo of the same age (Sauer 1935a). In all of the figures the (future) ventricular surface is at the bottom and the (future) meningeal surface at the top. It has already been mentioned (Plate 32) that (Fig. B) shows the appearance of the cells at one focal level only, and that Sauer (1935b) was able to trace most of the elements depicted throughout the width of the neural wall by focusing.

It is highly remarkable that the undifferentiated matrix cells in the neural tube retain during development their contacts with the ventricular and meningeal surfaces and, due to the increase in thickness of the neural wall, become transformed into greatly stretched elements on which a central primitive process (cpp), a peripheral primitive process (ppp), and a soma (s) can be distinguished (c). It should be reemphasized that these elongated matrix cells strictly adhere in position to the radii or vectors of the natural coordinate system, according to local degrees of transverse and longitudinal curvature of the neural wall.

8. The fundamental morphological units of the CNS

8.3 These FMUs are radial entities and developmental modules

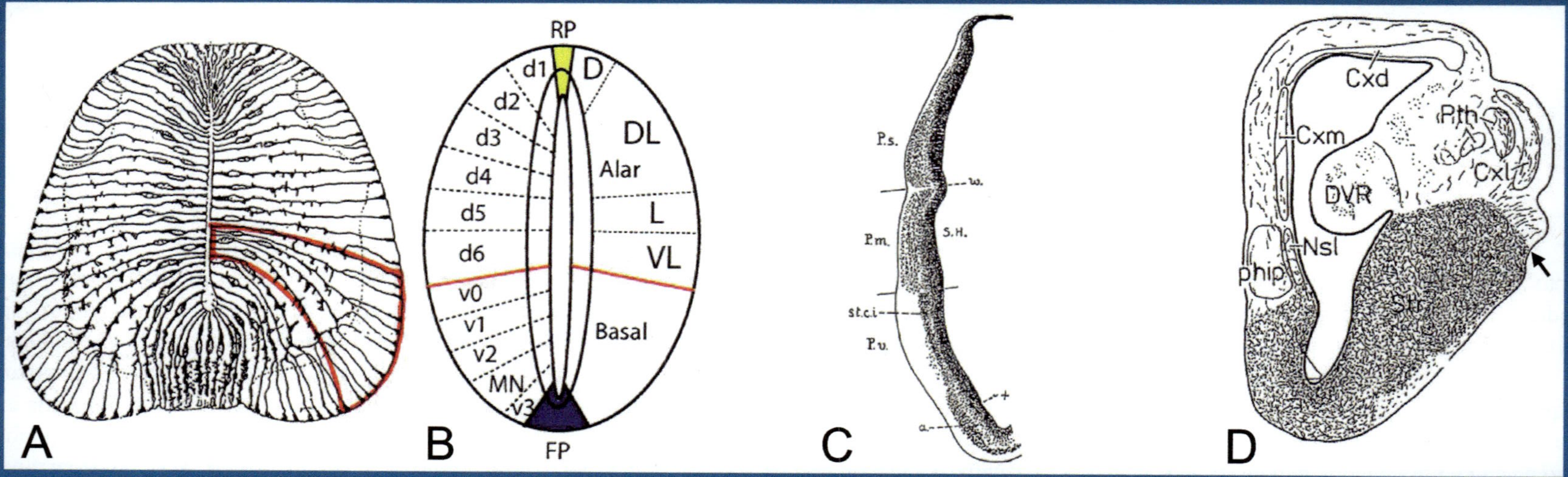

A. Transverse section through the spinal cord of a 4-days-old chick embryo showing the radial orientation of neuroepithelial cells (Ramón y Cajal 1909). The outline of a (hypothetical) FMU is marked with a red line

B. Schematic transverse section through the spinal cord of a mammalian embryo (Puelles 2013). The alar and basal regions are both subdivided into a number of microzones.

C. Horizontal hemisection through the caudal prosencephalon of a 4-day-old chick embryo (Rendahl 1924). The boundary between te areas P.s and P.m is marked by a cell-free zone; the areas P.m and P.v show considerable cytoarchitectonic differences.

D. Transverse hemisection through the telencephalon of a turtle, stained with an antibody against dopamine (Smeets 1988). The dopaminergic fibres are strictly confined to the striatal domain, thus identifying the striopallial boundary (arrow). Acetylcholinesterase histochemistry yields similar results

72

8.3 The FMUs Are Radial Entities and Developmental Modules

Plate 72

Plate 72 illustrates some important features of FMUs defining their nature and developmental properties.

1. The boundaries of the FMUs

These can be completely defined in terms of the natural coordinate system. All of these entities have:

a. their own specific area of ventricular surface;
b. their own specific area of meningeal surface, and
c. border zones which are radially or vectorially oriented throughout their extent (Plate 72, Fig. A).

2. The morphogenesis of the FMUs

FMUs, once identified at early embryonic stages, can generally be traced throughout development, provided that adequate identification techniques are used (see below). Early recognizable FMUs include the roof, alar, basal and floor domains of the 19 brain neuromeres, i.e. 76 primary supraspinal FMUs- and of the (varying numbers of) myelomeres (spinal neuromeres). During further development these initial FMUs, or at least the larger alar and basal ones, quite often split up into smaller radial entities, generally both anteroposteriorly and dorsoventrally. Thus, both the basal and alar domains of the myelomeres are currently known to differentiate into a number of ventrodorsally arranged microzones (Plate 72, Fig. B), and we have seen that the pretectum of tetrapods, i.e. the alar domain of the first prosomere, splits up into three rostrocaudally arranged sectors, and that subsequently each of these sectors differentiates into three or more radial subsectors (Plates 69, 73) These subsequent subdivisions of the primary FMUs may be provisionally designated as ‘secondary’, ‘tertiary’ and ‘final’ FMUs.

3. The identification and delineation of the FMUs

Initially, the FMUs were principally studied in embryonic and larval histological material stained with routine techniques, such as haematoxilin-eosin, haemalum and cresyl violet stainings (e.g. Palmgren 1921; Rendahl 1924; Bergquist 1932). Important criteria for the classic identificaton and delineation of the FMUs included the presence of radially oriented cell-free or cell-poor zones and cytoarchitectonic differences, such as differences in cell type, cell density and lamination patterns (Plate 72, Fig. C). Quite often these architectonic differences reflect differences in the phase of development (heterochrony; see below).

More recently, the application of histochemical and immunohistochemical techniques for the intracellular localization of neurotransmitters, the synthesizing and degrading enzymes of these substances, and other more or less selective neuroactive compounds, has contributed substantially to the delineation of FMUs. The definitive determination of the endlessly debated and disputed localization of the pallio-subpallial boundary in the forebrain of reptiles and birds represents a striking case in point (Plate 72, Fig. D).

During recent decades, the chemoarchitectural approach, although still valid, has been largely superseded by genoarchitectonics, a new subdiscipline which has become possible due to the availability of extensive genomic sequence data in an increasing number of vertebrate species, and of standardized in situ probe preparation and hybridization protocols. These techniques render it possible to visualize regulatory genes that are expressed in particular cells of the developing CNS. The differential expression patterns of these genes reveal the presence of stable and discrete territories, the primary FMUs, which during further development split up into smaller radial units.

The ‘new neuromorphology’ announced in the title of this course, is largely based on the systematic study of gene expression during the development of the CNS in a number of representative vertebrates, within the frame of a rigorous topological approach.

4. The nature of the FMUs: Developmental modules

As will be detailed below, the FMUs are progenitor centres where regionally differentiated matrix cells proliferate, and postmitotic neuroblasts and glioblasts migrate and differentiate into specific types of neurons or glia cells. They can be characterized as brain modules, i.e. brain entities which act at the same time, or successively, as *compartments* and *components*. The *compartment aspect* may be illustrated by the fact that the border zones of FMUs may act as cell lineage restrictors. They prevent (during a certain developmental period, and via characteristic adhesivity properties) the cells produced in one FMU from Intermingling with those of adjacent FMUs (Lumsden 1990; Figdor and Stern 1993; Fraser 1993). The *component aspect* appears from the fact that the successive primary FMUs in certain regions, particularly the spinal cord and rhombencephalon, show strikingly corresponding developmental programs, in so far as they respond to shared though regionally diversified, causal mechanisms. Shared dorsoventral patterning produces comparable (minimally different) cell contingents, which occupy mutually corresponding (dorsoventral) positions in neighbouring neuromeres, and which during further development become incorporated in di-, tri-, or multi-segmental nuclear complexes, such as the sensory descending trigeminal column. Thus, it has been demonstrated that the alar domain of the chick and mouse cryptorhombomere 8 ("r8") participates in the formation of one bi-, two tri - and no less than seven multisegmental cell columns (Cambronero and Puelles 2000; Marin et al. 2008; Tomas-Roca et al. 2014).

5. The genomic basis of the formation and differentiation of the FMUs

We have seen that the transverse and longitudinal gene expression patterns which create and delimit the primary FMUs are gradually produced by anteroposterior and dorsoventral inductive patterning mechanisms widely conserved in vertebrates (Plate 24). The developmental landscape implies gradual harnessing of the initially highly *dynamic* patterns of activity of a few early regulatory neural genes by a network of cell-cell interactions and secreted morphogen products. This leads progressively to regionalization (subdivision of early larger fields into smaller ones) and more selective and stable states of gene expression at each differentiated position of the neural plate and tube. Such pre-final and final states imply the read-out of numerous other sorts of transcription factors and differentiation proteins in locally characteristic combinations, ultimately reaching the equilibrated (stabile) composition and selective molecular profiles of individual FMUs. A FMU is thus a stable, uniquely molecularly specified and well-delimited progenitor domain capable of producing over time one or several characteristic derived cell types.

Molecular regionalization
of the pretectum
(alar region of prosomere 1)
of the chick
(Ferran et al. 2007)

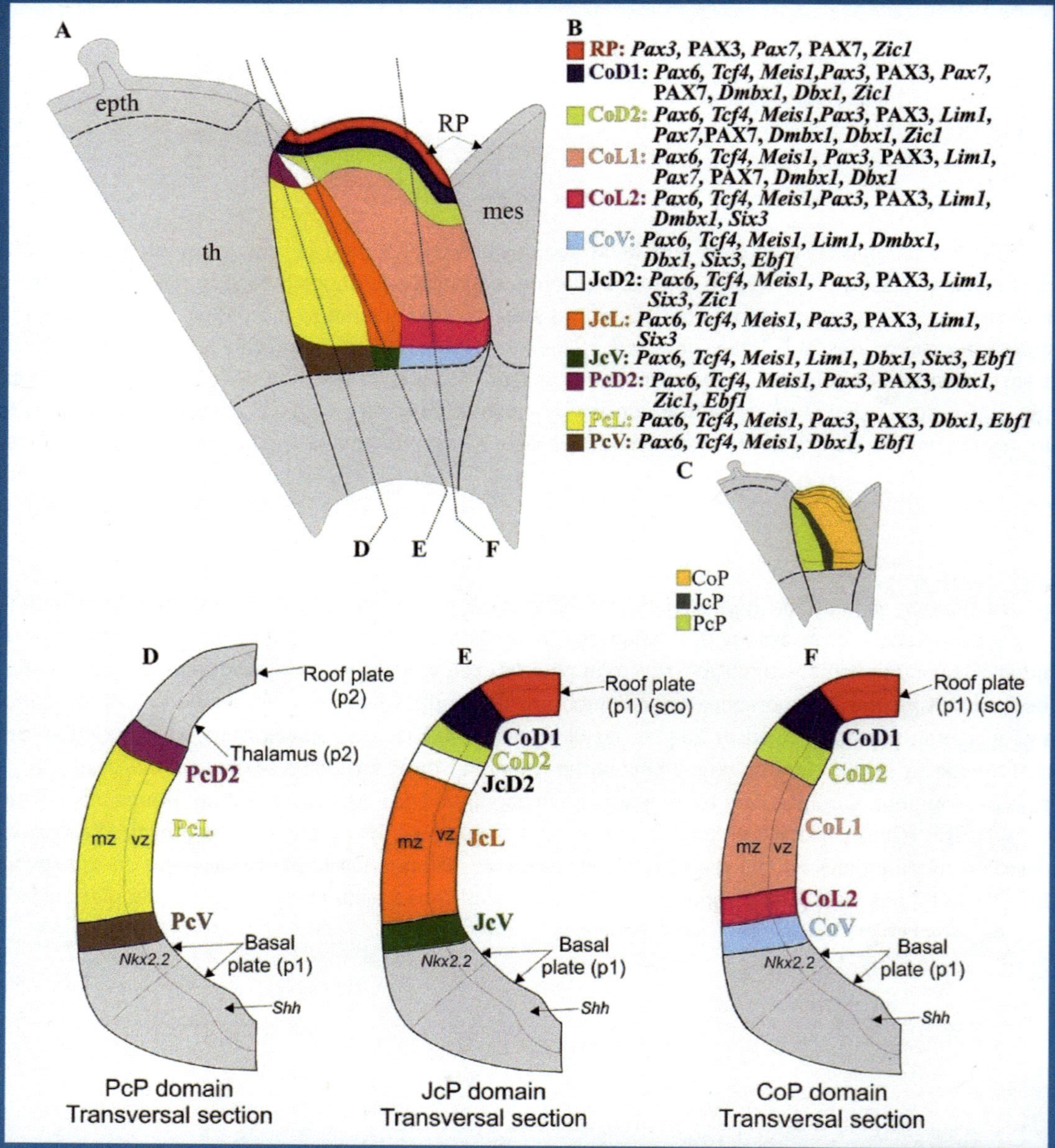

Note that all entities
delineated represent
radial units

73

Plate 73

Let us return to our model system, the avian pretectum.

Plate 73, Fig. A shows once more the state of differentiation of the pretectum of a chick embryo of about 60 h incubation (HH17–18). It will be seen that all of the three anteroposteriorly arranged secondary FMUs (PcP, JcP, CoP) present in an embryo of about 50 h of incubation (HH13–14; Fig. C) have differentiated into a number of dorsoventrally arranged subdomains. The letter codes used for these subdomains indicate their position within their respective secondary FMUs; thus, PcV means: ventral subdomain of the precommissural pretectum, and CoD2 is dorsal 2 subdomain of the commissural pretectum. All of these subdomains are characterized by unique combinations of expressed transcription factors. The colour-code used for the various subdomains in Fig. A and Figs. D–F refers to their molecular specification. This is explained in Fig. B, where the different molecular markers detected in each subdomain are listed next to their colour codes and abbreviations.

Figs. D–F present schematic transverse hemisections through the three AP pretectal domains, showing the DV pattern characteristic for each. The levels and orientations of these hemisections are indicated in Fig. A. Note that all of the subdomains identified represent molecularly differentiable *radial units,* which subsequently produce characteristic derivatives.

8.4 Developmental processes within the FMUs
8.4.1 Proliferation of matrix cells

1. Structure of neural plate

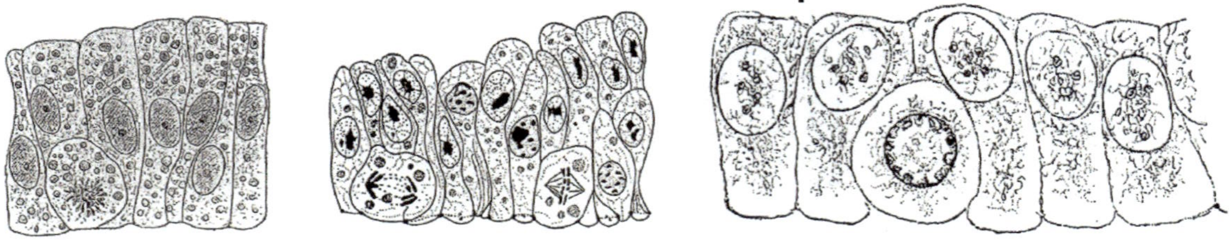

shark (His, 1889) chick (Sauer 1935b) rabbit (His 1889)

2. Proliferation within the (early) neural tube

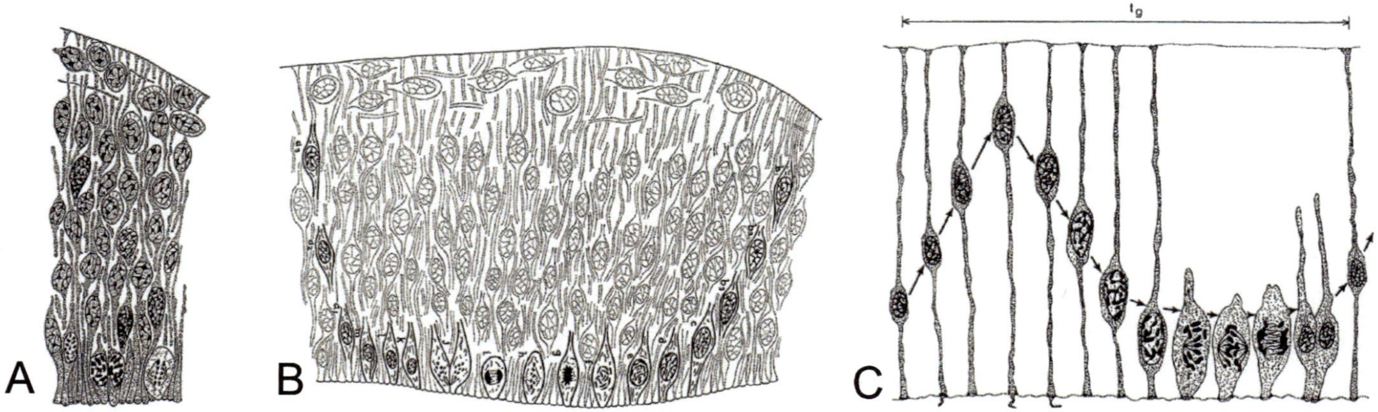

A pig embryo 10mm (Sauer 1935b)

B kinetics of nuclei of matrix cells and mitoses (Sauer 1935a)

C a neuroepithelial cell during the various phases of the mitotic cycle (Jacobson 1991)

74

8.4 Developmental Processes Within the FMUs

The developmental processes occurring within the FMUs include: (1) proliferation of undifferentiated matrix cells; (2) neurogenesis and gliogenesis; (3) migration of postmitotic neuroblasts and the formation of embryonic layers and (4) the differentiation of cell masses and cortices. These processes will now be discussed.

8.4.1 Proliferation of Matrix Cells

Plate 74

The pictures forming the upper row in Plate 74 have been discussed before. They show that the neural plate consists of columnar epithelial cells. In all of the three pictures, one or more cells are in mitosis, indicating that the neural plate is formed by a proliferating epithelium. Sections through the early neural tube yield similar pictures (e.g. Plate 74, Fig. 2A) and warrant a similar conclusion. However, it is important to note that the mitotic figures are by no means evenly distributed over the thickness of the neural wall. Rather, they are confined to the innermost zone, and it has been shown that mitoses are also crowded in the FMUs which hence may be interpreted as centres of proliferation (Bergquist 1932; Bergquist and Källén 1954). Sauer (1935a, 1936) advanced the hypothesis that the wall of the early neural tube consists of a single type of epithelial cell in various stages of the mitotic cycle. The nuclei of the resting cells reside in the superficial zone of the wall, while those of cells that are going to divide are translocated toward the ventricular surface. At the end of this nuclear displacement phase, the peripheral process of the cell loses its contact with the external or meningeal surface and retracts due to loss of adhesion and the depolarization of its cytoplasmic microtubular skeleton. Next, the cell rounds up and divides into two daughter cells. The neuroepithelial cells are uniformly attached to each other in the zone directly bordering the ventricular lumen by a desmosomal terminal bar network, and this together with the rounding up of the cells, accounts according to Sauer for the typical juxtaventricular position of the mitotic figures. Each daughter cell produces a new peripheral offshoot via a newly developed cytoskeleton and its nucleus moves away from the ventricle, probably pushed outwards by the pressure of other cells rounding up (Plate 74, Fig. 2B, C).

8.4 Developmental processes within the FMUs

8.4.1 Proliferation of matrix cells

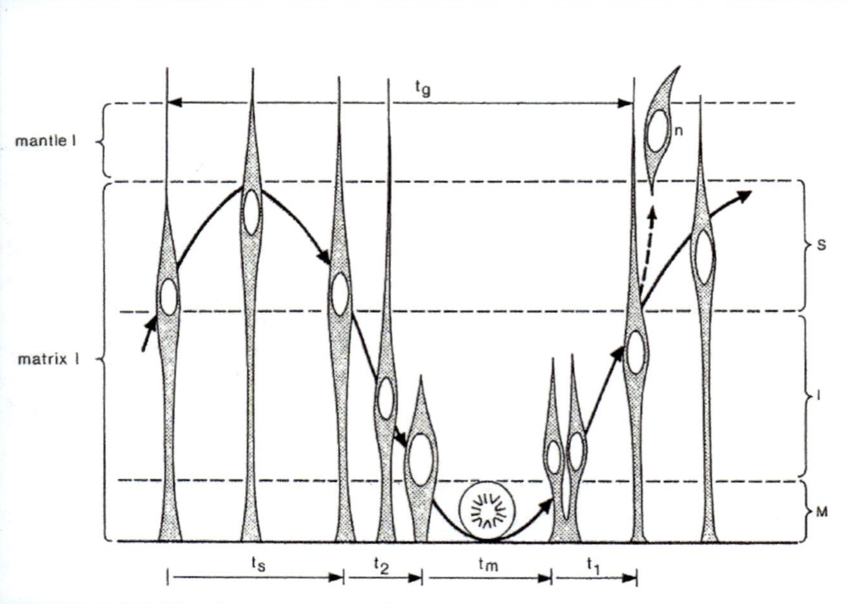

1. "Elevator movements" of a matrix cell
Based on autoradiography studies
(modified from Fujita 1966)
I, intermediate zone; M, mitotic zone;
S, zone of DNA synthesis; tg, generation time of matrix cell; tm, mitotic time;
ts, DNA synthesis time; t1, t2, postmitotic and premitotic resting time

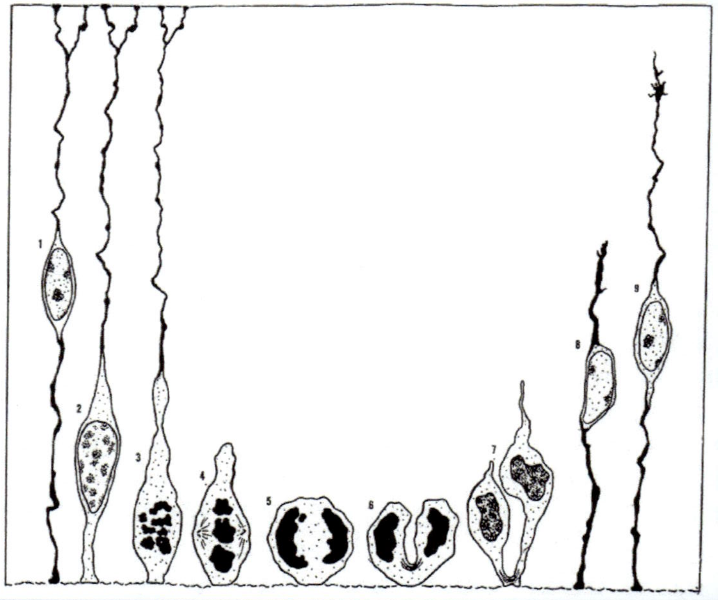

2. Generation cycle of matrix cells
In the telencephalon of
13 day - mouse - embryos
Based on Golgi and E.M. material.
(Hinds and Ruffett 1971)

75

Plate 75

Sauer's brilliant hypothesis on the developmental events in the early neural tube, just summarized, has been corroborated experimentally by numerous authors, including Fujita (1963, 1966) and Hinds and Ruffett (1977).

Fujita (1963, 1966) studied the early histogenesis of the neural tube in chick embryos with tritiated thymidine autoradiography combined with electron microscopy. He confirmed that the wall of the early neural tube consists of a homogeneous population of undifferentiated matrix cells, and that these cells perform the vertical displacements described by Sauer, which he designated as 'elevator movements'. He also found that the mode of cell production indicated by Sauer proceeds similarly in later stages, in which the three primordial layers of the neural tube, matrix layer, mantle layer, and marginal layer (in sequence from ventricular to meningeal surfaces, see below) have already become manifest (Plate 75, Fig. 1). Fujita subdivided the matrix layer into three zones: the M or mitotic, I or intermediate, and S or synthetic zones. He demonstrated that the nuclei of the matrix cells at the phase of DNA synthesis (t_s) are located in the superficial half of the matrix layer (S zone). When the nuclei have finished DNA synthesis, they descend during a postsynthetic or premitotic period, t_2, through the I zone to enter the deep juxtaventricular M zone. The matrix cells divide there and, after the time needed for mitosis, t_m, the nuclei of the daughter cells pass into the I zone, where they spend the postmitotic and presynthetic period, t_1. Finally, once the kinetic cycling mechanism triggers the beginning of a new cycle, they re-enter the S zone where genomic DNA starts to be copied.

Hinds and Ruffett (1977), who studied Golgi and E. M. material, found that the generation cycle of the matrix cells in the telencephalon of mouse embryos fully meets the descriptions of Sauer and Fujita (Plate 75, Fig. 2). Note that in both figures the ventricular surface is at the bottom.

8.4 Developmental processes within the FMUs

8.4.1 Proliferation of matrix cells

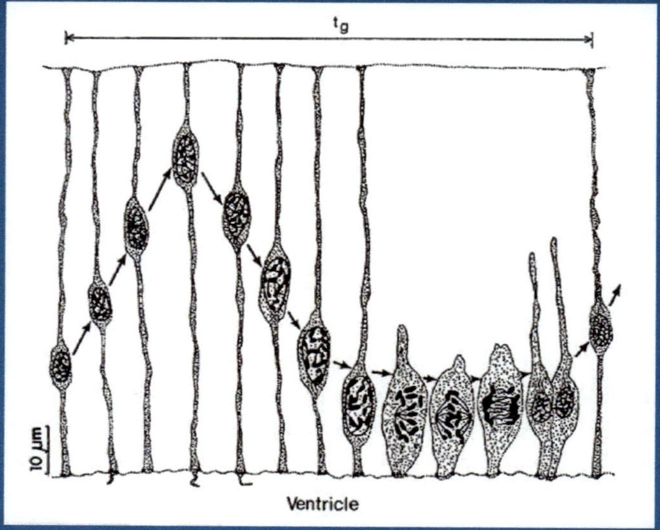

- The Natural Coordinate System is tangible and has taken the shape of a Cartesian coordinate system
- The ventricular surface represents the zero plane
- A detailed timetable prevents crowding
- Pseudostratified epithelium ⟶ hyperepithelium
- The nuclei and the regrowing processes remain within the confines of their FMU

76

Plate 76

Here follow some comments on the remarkable developmental events just discussed, illustrated by a figure shown before. In this figure (Plate 76), the cellular underpinning of the natural coordinate system is shown in its simplest form, viz. that of an orthogonal, Cartesian coordinate system. We have designated the ventricular surface of the neural tube as the zero plane of the natural coordinate system, because it is there that all newly produced cells make their appearance. It will be clear that the remarkable elevator movements of the nuclei of the matrix cells require a precise timetable to avoid crowding. A cyclic sequence of cyclin proteins apparently guides the neuroepithelial cells through the phases of the cell cycle. The molecular profile of each FMU has direct or indirect effects on the cyclin pattern, potentially leading to differential temporal regulation of the cell cycle within each of these units. Synchronous arrival of too many nuclei at one and the same level would lead to a serious 'Stau' or traffic jam. Because the neuroepithelium, by strictly adhering to its radial organization (see below), may attain a considerable thickness (remarkably, 10 cm or more!) we feel justified to state that this neuroepithelium becomes transformed in some places into a *hyperepithelium*. Finally, the rigorous radial arrangement of the matrix cells and the fact that postmitotic cells immediately and strictly restore this arrangement, brings about that all of these elements remain confined to their own FMU.

8.4 Developmental processes within the FMUs

8.4.2 Formations of embryonic layers

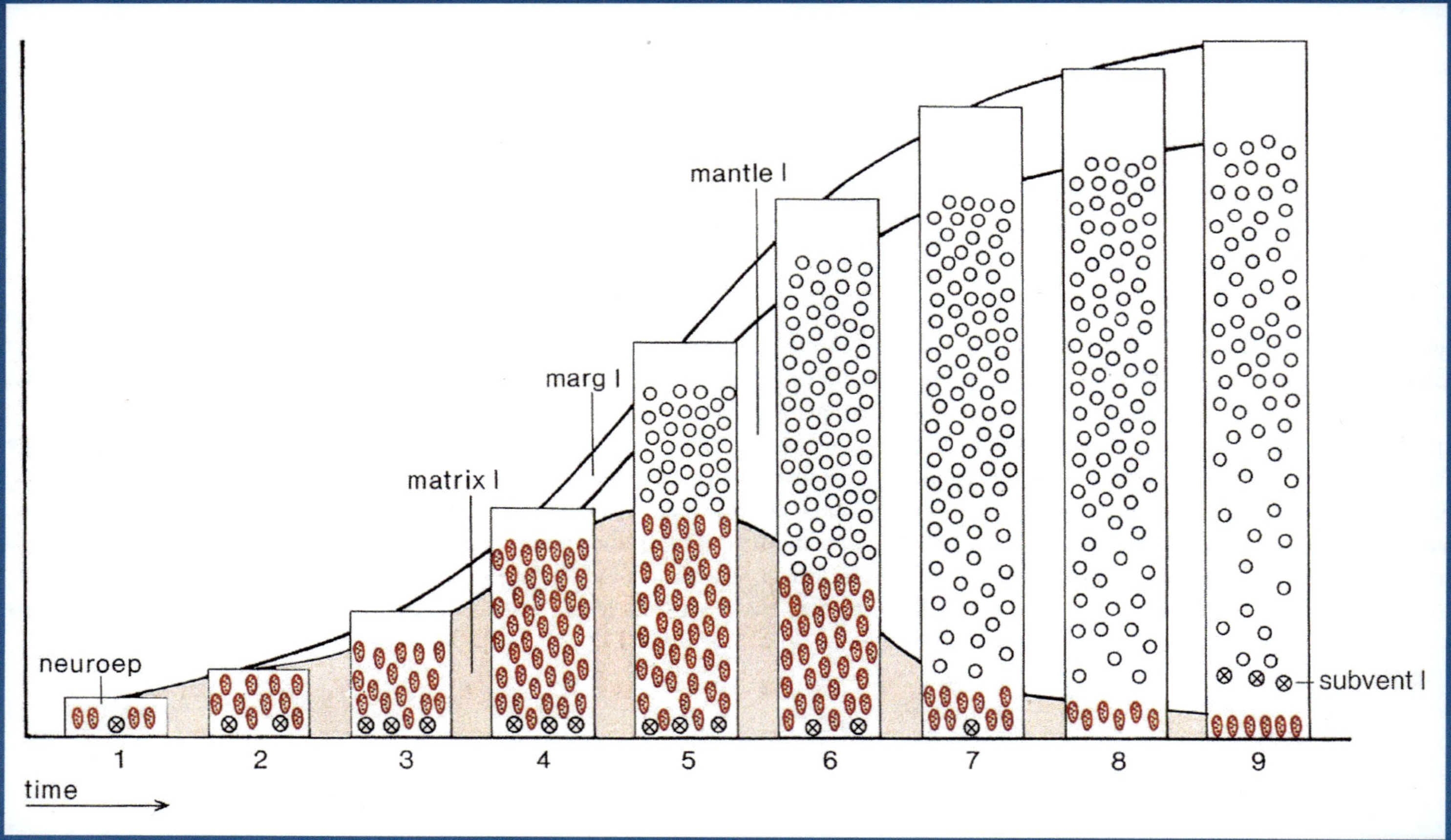

neuroepithelium

marginal layer
mantle layer
matrix layer

77

8.4.2 Formation of Embryonic Layers

Plate 77

Plate 77 presents a graphic reconstruction of the histogenesis of the CNS. Nine consecutive phases in the development of a small segment of the neural plate (1) and neural tube (2–9) are placed along the abscissa, whereas the changing thicknesses of the neural wall and of the individual layers are indicated along the ordinate.

We have seen that the neural plate and very early neural tube consists of a simple (later pseudostratified) proliferating columnar epithelium (1–2). At a certain developmental stage the nuclei of the elongated neuroepithelial cells withdraw from the most superficial zone of the neural tube and the wall becomes divisible into an outer, anuclear and an inner, nuclear zone (3–4). The outer zone, or *marginal layer*, consists initially exclusively of the external protoplasmic processes of the neuroepithelial cells, but is soon invaded by the axonal processes of maturing neuroblasts. The inner zone is termed the *matrix layer* (Kahle 1951; Fujita 1963, 1966). It contains the densely crowded nuclei of a morphologically homogeneous population of ventricular cells, all elements of which participate in the proliferation process. The matrix cells are the precursors (progenitors) of all neuronal and macroglial elements of the CNS.

During a certain early period, the matrix layer represents a purely proliferative compartment. Mother cells produce more mother cells, causing the surface area as well as the thickness of the tube to increase steadily (2–4). This period of symmetrical division of germinal cells is followed by a period of asymmetrical division, in which one of the daughter cells resulting from each mitosis differs from both its mother cell and its sister cell in that it withdraws from the mitotic cycle and migrates out from the matrix layer (Plate 75, Fig. 1: n). These postmitotic elements or neuroblasts that move out of the matrix layer then form a third compartment, the *mantle layer*, which is situated between the matrix and marginal layers (5). The opposite decisions of some postmitotic pairs of elements, giving rise to one postmitotic and one proliferative daughter cell seem due to the probabilistic nature of the transition at the end of the t1 phase of the cycle, with initially rather low probability of the neuronal fate being selected. In a later stage, the neurogenetic probability increases significantly, so that both daughter cells resulting from a mitosis may become postmitotic neuronal elements.

When the first neuronal elements appear in the matrix layer (distinguishable by various molecular markers, coded by differentiation genes, and the loss of some matrix-specfic markers), the period of pure proliferation has come to an end; proliferation will now overlap with neurogenesis. As more and more dividing matrix cells begin to generate postmitotic daughter cells, the mantle layer increases rapidly in thickness. During this new period, matrix cells and newborn neuronal elements coexist in the matrix layer, though the neuronal cells rarely remain within the latter. Within the mantle layer, which further expands at the expense of the matrix layer, since less and less cells select to cycling (5–8), the cell masses of the mature CNS are gradually created by the migration and selective adhesive aggregation of neuroblasts.

All parts of the CNS pass at some time point through a phase in which the three classical matrix, mantle and marginal layers can be clearly observed, but the developmental processes just discussed do not take place synchronously throughout the neuraxis. They are temporo-spatially patterned, this phenomenon being one of the consequences of the differential molecular profiles of the FMUs, which have effects on the probabilistic variables involved. This is beautifully illustrated by the work of Kahle (1951) on the development of the matrix layer in different parts of the human CNS.

8.4 Developmental processes within the FMUs

8.4.2 Formations of embryonic layers

Development of matrix layer in various parts of the human brain during the first four intrauterine months

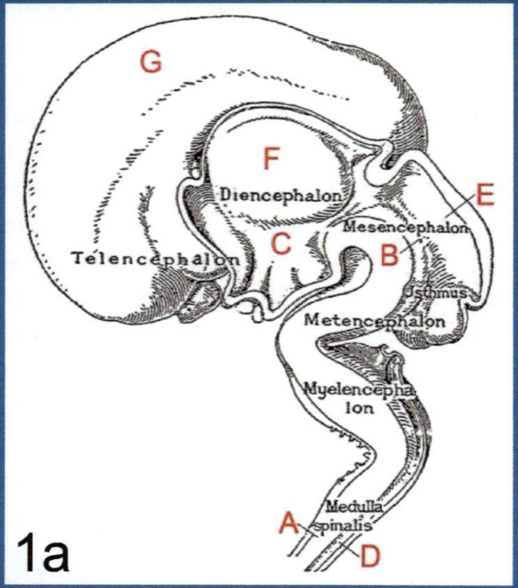

1a

(Kahle 1951)

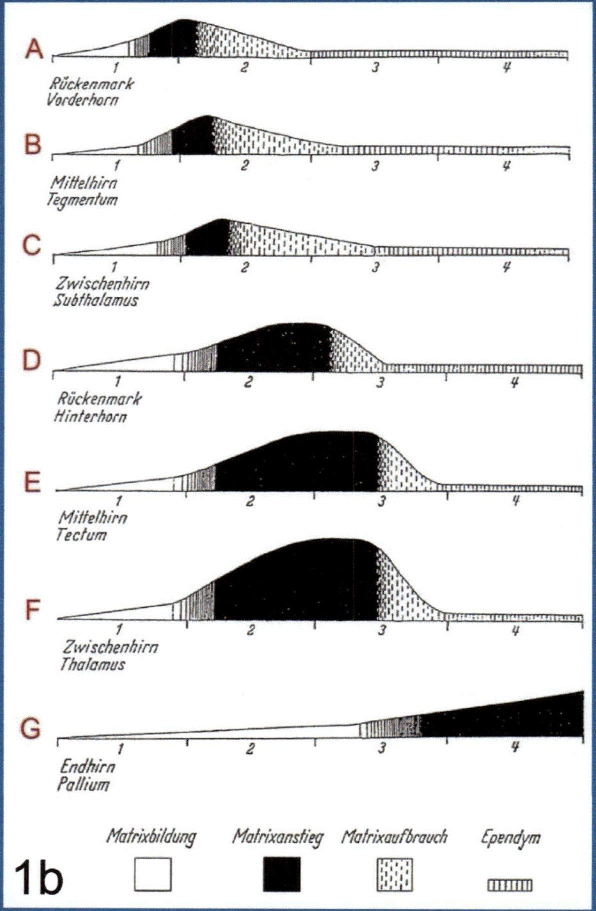

1b

Heterochrony

2. Transverse sections through spinal cord

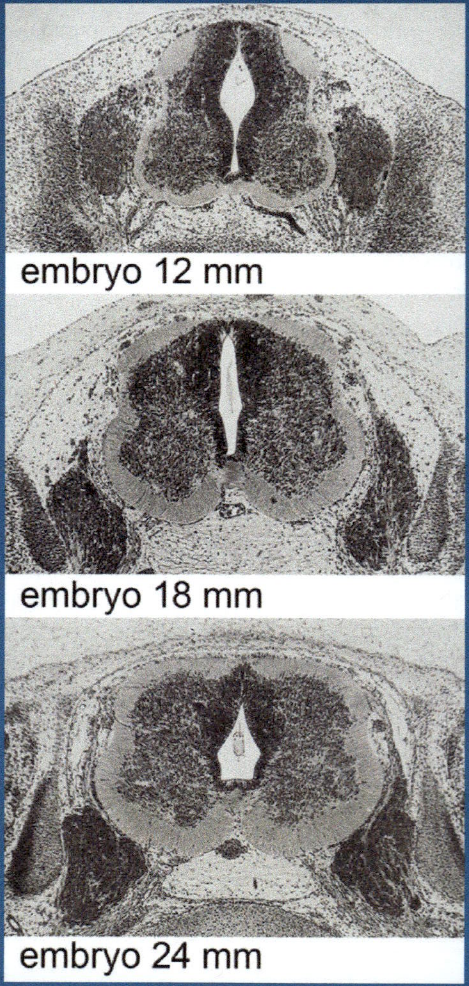

78

Plate 78

Kahle (1951) investigated the development of the matrix layer in seven different regions of the human neuraxis. These are indicated in Plate 78, Fig. 1a, which shows a medial view of the bisected brain of an embryo of about ten weeks' post conception. The waxing and waning of the matrix layer in the various regions during the first four intrauterine months is shown in Plate 78, Fig. 1b. Kahle subdivided the development of the matrix into four phases: formation (Matrixbildung), increase (Matrixanstieg), decrease (Matrixaufbrauch), and complete depletion and formation of ependyma (Ependym). It will be seen that the regions investigated may show considerable and characteristic heterochronic differences in the time course of matrix layer development. Thus, the unfolding of the matrix layer starts earlier and lasts shorter in the following regions: the spinal anterior horn (A), the midbrain basal plate /tegmentum (B), and the subthalamic region, a basal part of the hypothalamus (C) than in the spinal dorsal horn (D), the tectum of the midbrain (E), and the dorsal thalamus (F), all of which are alar domains. In the pallium (G), a singular alar domain that grows protractedly, the building up of the matrix starts late and proceeds for a very long time.

Plate 78, Fig. 2 shows transverse sections through the spinal cord of human embryos of 12, 18, and 24 mm crown-rump length. In all of these sections, the matrix layer is much wider in the dorsal (alar) region than in the ventral (basal) region because the basal motoneurons are generated very early and the corresponding matrix, being precociously depleted, transforms into ependyma.

Heterochrony (different timing in Greek) is very frequently observed in the CNS. It refers to the pattern by which different parts of a structure are in different phases of development at a given time point. The figures shown on this plate illustrate this feature. It quite often occurs that a given FMU is still in the two-layers-phase (matrix + marginal layer; see Plate 77, 3), whereas in an adjacent FMU, a distinct mantle layer is already manifest (Plate 77, 6). It will be clear that such heterochronies greatly facilitated the identification and delineation of FMUs before the advent of the geno-architectonic methods.

8.4 Developmental processes within the FMUs

8.4.2 Formation of embryonic layers

Proliferation; development of matrix layer

1

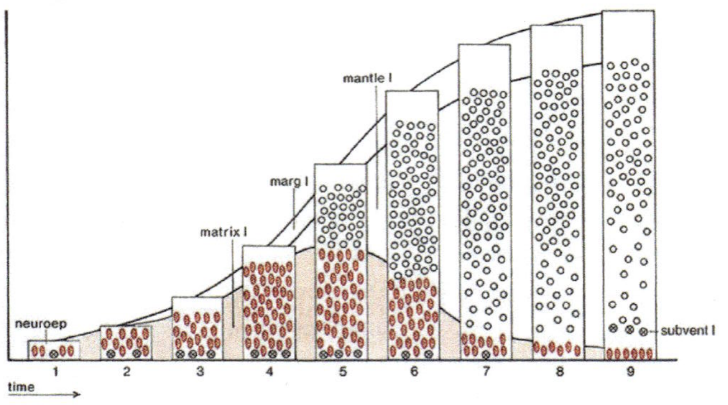

Factors involved in the production of extraordinarily large numbers of matrix cells - neuroblasts - neurons

a. thickening of matrix

b. lengthening of matrix phase

c. expansion of ventricular surface

formation of brain vesicles

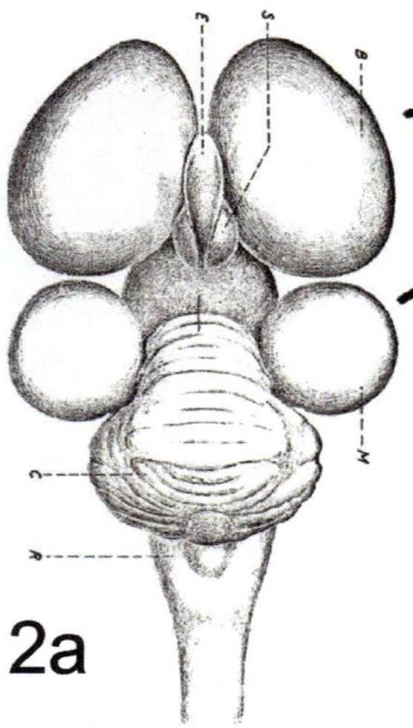

2a

ostrich embryo (Krabbe 1956)

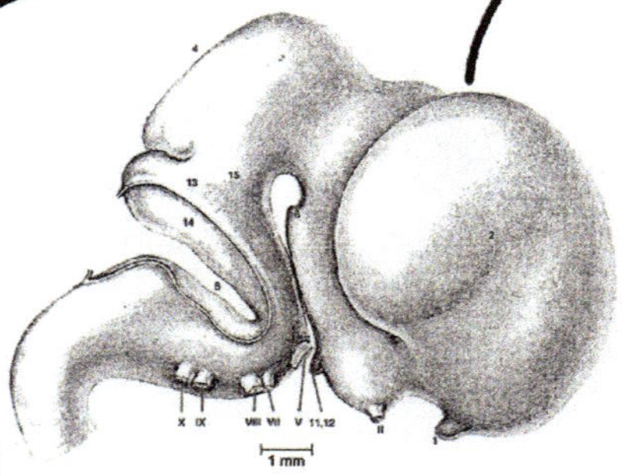

2b

B, 8 wks

human embryo

79

Plate 79

Here we briefly touch upon the mechanisms involved in the local production of extraordinarily large numbers of neurons, as occurs in, among other structures, the alar regions of mesomere1 (tectum of the midbrain) and hypothalamo-telencephalic prosomere1 (neocortex). Because neuroblasts and mature neurons do not have the capacity to divide, it is the proliferative capacity of the matrix layer, which determines the number of neurons to be produced in a given region. This is regulated molecularly through the production of various mitogenic factors, which promote proliferation and delay neurogenesis (decreasing its probability). The longer neurogenesis is postponed, the more neurons are produced finally, since the number of matrix progenitors increases exponentially. Comparative studies on the histogenesis of the neural wall in various regions of the brain show that the productive capacity of the matrix zone may be advanced in the following three ways:

(a) local thickening of the matrix layer (radial growth);
(b) lengthening of the phase during which an active matrix is present, and
(c) expansion of the neuroepithelial surface (tangential growth), including the corresponding matrix layer.

As regards a: Widening of the matrix layer is coupled of course with a local increase of the number of matrix cells available for proliferation (Plate 79, Fig. 1). This widening implies, however, that the number of vertically stacked nuclei (pseudostratification) increases, which hamper the elevator movements of the nuclei involved in the proliferation process and eventually saturate the ventricular surface with mitotic figures, a phenomenon designated as 'choking' (Smart 1985).

As regards b: We have seen already (Plate 78, Fig. 1b: G) that this mode of increasing the proliferative capacity is prominent in the mammalian pallium.

As regards c: It will be clear that the problems related to a local thickening of the matrix layer, particularly 'choking' (see a), may be compensated for by local increase of the ventricular surface area. Such local expansions correspond to the formation of thin-walled vesicular structures, such as the avian 'optic lobe' (Plate 79, Fig. 2a), and the avian and mammalian cerebral hemispheres (Plate 79, Fig. 2a, b).

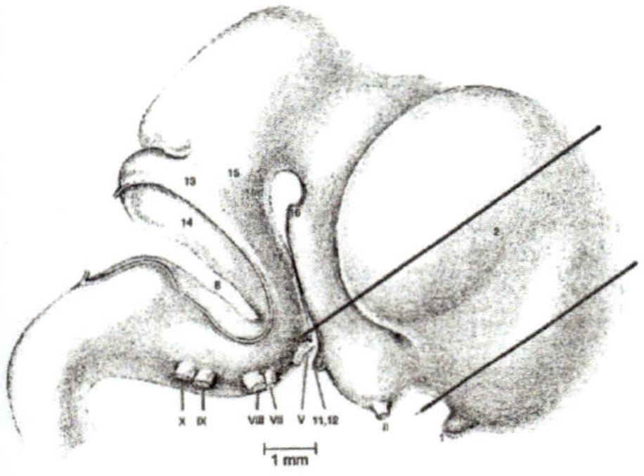

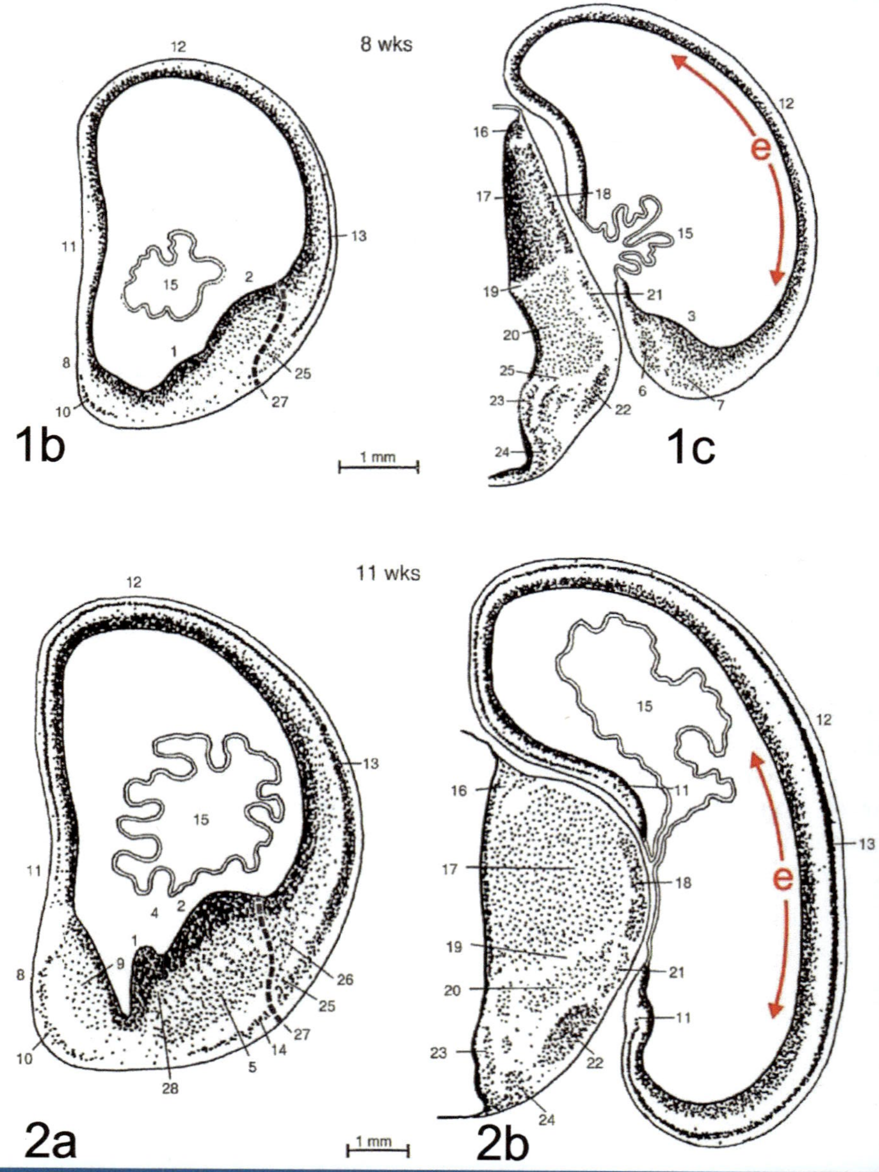

The telencephalic vesicle in human embryos

Note:

Expansion of matrix zone

80

Plate 80

Plate 80 shows cross sections through the rostral part of the brains of human embryos of 8 weeks (Fig. 1) and 11 weeks (Fig. 2) gestation. The expansion (e) of the thin-walled pallial domain of the telencephalic vesicles is indicated by red arrows. It is this expansion, and the concomitant increase in surface area of the matrix zone, which accommodates the production of the enormous number of neurons required for the construction of the cerebral cortex, without choking.

We should like to draw attention to the striking phenomenon of *heterochrony* in the medial part of the section shown in Fig. 1c, where four consecutive FMUs are visible: the thalamus (17), the prethalamus (20), and the basal regions of the peduncular (23) and terminal hypothalamus (24). Note that in the thalamic region a wide, dense matrix zone is present, whereas in the prethalamus a narrow matrix zone is flanked by a wide, apparently undifferentiated, mantle layer. In the peduncular and terminal hypothalamic sectors signs of lamination are manifest (the superficial primordium of the migrated subthalamic nucleus (22) is clearly visible). In the terminal domain a narrow matrix zone can still be recognized, but in the peduncular one this zone seems to be already completely exhausted. It is also of note that the interneuromeric boundaries between the dorsal thalamus and the prethalamus, as well as that between the prethalamus and the peduncular hypothalamus, are marked by distinct cell-free zones (19, 35).

Before passing on to the phenomenon of radial migration of neuroblasts, some remarks should be made on additional or alternative matrix zones and on the ontogeny of neuroglia. Apart from the ubiquitous ventricular matrix, two other proliferative compartments, i.e. the subventricular zone and the external germinal layer have been observed in the developing CNS.

The *subventricular zone (SVZ)*, sometimes also denoted as the subependymal layer or cell plate has only been described in the lateral (pallial) and basal (subpallial) walls of the mammalian telencephalon (Smart 1961; Fujita 1966; Blakemore 1969; Smart and Sturrock 1980). This layer develops at the junction of the matrix zone and the mantle layer. Most authors report that this layer appears relatively late, at a moment when standard ventricular neurogenesis is largely or entirely completed (Plate 77: 9). However, according to Rakic (1974), the subventricular zone can be recognized in the pallial wall of the rhesus monkey as early as at 45 days of gestation (the total duration of gestation is about 165 days in this animal). This agrees with the notion of Smart (1985) that its appearance occurs at the height of matrix proliferation, as a result of choking. This layer is thus formed by interstitial (free) progenitor cells derived from the matrix ones, but in contrast, the subventricular elements lack an epithelial organization, having lost the desmosomal attachments at the ventricular surface, and do not exhibit interkinetic nuclear displacements while they divide in situ. However, recent data (see e.g. Hansen et al. 2013) suggest that the interphasic subventricular cells continue to have peripheral cytoplasmic processes that reach the pial surface (these may be crucial for guidance of newborn neurons into the cortical plate). The cortical field of primates is held to develop separate inner and outer subventricular zones (ISVZ, OSVZ), the latter being best developed in humans (Pilz et al. 2013). The SVZ persists after birth, and in vestigial manner, into adult life and even senescence.

The subventricular layer gives rise to special classes of neurons and to all types of macroglial elements, with the possible exception of ependymal cells. A particularly strongly developed subventricular zone can be observed in the mammalian striatal and pallidal primordia.

Another extra proliferative compartment, i.e. the *external germinal layer*, is confined to the cerebellum. This layer, also referred to as the external granular layer, develops from the ventricular matrix in the rostral, cerebellar part of the rhombic lip. The *rhombic lip* is a thickened germinal zone in the dorsalmost part of the rhombencephalic alar plate, adjacent to the attachment of the choroidal roof of the fourth ventricle (Plate 33, Fig. e). Whereas cells migrate out tangentially from the whole rhombic lip, only those issuing from the rostral part of the lip, which is associated with the cerebellar plate (isthmus and rhombomere 1), are capable of massive, long-lasting proliferation, which ends only postnatally. The resulting subpial layer spreads by tangential marginal migration of its elements over the entire external surface of the cerebellar primordium; these superficial cells continue to proliferate, giving rise to enormous numbers of postmitotic cells. The elements born within this transient and ectopic germinal zone ultimately migrate radially inward to form the granular layer of the cerebellar cortex.

As regards gliogenesis, Fujita (1966) held that the germinal cells which constitute the ventricular matrix layer first give rise to neuroblasts and that only later, after completion of the local neurogenetic program, the remaining germinal cells transform into astroglial glioblasts, or differentiate into ependymal cells. In contrast, oligodendoglial cells are produced selectively in some FMUs, invading from there the remaining territories.

Studies using Golgi and E.M. techniques have shown, surprisingly, that radial glial cells already exist during late stages of neurogenesis (Rakic 1971, 1972; Schmechel and Rakic 1979). This possibly can be explained as resulting from clones that finish their neurogenetic program earlier than others. As the name implies, radial glial cells are a specialized non-neuronal epithelial cell class characterized by an elongated offshoot that radially spans the entire wall of the neuraxis, and a specific set of differentiation markers that distinguish them from matrix cells, allowing ascription to the astroglial lineage.

Radial glial cells fulfil a variety of roles in the CNS. It has been demonstrated that these elements are mitotically active throughout the development of the CNS (Misson et al. 1988; Hartfuss et al. 2001), and that they (a) guide the radial migrations of neuroblasts (Rakic 1972; Hatten 1999; see below); (b) form canal-like compartments which determine the growth and orientation of outgrowing axons (Singer et al. 1979; Nordlander 1987); (c) may transform into astrocytes (Schmechel and Rakic 1979), and (d), according to some authors, are even able to produce neurons (Noctor et al. 2001; Weissman et al. 2003: Anthony et al. 2004; Hansen et al. 2010).

8.4 Developmental processes within the FMUs
8.4.3 Radial migrations of neuroblasts

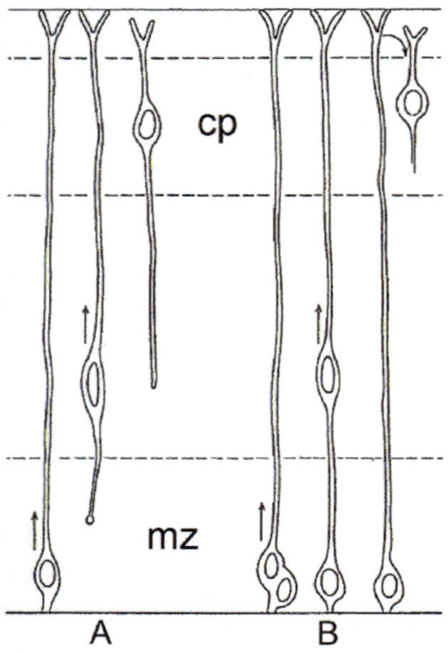

1. Migration of neuroblasts from the pallial matrix zone (mz) to the cortical plate (cp), according to (A) Morest (1970), and (B) Berry and Rogers (1965)

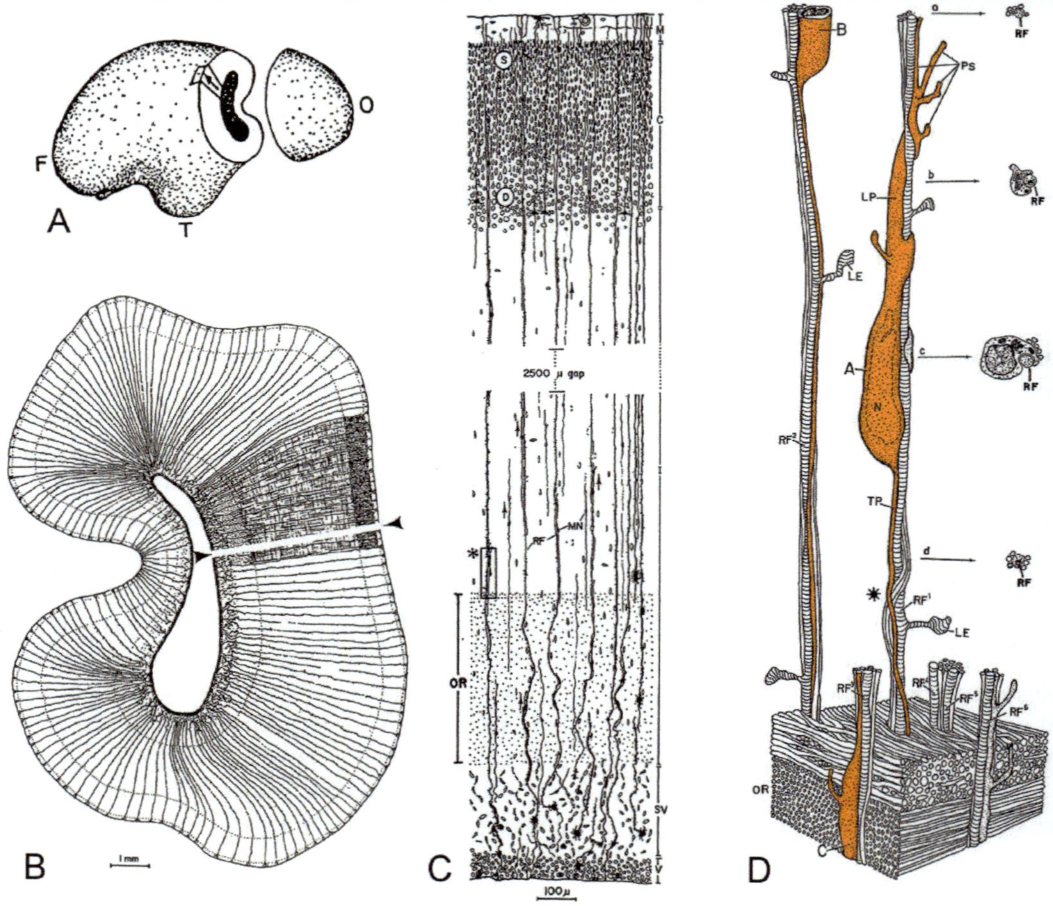

2. Migration of neuroblasts along ependymal gliocytes in the occipital part of the tetencephalon of fetal monkeys (Rakic 1972)

81

8.4.3 Radial Migration of Neuroblasts

Plate 81

After their formation, young neuroblasts are expelled passively out of the matrix zone and then migrate away from their origins in the proliferative zones to their final positions in cell assemblies of the mantle layer. Particularly in the larger brains of amniotes there may be a considerable distance between the deep *locus* where a given neuron was generated and the position it occupies in the mature brain. The intervening migration process involves active navigational movements of the neuroblasts through the surrounding cellular environment, using various guidance mechanisms to their final destination, before they settle to establish their specific synaptic connections. Generally, the migrations just referred to, occur in a strictly radial direction. Several mechanisms have been proposed for these radial migrations, three of which will be briefly discussed and are shown in the plate.

Morest (1970) studied the early development of neurons in the telencephalon of pouch young opossums in Golgi material. He concluded that (a) neuroblasts maintain their central and peripheral primitive processes that extend to the ventricular and pial surfaces, respectively; (b) not the cells as such, but rather their nuclei and surrounding protoplasm move through their peripheral primitive process outward, and (c) this translocation is accompanied or followed by a loss of the central primitive process, and the cell later detaches also from the pial surface (Plate 81, Fig. 1A).

Berry and Rogers (1965), who studied the development of the cortex in the rat, suggested a similar mechanism. They observed in their Golgi preparations 'binucleate' elements maintaining their attachment to both the ventricular and pial surfaces and suggested that following the nuclear division one of the daughter nuclei travels peripherally through the primitive process, and that the cytoplasmic division only occurs later, after the migrating nucleus has attained its destination in the cortical plate (Plate 81, Fig. 1B). This mechanism has not been corroborated in more recent studies.

Rakic (1972) studied the development of the occipital cortex in rhesus monkey fetuses. He proposed that radially oriented glial fibres, which span the full thickness of the pallial wall, provide crucial contact guidance paths for the freely migrating neuroblasts.

The investigations of Rakic are summarized in Plate 81. Figure 2. A shows the cerebrum of a 97-day Rhesus monkey fetus in which the levels of the sections studied are indicated. In Fig. 2B such a section, impregnated with the Golgi technique, is depicted. It beautifully shows the arrangement of the radial glial fibres. Figure 2C presents a more detailed drawing of the cerebral wall in the area indicated by the white strip (arrowheads) in Fig. 2B (middle 2500 μ of the intermediate zone is omitted). The small rectangle, marked with an asterisk, shows the position of the reconstruction in Fig. 2D. Abbreviations: C, cortical plate; D, deep cortical cells; S, superficial cortical cells; I, intermediate zone; M, marginal layer; MN, migrating cell; OR, optic radiation; RF, radial fibre; S, superficial cortical cells; SV, subventricular layer; V, ventricular zone or matrix. D shows a three-dimensional reconstruction of the relationships between migrating cells (A, B, C) and radial fibres (RF), based on serial E. M. images.

It should be emphasized that perikaryal translocation (Plate 81, Fig. 1A) and neuroblast migration along radially disposed glial fibres (Plate 81, Fig. 2) are not mutually exclusive. It might well be that the former mode of displacement is operative during very early stages of neurogenesis, whereas the latter process prevails during later development, particularly in regions where considerable distances have to be bridged (Puelles and Bendala 1978).

8.4 Developmental processes within the FMUs

8.4.3 Radial migrations of neuroblasts

Radial and deflected radial migration of neuroblasts along ependymal gliocytes in the telencephalon of mouse embryos (Misson et al. 1991)

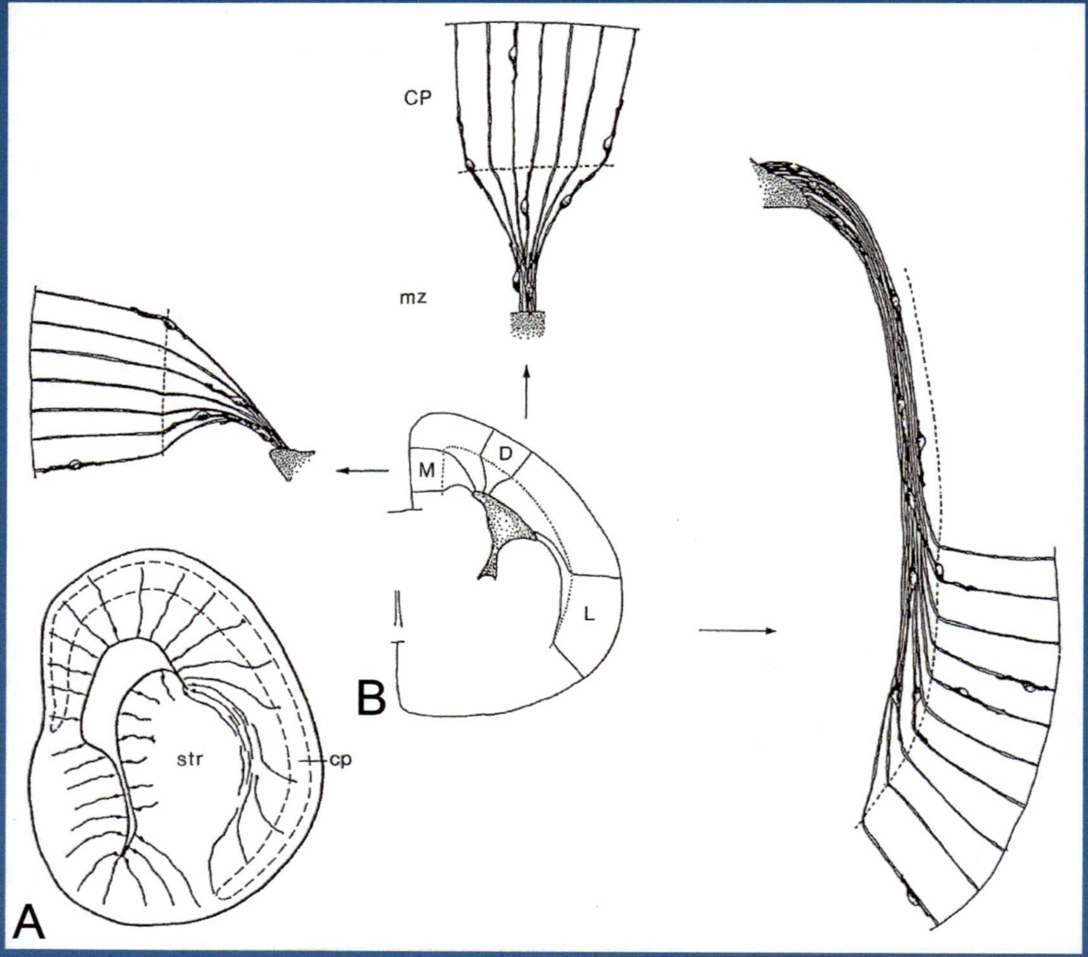

82

Plate 82

During morphogenesis, the walls of the neural tube may be considerably deformed. The vectors interconnecting its ventricular and pial surfaces are important indicators of the deformations that have taken place. Smart and Sturrock (1979) studied the development of the telencephalon in the mouse in Golgi material. They observed that early in development the telencephalic matrix cells and ependymal gliocytes show a radial orientation. However, later in development the ependymal fibres in the striatal anlage apparently lose contact with the pial surface, while those emanating from the lateral part of the pallial ventricular surface are deflected ventrally in association with the tangential expansion of the cortical plate. Though the latter ependymal processes seem to pursue curving courses, which circumscribe the striatal primordium, they remain radially oriented in natural coordinate terms (Plate 82, Fig. A). Given the close association between the ependymal glial fibres and the migration of neuroblasts from the ventricular matrix to the cortical plate, it may be expected that the glial processes not only reveal the formal changes in the hemisphere wall, but also indicate the route along which migrating cells attain the ventrolateral part of the cortical plate, which forms the anlage of the prepiriform cortex and the adjacent claustro-insular complex (Puelles 2014). This presumption has been substantiated by the work of Misson et al. (1991), who studied the migration of neuroblasts in relation to the radial glial fibres in the murine pallium. They found that during embryonic days 16–18 the leading processes and the somata of migrating neuroblasts are close to and aligned with adjacent radial glial fibres. As regards these neuroblast-glia relations, Misson et al. (1991) distinguished three different patterns (Plate 82, Fig. B), namely *radial divergent*, observed in the dorsal pallial region, *curving divergent*, occurring in the medial pallial region, and *curving convergent, then divergent*, prevailing in the most lateral pallial region. Note that the glial fibres emanating from the lateral angle of the lateral ventricle apparently cover a considerable distance oriented tangentially relative to the wall of the developing cerebral hemisphere, and the same holds true for the neuroblasts migrating along these fibres. However, it will be clear that the course of these glial fibres and the direction of this migration, though apparently tangential, is *topologically* radial. In order to distinguish this mode of displacement from the true tangential migrations, to be discussed in Sect. 10, it is designated here as *deflected radial migration*.

It stands to reason that all of the neuroblasts involved in the radial migrations just discussed, remain categorically within the confines of their own respective FMUs, namely the different pallial sectors.

8.4 Developmental processes within the FMUs
8.4.4 Differentiation of cell masses and cortices

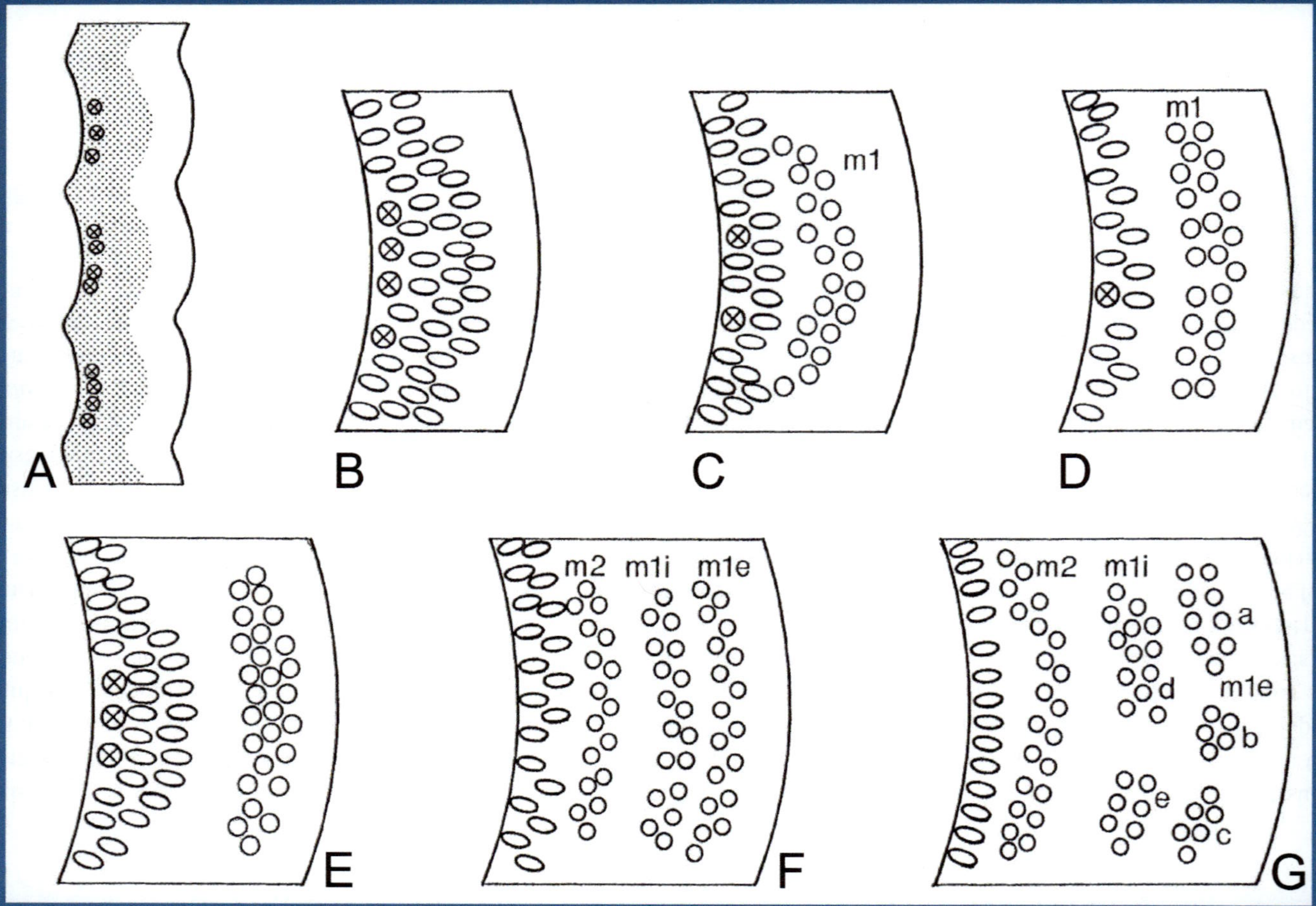

Formation of migration layers, which are the products of waves of mitotic activity.
The migration layers split up into cell masses or nuclei
(Based on Bergquist and Källén 1954)

83

8.4.4 Differentation of Cell Masses and Cortices

Plate 83

Plate 83 summarizes the ideas of Bergquist and Källén (Bergquist 1932, 1953, 1954a, b; Källén 1951a–d, 1952, 1962, 1965; Bergquist and Källén 1954). According to these authors, the formations designated here as FMUs manifest themselves initially as centres of proliferation (A), and later as *migration areas*, i.e. cushion-like thickenings of the embryonic mantle layer, produced by local accumulations of neurons due to the proliferative activity in the ventricular matrix (B). According to the observations of Bergquist and Källén, the following events may take place within a given migration area:

1. There is a continuous radial migration and the cells migrate over only short distances. The resultant cell mass remains in direct contact with or at least close to the ventricular matrix zone or the ependyma (B). This is the prevalent situation in many parts of the brains of anamniotes.
2. Migrating cells detach themselves from the ventricular zone (usually due to passing tangential fibres or differential adhesive properties) and form a separate migration layer in the mantle zone (C, D: m1).
3. New consecutive waves of local proliferative activity (E) lead to the formation of one or more additional migration layers (F: m2). In certain prosencephalic areas of amniotes no less than four consecutive waves of proliferation and four resultant migration layers have been observed.
4. The primary migration layers may secondarily split up into two (F: m1i, m1e) or sometimes three sublayers. This is usually due to a mixture of differential adhesivity and arrival of afferents that selectively target one cell population, or a perticular somatodendritic domain.
5. Each of the layers or sublayers within a migration area may fractionate into two or more separate compact cell masses, losing the appearance of a layer and giving rise to various nuclei (G: a–e).

Bergquist and Källén (1954, p. 641) emphasized that the *homologization* of brain nuclei should be based on analysis of their respective developmental histories. For each given cell mass the following questions should be answered: (1) from which migration area does this nucleus develop?; (2) within which migration layer is it formed?, and (3) within which part of the migration layer does it develop?

We essentially agree with this approach, which, *mutatis mutandis*, forms the conceptual basis of the *new neuromorphology* set forth in this course. However, genoarchitectonic studies so far suggest that the clearly conservative property of brain cell populations is their origin within a given FMU; molecularly corresponding cell types of different vertebrates may become incorporated into different layers or sublayers (see Sect. 10).

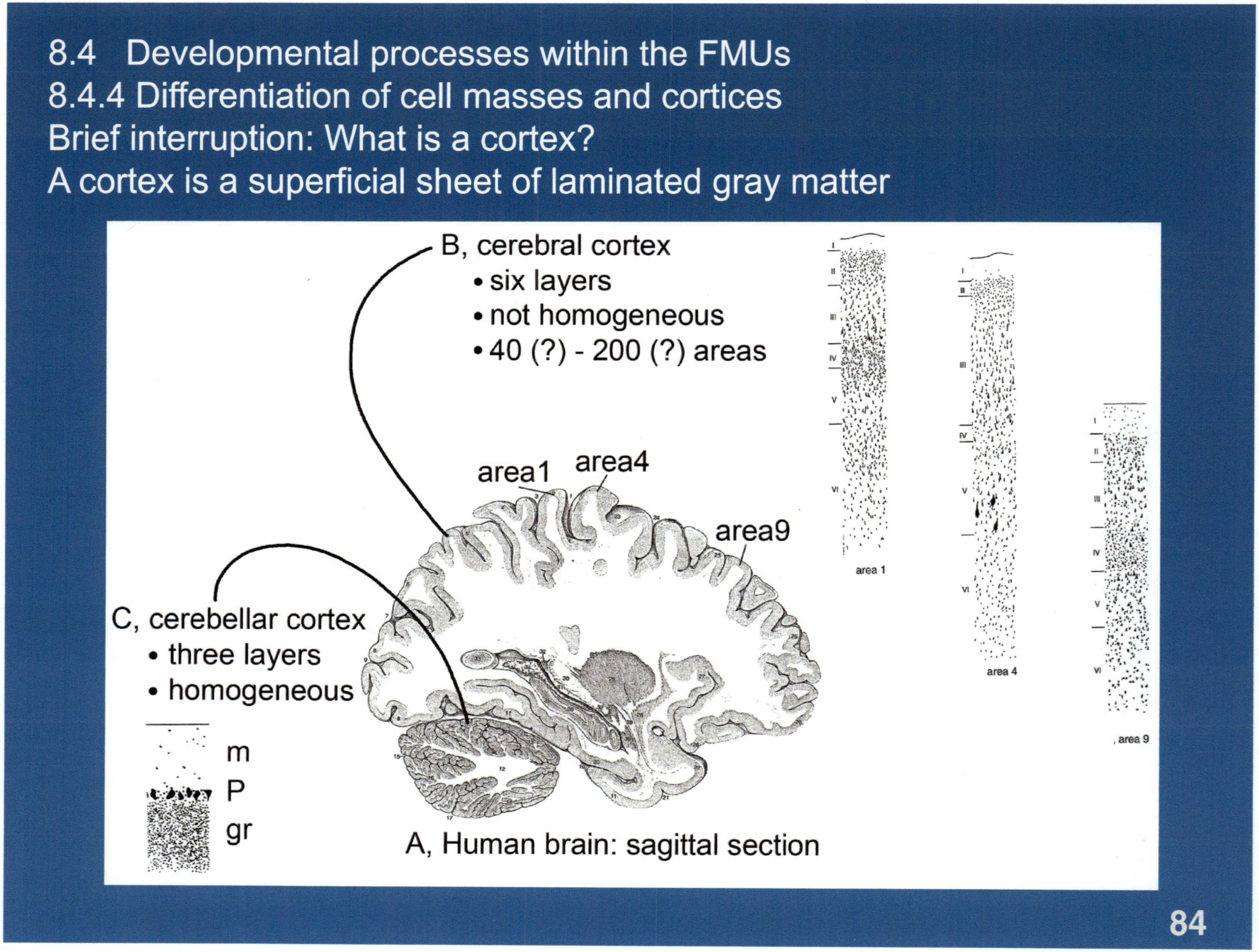
8.4 Developmental processes within the FMUs
8.4.4 Differentiation of cell masses and cortices
Brief interruption: What is a cortex?
A cortex is a superficial sheet of laminated gray matter
B, cerebral cortex
• six layers
• not homogeneous
• 40 (?) - 200 (?) areas
area1
area4
area9
area 1
area 4
area 9
I
II
III
IV
V
VI
C, cerebellar cortex
• three layers
• homogeneous
m
P
gr
A, Human brain: sagittal section
84

Plate 84

We want to say something about the development of the neocortex, but this discussion should of course be preceded by answering the question: what is a cortex? The answer to this question may be formulated as follows:

A cortex is a superficial sheet of laminated grey matter, whose neurons of diverse typology interact functionally with each other.

Plate 84, Fig. A shows a sagittal section through the human brain, in which the two largest cortices of that brain, the *cerebral cortex* and the *cerebellar cortex* are clearly visible. The convoluted cerebral cortex covers almost the entire surface of the cerebral hemispheres. The even more strongly convoluted cerebellar cortex does the same in the cerebellum. Both of these cortices are far removed from their respective matrix zones, and both are the product of highly complex, composite migration processes.

The two cortices differ considerably in structure. Most of the cerebral cortex, the so-called neocortex, consists of six layers (numbered I–VI) and is not homogeneous across many structurally and functionally differentiated areas. These layers can be distinguished more or less clearly in the three strips of cortex shown in Fig. B, taken from specific cortical areas, the respective positions of which are indicated in Fig. A. The three strips depicted are drawn from Nissl preparations, which show only and selectively the neuronal perikarya. It will be seen that these strips, and hence the areas from which they are taken, show clear structural differences. Thus, area 4, which represents the primary motor cortex, has a very feebly developed layer IV, and its Vth layer contains the giant pyramidal cells of Betz, which occur nowhere else in the cortex. The study of the structural differentiation of the cortex with the aid of Nissl preparations is known as cytoarchitectonics, and a complementary approach, which focuses on the density and orientation of myelinated fibres in the cortex, is termed myeloarchitectonics. There is no unanimity as regards the number of architectonic fields in the human cortex. Brodmann (1909) distinguished 44 cytoarchiterctonic fields in this cortex, but according to Vogt and Vogt (1919), who combined cyto- and myeloarchitectonics, the number of architectonic fields is much higher and may well amount to 200. There is converging quantitative cytoarchitectonic, receptor architectonic and functional evidence that the number of cortical fields as estimated by the Vogts is essentially correct (Nieuwenhuys 2013; Nieuwenhuys et al. 2015). Comparative studies in different mammalian species indicate that the number of distinct areas generally increases with the total surface area of the cortex, suggesting a size threshold mechanism. It is of note that all cortical areas show a distinct radial organization (see Plate 85), and that their boundaries fit perfectly in the natural coordinate system.

The cerebellar cortex (Plate 84, Fig. C) has only three layers: molecular layer (m), layer of Purkinje cells (P) and granular layer (gr), and shows, contrary to the cerebral cortex, a homogeneous stucture throughout. Any areal functional specialization in the cerebellar cortex is due to differential origins of the afferents.

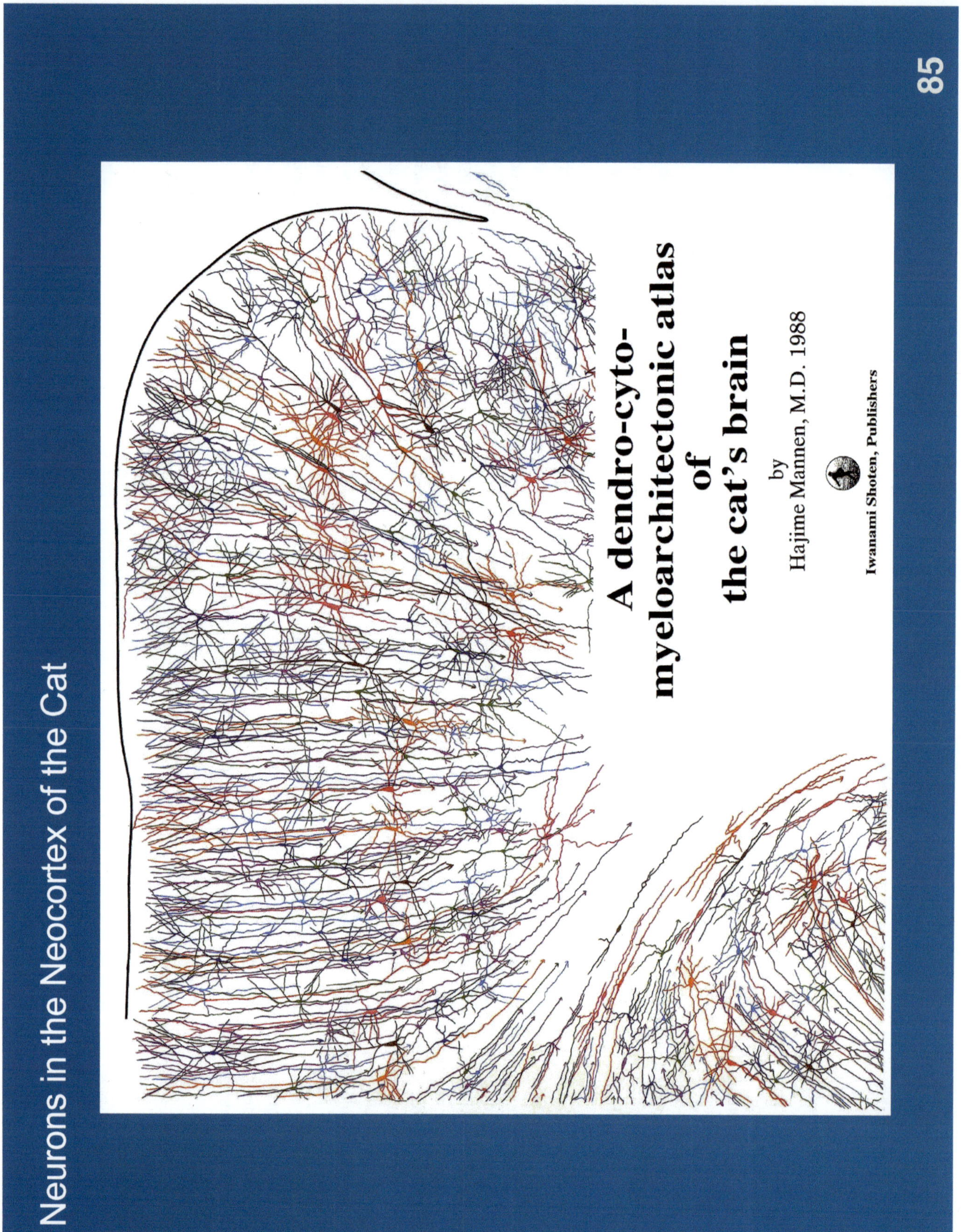
Neurons in the Neocortex of the Cat
85
A dendro-cyto-
myeloarchitectonic atlas
of
the cat's brain
by
Hajime Mannen, M.D. 1988
Iwanami Shoten, Publishers

Plate 85

Plate 85, taken from an astonishingly detailed atlas of the cat's brain, shows the overall radial orientation of the neurons in the cerebral cortex due to the prevalence (over 70 %) of pyramidal cells displaying a large radially oriented apical dendrite.

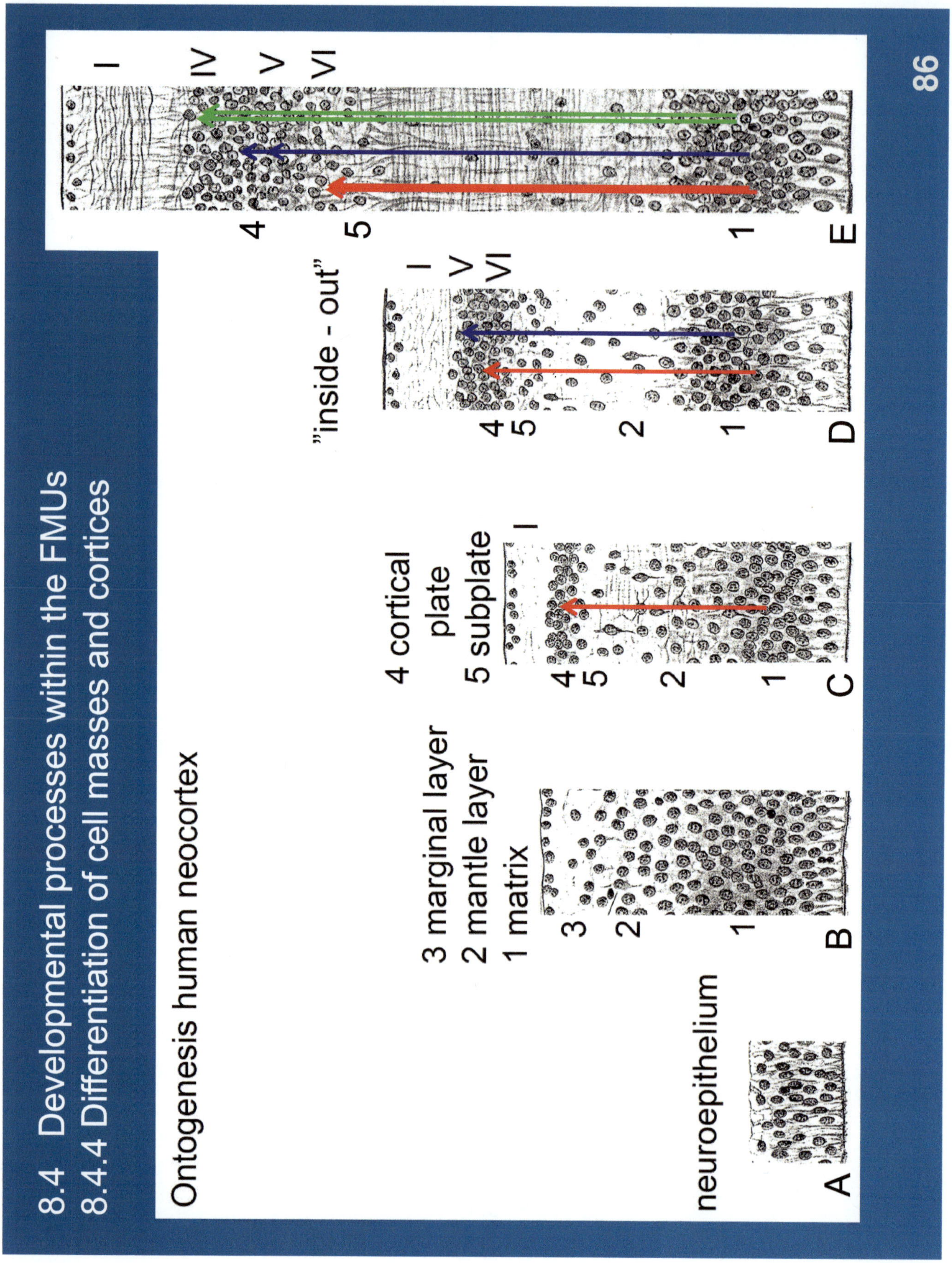
8.4 Developmental processes within the FMUs
8.4.4 Differentiation of cell masses and cortices
Ontogenesis human neocortex
neuroepithelium
A
3 marginal layer
2 mantle layer
1 matrix
3
2
1
B
4 cortical plate
5 subplate
I
4
5
2
1
C
"inside - out"
I
V
VI
4
5
2
1
D
I
IV
V
VI
4
5
1
E
86

Plate 86

Plate 86 shows some aspects of the histogenesis of the human neocortex, which develops in the pallial part of the cerebral hemispheres.

It has already been mentioned that when fully formed, this structure consists of six layers, numbered I–VI. In the adult brain it is separated from the lateral ventricle by a wide fibre zone (white matter).

The early development of the pallium does not differ from that observed in other parts of he neuraxis. It consists initially of a pseudostratified epithelium (Fig. A), and somewhat later, the three classical layers, matrix, mantle, and marginal, can be distinguished (Fig. B). During the sixth week of gestation radially migrating neuroblasts start to settle in the marginal layer, where they form a compact cortical plate (Fig. C; see also Plate 80, Fig. 2: 13). This cortical plate divides the previously existing more widely spaced primordial cortical population into an outer *zonal layer* and an inner *subplate* (Marin-Padilla 1988). The subpial zonal layer is transformed into the first cortical layer (I). Just like the cortical plate, the subplate is a transient structure. Its neurons apparently play an important role in the guidance of thalamic afferents, arriving via the deep white matter into the developing cortex (Shatz 1992). The neuroblasts that first arrive in the cortical plate later form layer VI (D), they are followed sequentially by those destined for the more superficial layers V–II (D, E). The latter migrate through the subplate and the earlier formed layers to settle more peripherally in an 'inside-out' fashion. At some point the cortical subventricular layer with its inner and outer components starts to produce additional cortical cells, overlapping the gradual depletion of the ventricular matrix layer (not shown).

The massive radial migration of neuroblasts to the developing cortex lasts until about the 26th week of gestation (Marin-Padilla 1992). After that period, the pallial mantle layer is gradually transformed into a fibre-rich intermediate zone, consisting of cortical afferents and efferents, plus cortico-cortical connections, and few interstitial neurons.

It has been demonstrated that, apart from the pallial ventricular matrix, the ventricular and subventricular zones (VZ, SVZ) of the primordial basal ganglia including the preoptic region contribute substantially to the formation of the cortex (Marín and Rubenstein 2001; Hansen et al. 2013; Ma et al. 2013). Whereas the elements produced in the pallial ventricular matrix and SVZ differentiate after a radial migration into glutamatergic pyramidal neurons, the diverse subpallial FMUs give rise to a diversity of GABA-releasing local circuit neuron types which migrate tangentally to the cortex, jointly building up over 20 % of its total cellurarity (see Sect. 10).

The SVZ extends from the anlagen of the basal ganglia into the pallium. Smart et al. (2002) studied the ontogeny of the primary visual cortex in the rhesus monkey. They observed that in this region a particularly strongly developed SVZ is present which, in terms of size and duration of activity, clearly exceeds the relatively early declining ventricular matrix. It could not be established which neuron type(s) develop(s) from this 'hypertrophic' SVZ.

Hansen et al. (2010) found that the SVZ in the human pallium has an expanded outer zone, containing numerous modified radial glial cells. These elements are each provided with a long basal process, which however does not contact the ventricular surface. It could be demonstrated that these elements can undergo proliferative divisions and self-renewing asymmetric divisions to generate neuronal progenitor cells and accordingly, neurons.

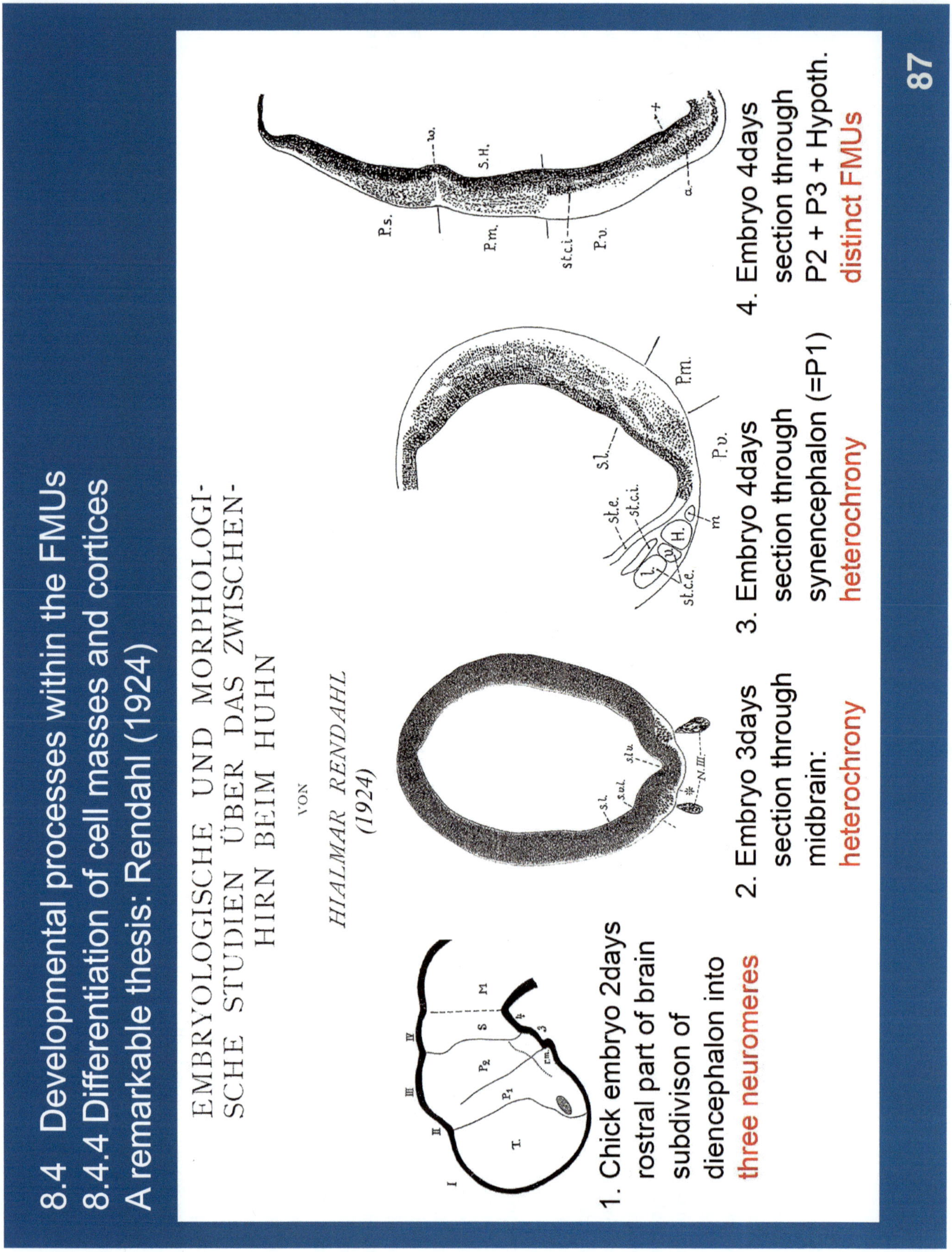
8.4 Developmental processes within the FMUs
8.4.4 Differentiation of cell masses and cortices
A remarkable thesis: Rendahl (1924)
EMBRYOLOGISCHE UND MORPHOLOGISCHE STUDIEN ÜBER DAS ZWISCHENHIRN BEIM HUHN
VON
HIALMAR RENDAHL
(1924)
1. Chick embryo 2days rostral part of brain subdivison of diencephalon into three neuromeres
2. Embryo 3days section through midbrain: heterochrony
3. Embryo 4days section through synencephalon (=P1) heterochrony
4. Embryo 4days section through P2 + P3 + Hypoth. distinct FMUs
P.s.
P.m.
P.v.
S.H.
st.c.i.
87

Plate 87

Before returning to our model system, the pretectum of the chick, we should like to show some pictures taken from Hialmar Rendahl's (1924) excellent thesis on the development of the diencephalon of the chick, which appeared 92 (!!) years ago. This author used exclusively cytoarchitectonic criteria

Plate 87, Fig. 1 shows that Rendahl subdivided the diencephalon of the early embryonic chick into three neuromeres, which he designated as synencephalon (S), parencephalon posterius (P2), and parencephalon anterius (P1). This subdivision closely corresponds to that of the current prosomeric model (p1–p3). The only major difference is that the basal portion of Rendahl's P1 extends further rostrally than the corresponding part of prosomere 3, partly invading the hypothalamus.

Plate 87, Fig. 2 presents a section through the midbrain of an embryo of 3 days' incubation. Note that the entire dorsal (alar) region is still in the neuroepithelial state, whereas in the ventral (basal) region a well-differentated oculomotor nucleus can already be already distinguished in the mantle layer.

Plaee 87, Fig. 3 presents a section through the synencephalon (prosomere 1), and a further example of heterochrony. In the dorsal region of this section, which represents the pretectum, a matrix zone is flanked by an as yet undifferentiated mantle layer; however, in the mantle layer of the ventral (tegmental) region some (pro) nuclei can be already clearly distinguished.

Plate 87, Fig. 4 shows some particularly sharply delimited FMUs in a horizontal section through the two parencephalic neuromeric domains (prosomeres 2 and 3) and the hypothalamus. The latter shows less abundant mantle cells within its peduncular portion (hp1), as opposed to the terminal portion (hp2).

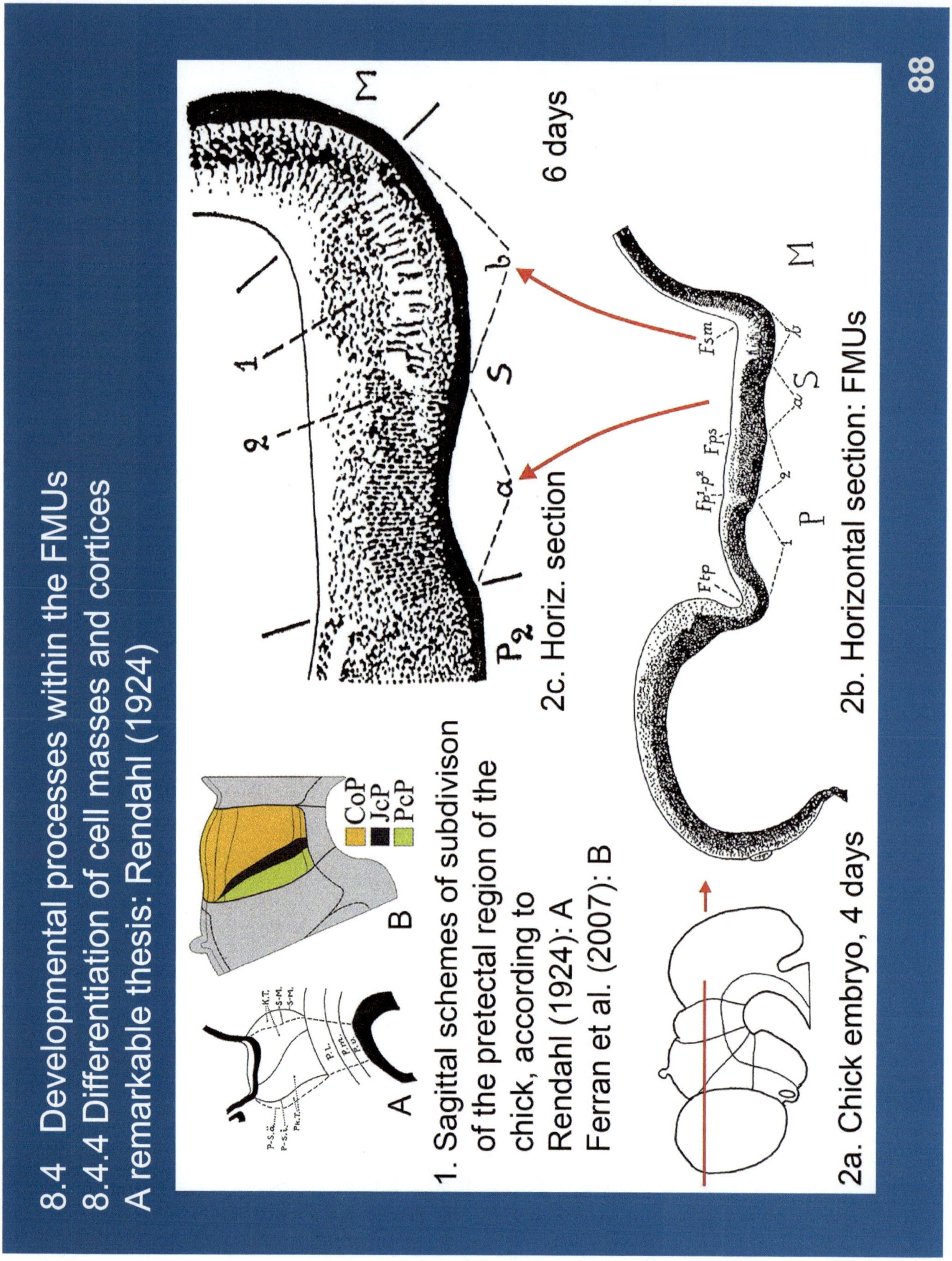
88
8.4 Developmental processes within the FMUs
8.4.4 Differentiation of cell masses and cortices
A remarkable thesis: Rendahl (1924)
A
B
CoP
JcP
PcP
1. Sagital schemes of subdivison of the pretectal region of the chick, according to Rendahl (1924): A
Ferran et al. (2007): B
2a. Chick embryo, 4 days
2b. Horizontal section: FMUs
2c. Horiz. section
6 days
P
S
M
Ftp
Fps
Fsm
P2
a
b

Plate 88

Plate 88 presents some additional pictures from the thesis of Rendahl (1924)

Fig, 1A shows that Rendahl's subdivision of the pretectum of the chick, which was based on haemalum and cresyl violet preparations, closely resembles the genoarchitectonic parcellation of Ferran et al. (2007; Fig. 1B). It appears that Rendahl only overlooked the narrow, strip-like JcP domain.

Figure 2 presents horizontal sections through the pretectal region and adjacent structures in chick embryos of 4 (Fig. 2a, b) and 6 days' incubation (Fig. 2c) Viewed in combination, they give an impression of the progressive differentiation of the pretectum during this period. Rendahl does in Fig. 2c identify a small partly separate nuclear primordium (tagged '2'), which corresponds to the JcP mantle zone.

8.4 Developmental processes within the FMUs
8.4.4 Differentiation of cell masses and cortices

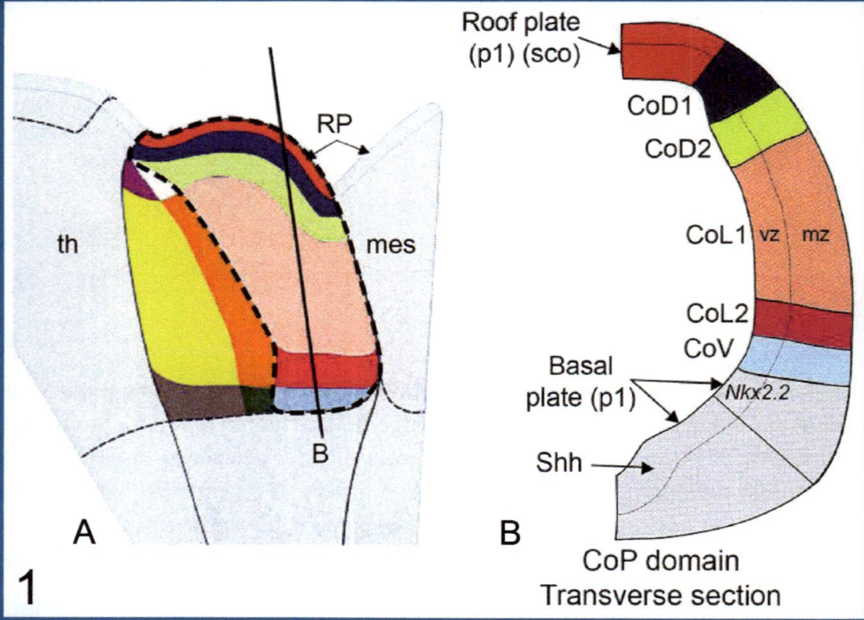

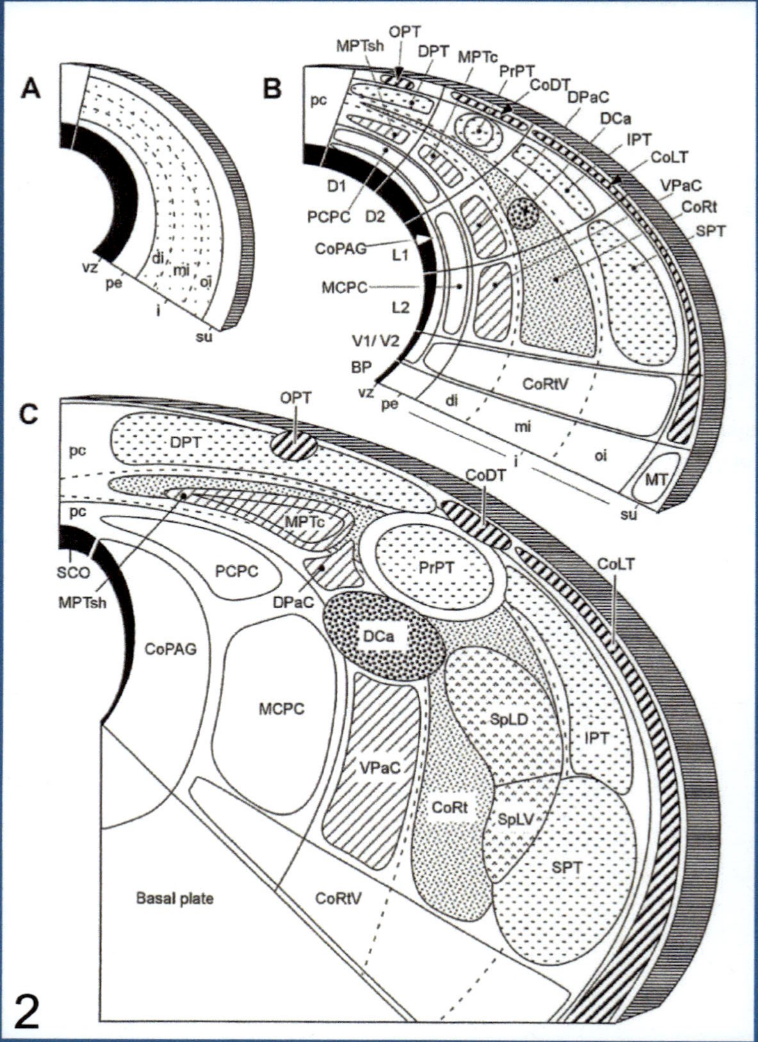

Once again:
The molecular differentiation of the pretectum of the chick (Ferran et al. 2009)

Further differentiation of the commissural domain

Plate 89

We return to our model system, the pretectum of the chick.

It may be recalled that:

1. The pretectum, i.e. the alar domain of the first prosomere (pr1) represents a FMU of the first order (FMU1).
2. During early development, this FMU1 splits up into three distinct anteroposterior domains: CoP, JcP, and PcP, which represent FMUs of the second order (FMU2).
3. Next, each of these anteroposterior domains is subdivided into a number of molecularly defined, dorsoventrally arranged subdomains, representing FMUs of the third order (FMU3). It is to be emphasized that these microzonal molecular specification patterns are established step by step during the second and third days of incubation, largely before neurogenesis and morphological differentiation (mantle layer development) begins.

In the following synopsis of the later development of the pretectum, we will confine ourselves mainly to the commissural pretectal domain PcP. In Plate 89, Fig. 1A the boundaries of this domain are marked with a heavy dashed line. Plate 89, Fig. 1B shows a diagrammatic transverse hemisection through PcP, in which the positions of its five molecularly defined, dorsoventrally arranged subdomains, CoD1, CoD2, CoL1, CoL2, and CoV, are indicated. Note that the meaning of the descriptors 'dorsal', 'lateral', 'ventral' refers exclusively to position within the alar plate, and not to the whole neuromere.

Neurodifferentiation begins with the radial migration of neuroblasts out from the matrix zone to form the mantle layer (Plate 77). In many parts of the developing CNS, including the pretectum, the migrated neuroblasts within the mantle layer aggregate in a number of layers or strata. We have seen that according to Bergquist and Källén (1954) the formation of these various migration layers is preceded by waves of increased proliferative activity in the matrix zone (Plate 83).

The strata and layers distinguished in the commissural mantle zone at 6–7 days' incubation are indicated in Plate 89, Fig. 2A. They include: (1) a periventricular stratum (pe), situated directly peripherally to the ventricular zone (pv) or matrix layer; (2) an intermediate stratum (i), and (3) a superficial stratum (su). Moreover, the intermediate stratum appeared to be divisible into separate deep (di), middle (mi), and outer (oi) layers. These strata and layers represent pronuclear formations, from each of which later on, one, two or several distinct nuclei may arise. The detailed genoarchitectonic analysis of Ferran et al. (2009) has shown that the formation of the various pretectal nuclei (or their direct ontogenetic precursors) principally occurs within the boundaries of the radially arranged, dorsoventral progenitor subdomains D1, D2, L1, L2, and V1/V2. This is shown in Plate 89, Fig. 2B, which diagrammatically illustrates the 'state of nuclear differentiation' at 9–10 days' incubation.

Plate 89, Fig. 2C presents a final scheme of commissural pretectal nuclei after 12 days' incubation. It appears that the initial differentiation pattern, determined by the intersection of the tangential strata/layers and the radial precursor subdomains as shown in Plate 89, Fig. 2b, is now partially obscured by displacement of cell masses. Three examples of these processes are instructive: (1) Nucleus designated as PrPT is ventrolaterally displaced relative to nucleus MPT. (2) Cell mass DCa originates from PcP, *i e.* the rostral or precommissural pretectal domain, but the ventral portion of this cell mass forms during development a caudally directed extension which encroaches upon the juxtacommissural (JcP) as well as the commissural (CoP) domains. (3) Due to the changes in position of the PrPt, MPT and DCa nuclei, the PPaC nucleus is deformed and comes to be lodged in the corner between the PrPT and DCa nuclei.

The detailed genoarchitectonic analyses of Ferran et al. (2007, 2009) have shown that the development of the PcP and JcP pretectal domains, closely resemble that of the CoP domain. Also in these domains, the positional relations of the various (pro) nuclei is determined by the intersection pattern of the tangential cell strata / layers and the radial precursor sectors.

The histogenesis of the avian pretectum appears to answer entirely to the scenario for the development of prosencephalic centres in amniotes articulated by Bergquist and Källén (1954, p. 654):

"From the different squares of this network (i.e. the FMUs of the present course; R.N., L.P.), a number of migration layers is formed by one or many successive migrations. These layers are finally divided into cell masses, representing nuclear anlagen. Within them the cell differentiation takes place. In later stages the picture, obtained in this way, is somewhat blurred by a spreading of the cells."

The new neuromorphology outlined in this course is based on the foundations laid by Bergquist and Källén and by the Swedish school of neuroembryology in general (for a review, see Nieuwenhuys 1998a). In general, it may be said that the numerous outstanding neuroembryological studies, produced by this school during the period from 1920 to 1970, attracted very little attention (e.g., Kuhlenbeck 1973; Swanson 2012) and sometimes received fully undeserved criticism (see e.g. Herrick 1933). Due to the considerable resolution power of the genoarchitectonic approach, the ideas of the Swedish school can now be explored and provided with a solid molecular underpinning.

Genoarchitectonic analysis of the brain may be defined as the practical application of gene expression patterns to dissect details of neural structural differentiation (cell populations, nuclei, and their boundaries) connecting these items causally with pre-existent molecular regionalization of the neuroepithelium (progenitor domains, FMUs) (Ferran et al. 2009). One of the fruits of this new and powerful approach is that 15 novel pretectal formations, which were not distinguished in any previous study of the avian pretectum, could be identified. This emerging new complexity of brain structure may be repellent to those who prefer a simplistic view of the brain. However, there can be no doubt that real progress lies in exploring the new frontier exactly as it appears in front of us (unsuspected levels of complexity may still exist). Moreover, as the history of neuroscience clearly proves, overall functional analysis will not provide satisfactory results unless every bit and piece of the brain is considered in our theories.

8.4 Developmental processes within the FMUs
8.4.4 Differentiation of cell masses and cortices

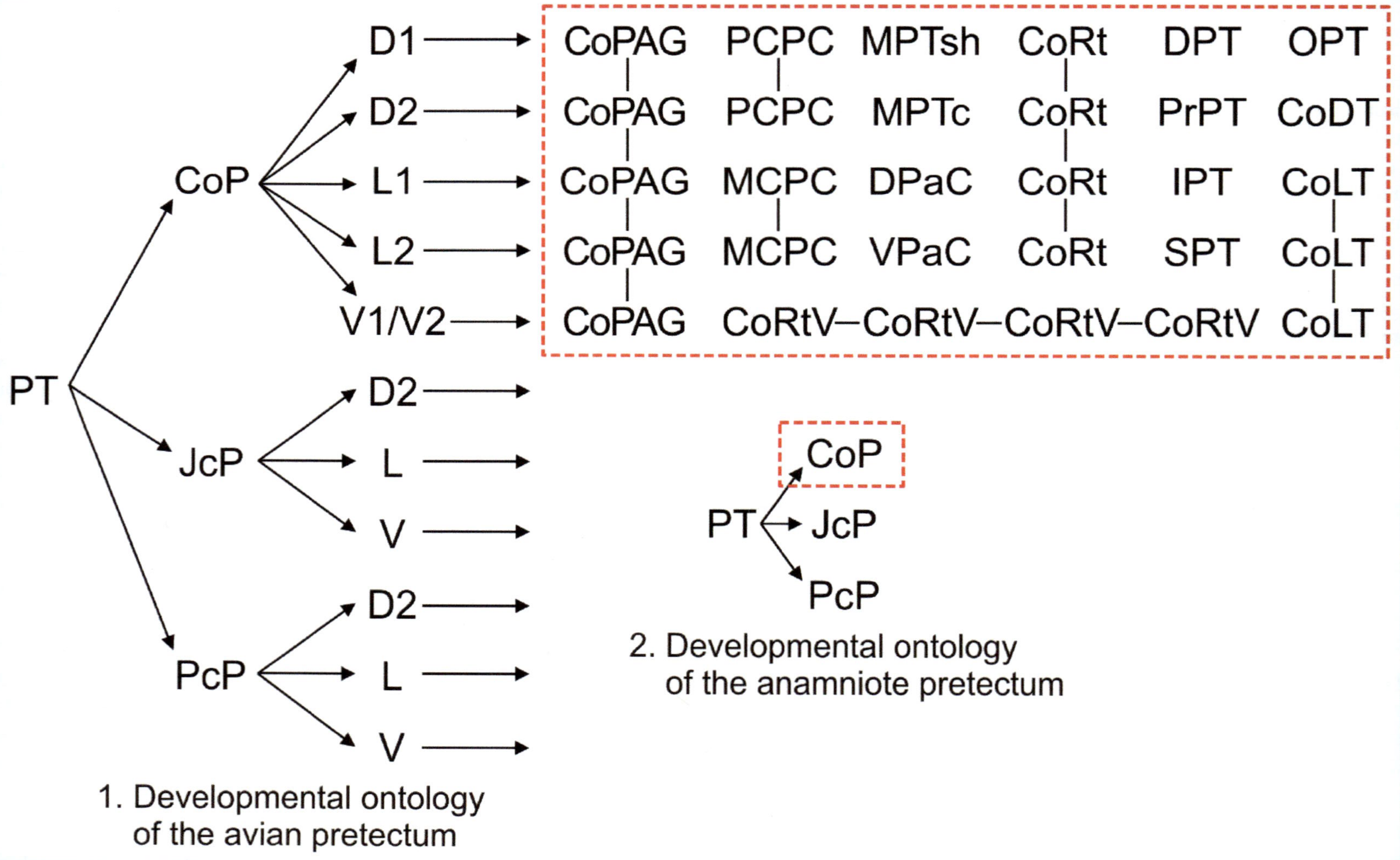

90

Plate 90

So, the molecular regionalization of the avian commissural pretectal domain can be summarized as follows:

1. Individualization of the alar region of prosomere 1 (PT), one of the 38 primary FMUs in the basal and alar plates of the avian brain.
2. Individualization of three anteroposteriorly arranged secondary FMUs, the domains PcP, JcP, and CoP, within PT.
3. Individualization of five dorsoventrally arranged tertiary FMUs, the subdomains CoD1, CoD2, CoL1, CoL2, and CoV (or D1, D2, L1, L2, and V1/V2).
4. Individualization of strata and layers within the mantle layer of each subdomain of PcP.
5. Individualization of (pro) nuclei within the individual strata/layers, relative to the borders of the dorsoventrally arranged tertiary FMUs.

This progressive molecular regionalization process or *ontology* of CoP is visualized in Plate 90, Fig. 1. Similar ontologies can be determined for the other avian pretectal domains, and should be laid out for all of the 37 remaining primary FMUs of the avian brain, as well as for all of the 38 primary FMUs of a number of representative other vertebrate species. This enormous amount of pending work forms an indispensable prerequisite for the development of our new neuromorphology, the central aim of which is (a) the establishment of sound, molecularly buttressed causal explanations of the complex structure of the brain, to aid in the analysis of neural systems, and (b) the elucidation of homologies, in order to harmonize the theory of brain evolution with genomic evolution. The ideas which Bergquist and Källén (1954) enunciated concerning the homologization of neural cell masses may be summarized in the following definition:

Cell masses in the brain of related vertebrates, which emerge from the same locus of the same stratum /layer of migrated neuroblasts within the same migration area (our FMU) are homologous.

Later in this course (Sect. 10) we will address the question whether, and if so, to what extent this definition needs to be revised in light of our new insights

An algorithm for the systematic comparison of the developmental ontologies of corresponding brain regions in related species is to be developed, though it is already clear that the natural coordinate system will be of paramount importance in this analysis. There is evidence indicating that the (early) development of corresponding FMUs in different species is governed by strong evolutionary conservation of gene expression patterns and genomic regulatory mechanisms. Hence, the noticeable similarity of early brain genoarchitectonic profiles across all vertebrate lineages studied so far. Thus, it has been established that the developmental ontology pattern visualized in Plate 90, Fig, 2, is valid for all tetrapods, and perhaps even for all vertebrates (Ferran et al. (2009), representing literally a Bauplan of the pretectum. It needs to be determined whether the regionalization of the pretectum in anamniotes does not proceed further than indicated in Plate 90, Fig. 2. If that happens to be the case, it may be stated that the anamniote CoP is the *field homologue* of all of the 16 avian pretectal nuclei surrounded by a red dashed line in Plate 90, Fig. 1 (Puelles and Medina 2002; Medina 2007). It stands to reason that in the comparison of developmental ontologies of particular FMUs in the brain of different amniotes much more complex situations may obtain, due to variation and specialization.

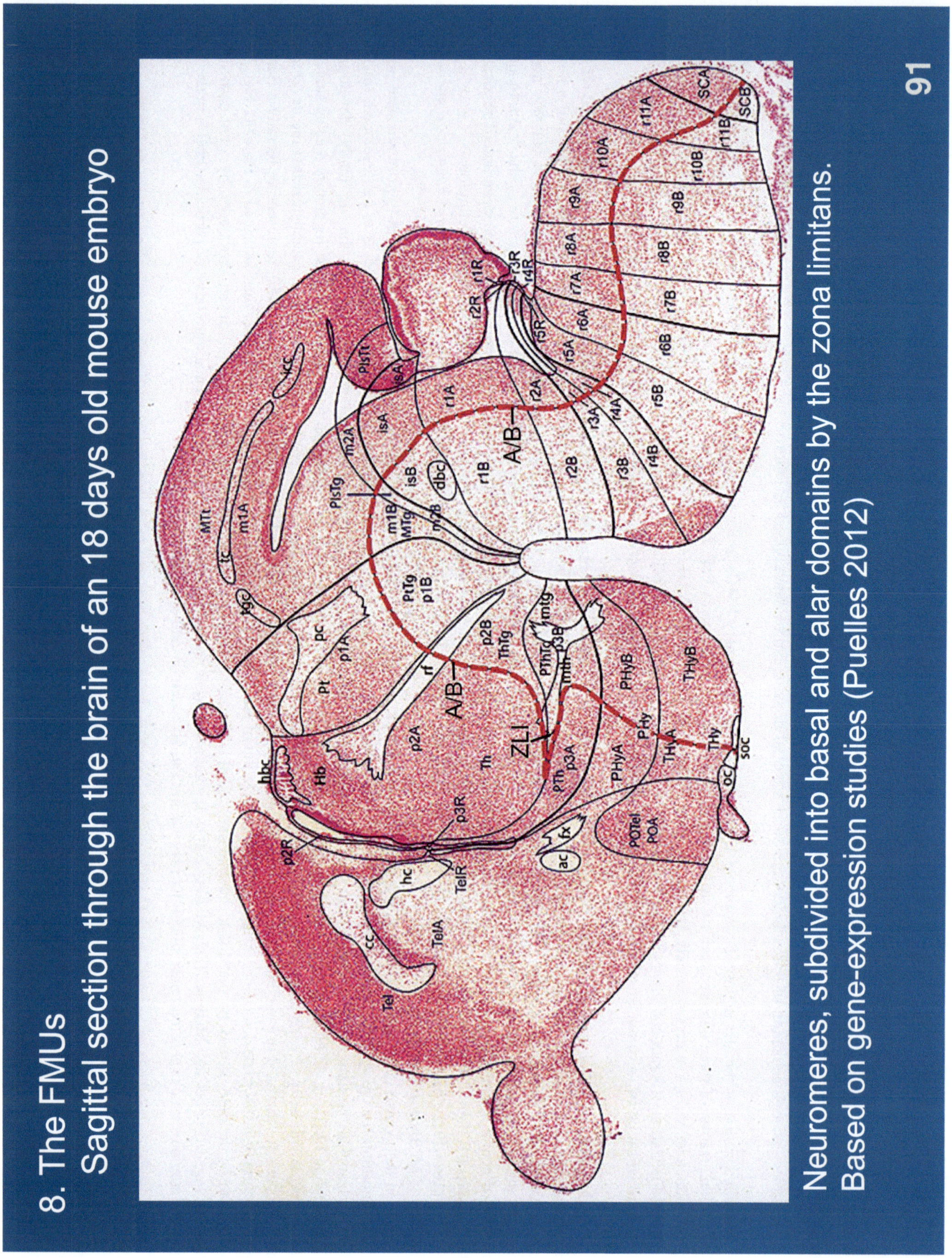
8. The FMUs
Sagittal section through the brain of an 18 days old mouse embryo
Neuromeres, subdivided into basal and alar domains by the zona limitans.
Based on gene-expression studies (Puelles 2012)
91
MTt
m1A
PIsTt
isA
IsA
m2A
PIsTg
isB
dbc
r1A
r1B
A/B
r2R
r1R
r3R
r4R
r5R
r2A
r2B
r3A
r3B
r4A
r4B
r5A
r5B
r6A
r6B
r7A
r7B
r8A
r8B
r9A
r9B
r10A
r10B
r11A
r11B
SCA
SCB
m1B
MTg
m2B
PtTg
p1B
pc
p1A
Pt
rf
p2B
Thtg
PThtg
p3B
mtg
mth
ZLI
p2A
Th
Hb
hbc
p3A
PTh
p3R
p2R
PHyB
PHy
PHyA
THyB
THyA
THy
soc
oc
POTel
POA
fx
ac
hc
TelR
TelA
cc
Tel

Plate 91

Plate 91 shows a sagittal section through the brain of an 18 days old mouse embryo. The positions of the 19 encephalomeres are indicated: 11 rhombomers (r1–r11), the isthmic neuromere (is), 2 mesomeres (m1, m2), 3 prosomeres (p1–p3), the peduncular hypothalamo-telencephalic neuromere (PHy), and the terminal hypothalamo-telencephalic neuromere (THy). The longitudinal alar-basal boundary (A/B; heavy dashed red line) subdivides all of the encephalomeres into basal (B) and alar (A) primary domains. The A/B boundary is deflected dorsally between the p2A and p3A domains, forming a transverse spike, called the zona limitans intrathalamica (ZLI). The positions of the following tracts and commissures are indicated: anterior commissure (ac), hippocampal commissure (hc); fornix tract (fx); habenular commissure (hbc); optic chiasma (oc); supraoptic commissure (soc) retroflex fasciculus (rf); mamillothalamic tract (mth); mamillotegmental tract (mtg); posterior commissure (pc); tectal gray commissure (tgc); tectal commissure (tc); intercollicular commissure (icc), and decussation of the brachium conjunctivum (dbc). Several of these fibre systems, including fx, oc, soc, rf, pc, and dbc, are clearly associated to neuromeric boundaries (see Sect. 11).

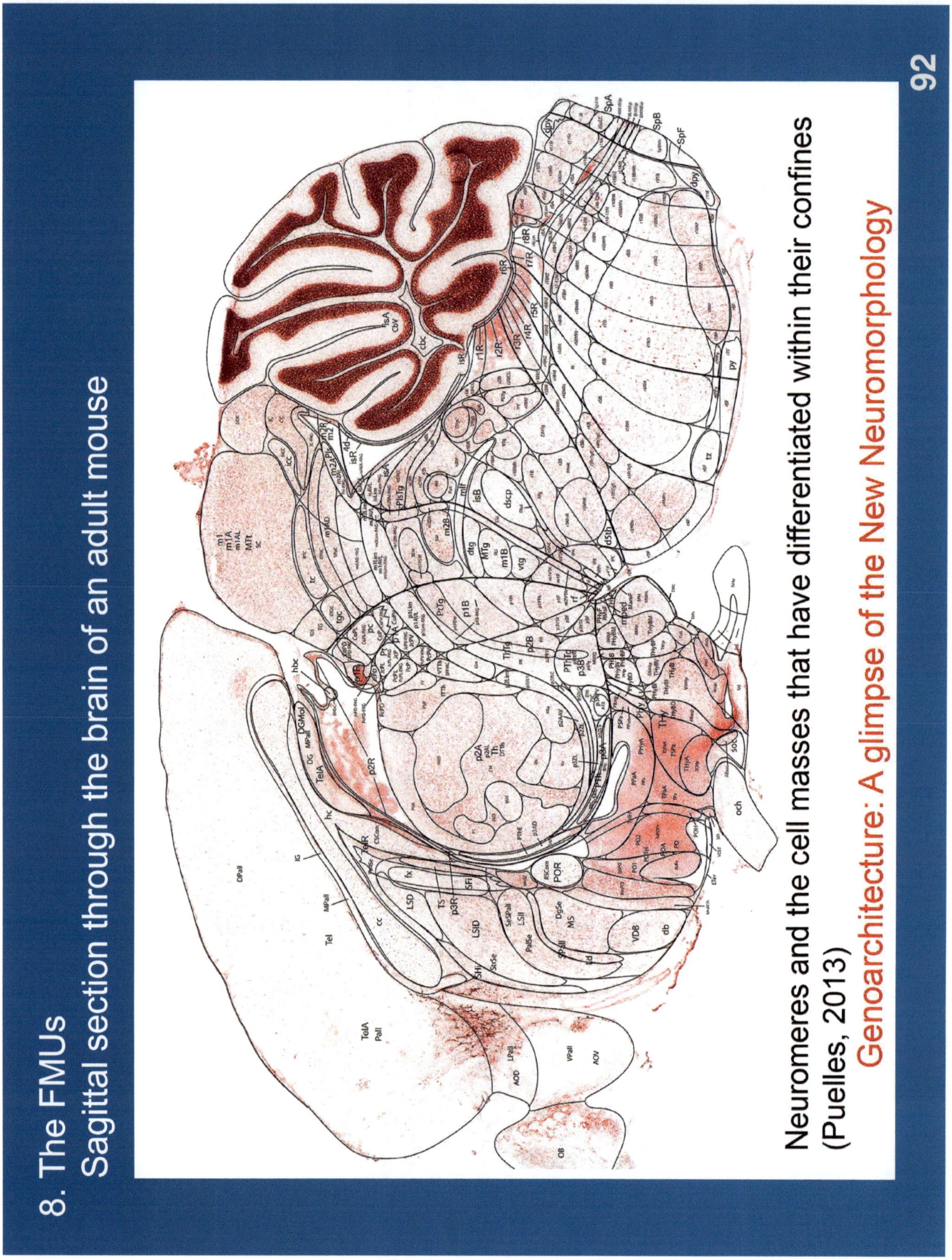
8. The FMUs
Sagittal section through the brain of an adult mouse
Neuromeres and the cell masses that have differentiated within their confines (Puelles, 2013)
Genoarchitecture: A glimpse of the New Neuromorphology
92

Plate 92

Neuroanatomical tradition usually deals with brain nuclei represented in given areas as if they were wholly independent one from another, lacking any particular describable order. They are conceived as "potatoes in a potato sack", making very difficult the understanding of tridimensional complexes of individual nuclei. However, progress in the analysis of the histogenesis of nuclei indicates that each of them is formed within particular mantle strata of radial histogenetic complexes, or fundamental morphological units (FMUs; Plate 92). Thus, in the first place, the nuclei that derive from the same radial complex are related to one another as relatively deeper or superficial entities along the ventriculo-pial dimension of the brain wall (irrespective that this dimension may be morphogenetically deformed, it remains traceable according to natural coordinates). Secondly, the radial histogenetic complexes in the mantle layer retain the mutual planar neighbourhood relationships originally held by the neuroepithelial progenitor domains from which they developed. Therefore, the derived nuclei occupying the same stratum in neighboring radial histogenetic domains also retain these invariant bidimensional (AP/DV) positional relationships. Neuroanatomic descriptions should be more useful if the radial and stratigraphic relationships of studied grisea and related connections were made clear.

8.5 Summary of FMUs

1. The central nervous system is composed of a fixed number of fundamental morphological units (FMUs) characterized by unique molecular profiles and fixed topological positions within the neural tube, irrespective of morphogenetic deformations.
2. During development and also in the adult, these FMUs are radially organized histogenetic entities, which extend from the ventricular surface to the meningeal surface of the neuraxis.
3. The earliest appearing or primary FMUs are the floor, basal, alar and roof domains of the various neuromeres.
4. The primary FMUs quite often split up during further patterning under influence of secondary organizers, into smaller radial entities, which may be provisionally designated as 'secondary', 'tertiary' and 'final' FMUs. The number of final brain FMUs lies in mammals in the hundreds.
5. All of the FMUs have their own specific gene expression profile, referred to as their 'molecular identity' or 'profile' (apparently involving minimally a combination of some 80–100 characteristic genes, many of which are individually shared with other FMUs, but in different combinations).
6. The FMUs are progenitor centres (domains) where matrix cell proliferate and differential types of postmitotic neuroblasts and glioblasts are produced; these migrate radially and finally differentiate into specific types of neurons or glia cells.
7. All FMUs pass through a stage in which three primordial layers, matrix layer, mantle layer, and marginal layer, can be distinguished. The adult brain structure contains only ependyma in place of the former (now depleted) matrix layer, and largely consists of derivatives of the mantle and marginal layers
8. In many parts of the CNS of amniotes, the neuroblasts, building the mantle layer aggregate in one or more layers. Each of the various cell masses emanating from a FMU develops generally from a specific locus in one of these layers.

9. The structural pattern, 'Bauplan' or 'blueprint' of the CNS

9.1 The overall pattern

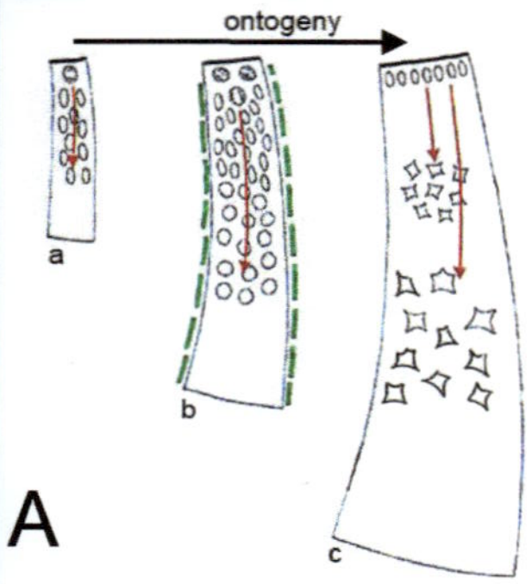

A

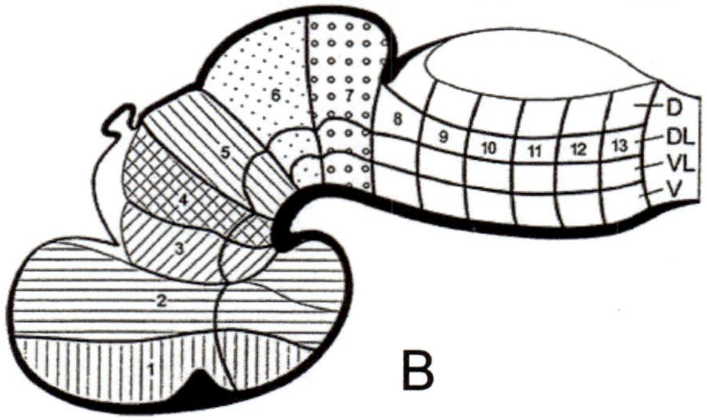

B

A. Diagrammatic repressentation of the development of a FMU.
B. The morphological plan of the vertebrate brain, according to (Bergquist and Källén 1954)
C. Primary FMUs in the CNS of a generic mammalian embryo (Puelles et al. 2012)

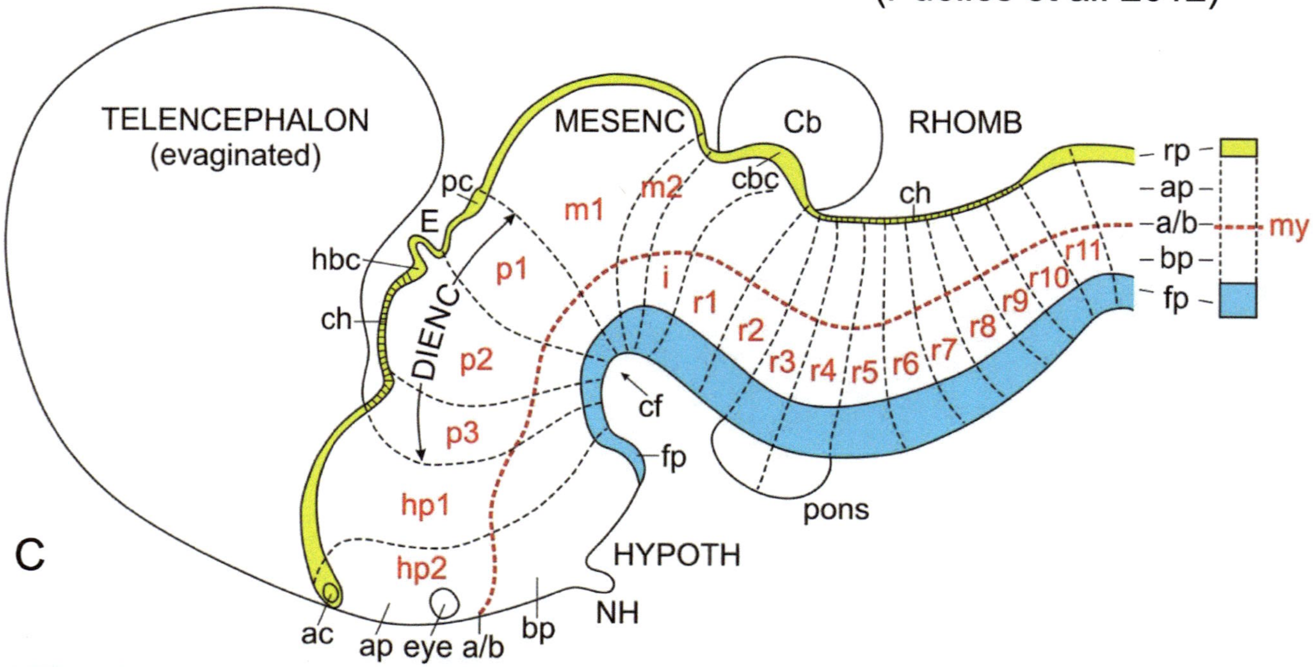

C

9 The Structural Pattern, ‘Bauplan’ or ‘Blueprint’ of the CNS

9.1 The Overall Structural Pattern

Plate 93

At the beginning of this new section it may be recalled that:

1. The CNS consists of fundamental morphological units (FMUs), which are centres of neuronal and glial proliferation, migration and differentiation (Plate 93, Fig. A).
2. All of these FMUs are radial entities, the ventricular/ependymal surfaces of which occupy a specific area of the ventricular surface of the neural tube.
3. The comparative neuroembryological studies of Bergquist and Källén (1954) have indicated that the FMUs (their ‘migration areas’) are spatially organized in a pattern resulting from the intersection of transversely oriented bands of high mitotic activity with longitudinal zones showing similar proliferative features. They found that this pattern shows a remarkable consistency throughout the vertebrate kingdom, and hence may be considered as representing the ‘morphological plan’ of the CNS of vertebrates (Plate 93, Fig. B).
4. Studies on the expression patterns of developmental regulatory genes and their products in embryonic brains of various vertebrate groups have revealed the presence of sharply delimited progenitor domains in the wall of the early neural tube, the spatial pattern of which shows a striking resemblance to the ‘morphological plan’ of Bergquist and Källén (Plate 93, Fig. C).

Plate 93, Fig. C is largely derived from the ‘updated prosomere model’ of Puelles et al. (2012); however, a strip representing a spinal neuromere or myelomere (my) has been added, and the ventricular surface of the rhombencephalon has been exaggerated and turned toward the beholder. By this manipulation, we obtained a flat map of the entire ventricular surface of the CNS except for that of the evaginated telencephalic hemispheres, on which the molecularly defined *primary FMUs* could be projected. The topologically longitudinal alar/basal boundary (red dashed curve) divides, as its name implies, all myelomeres and encephalomeres into a basal and an alar domain. Hence, all neuromeres comprise their own specific sectors of the floor, basal, alar, and roof plates.

The outline of Plate 93, Fig. C is based on a medial view of the bisected CNS of a generic mammalian embryo. There is, however, abundant evidence that the pattern of boundaries delimiting the various primary FMUs, as shown in the figure, is shared widely among vertebrates. Hence, Fig.C reflects our current ideas concerning the structural plan of the vertebrate CNS.

R. Nieuwenhuys and L. Puelles, *Towards a New Neuromorphology*,
DOI 10.1007/978-3-319-25693-1_9

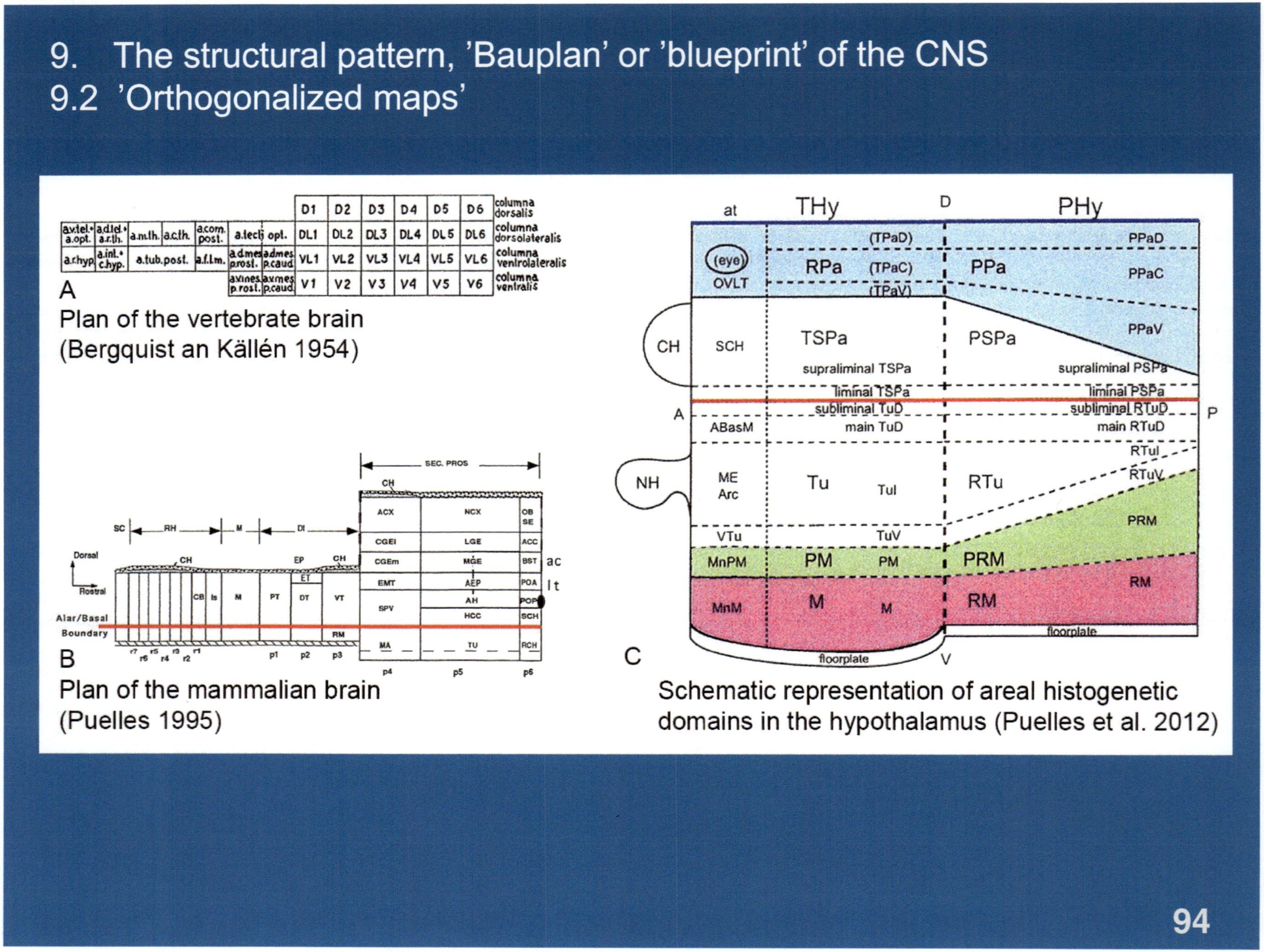
9. The structural pattern, 'Bauplan' or 'blueprint' of the CNS
9.2 'Orthogonalized maps'
A
Plan of the vertebrate brain
(Bergquist an Källén 1954)
D1 D2 D3 D4 D5 D6
DL1 DL2 DL3 DL4 DL5 DL6
VL1 VL2 VL3 VL4 VL5 VL6
V1 V2 V3 V4 V5 V6
columna dorsalis
columna dorsolateralis
columna ventrolateralis
columna ventralis
B
Plan of the mammalian brain
(Puelles 1995)
SEC. PROS
Dorsal
Rostral
Alar/Basal Boundary
ac
lt
C
Schematic representation of areal histogenetic domains in the hypothalamus (Puelles et al. 2012)
at
THy
D
PHy
A
P
V
CH
NH
OVLT
SCH
ABasM
ME
Arc
VTu
MnPM
MnM
RPa
TSPa
supraliminal TSPa
liminal TSPa
subliminal TuD
main TuD
Tu
TuI
TuV
PM
M
PPa
PPaD
PPaC
PPaV
PSPa
supraliminal PSPa
liminal PSPa
subliminal RTuD
main RTuD
RTu
RTuI
RTuV
PRM
RM
floorplate
94

9.2 'Orthogonalized' Maps

Plate 94

Several authors have published simplified flat maps of the ventricular surface of the CNS in which the FMUs are represented by rectangular fields. Bergquist and Källén (1954: Plate 94, Fig. A), and Puelles (1995: Plate 94, Fig. B) have summarized the ideas they held at that time on the basic structural pattern of the vertebrate brain in maps of this type. Both maps are obsolete and have now only historic (and heuristic) value. Note that the rostral end of the brain is to the left in Fig. A, but to the right in Fig. B.

Plate 94, Fig. C shows an example of an 'orthogonalized' representation of the principal histogenetic domains in the hypothalamus of the mouse, derived from Puelles et al. (2012). The longitudinal dimension is marked by the floor plate at the bottom, the alar/basal boundary is in the middle (horizontal red line), and the telencephalo-hypothalamic limit is at the top (blue line). The acroterminal territory (at) is represented on the left, with a symbolic optic chiasma (CH) and neurohypophysis (NH). The hypothalamo-diencephalic boundary (i.e. the boundary between the third diencephalic prosomere, p3, and the first hypothalamic prosomere, hp1) lies at the right edge of the diagram. The intrahypothalamic boundary (i.e. the boundary between the first and the second hypothalamic neuromeres, hp1, hp2) is marked with a vertical thick dashed line. This boundary separates the hp1-derived peduncular hypothalamus (PHy) from the hp2-derived terminal hypothalamus (THy). The basal domains of PHy and THy represent primary FMUs. The alar domains of PHy and THy form part of primary FMUs, which extend into the telencephalon (Plate 95, Fig. C). The four principal hypothalamic territories, basal and alar PHY, and basal and alar THy, can all be divided into a number of secondary FMUs (indicated by different colours), and each of these is further subdivisible into several tertiary FMUs (delineated with thin dashed lines). The detailed analysis of Puelles et al. (2012) has shown that within the hypothalamic anlage no less than 33 distinct progenitor areas can be distinguished (indicated on this diagram but not described here).

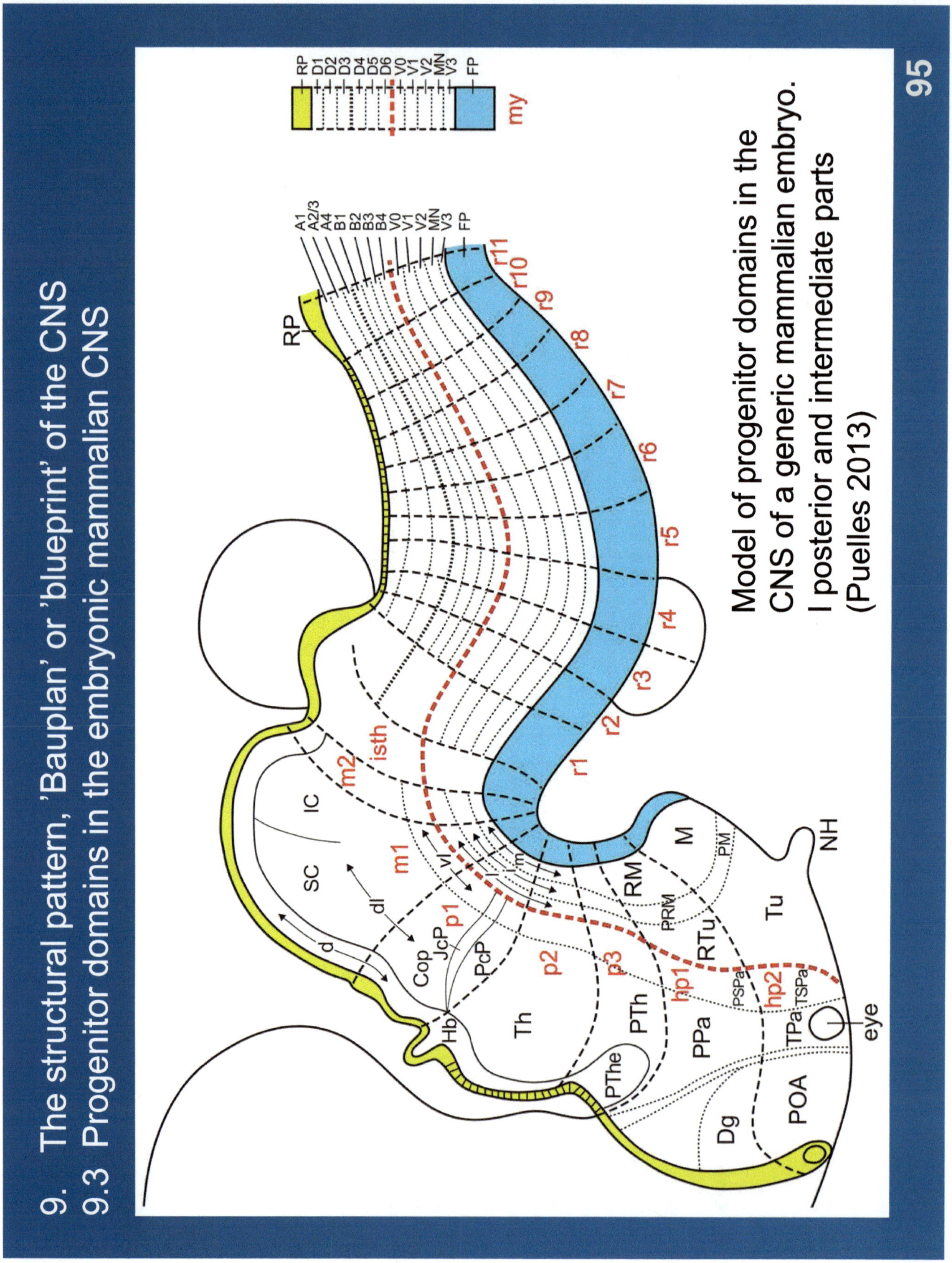
9. The structural pattern, 'Bauplan' or 'blueprint' of the CNS
9.3 Progenitor domains in the embryonic mammalian CNS
Model of progenitor domains in the CNS of a generic mammalian embryo.
I posterior and intermediate parts
(Puelles 2013)
95
RP
D1
D2
D3
D4
D5
D6
V0
V1
V2
MN
V3
FP
my
A1
A2/3
A4
B1
B2
B3
B4
r1
r2
r3
r4
r5
r6
r7
r8
r9
r10
r11
isth
m1
m2
IC
SC
d
dl
vl
vm
Cop
JcP
PcP
p1
p2
p3
Hb
Th
PTh
PThe
RM
PRM
PM
M
RTu
Tu
hp1
hp2
PSPa
TSPa
PPa
TPa
POA
Dg
eye
NH

9.3 A More Detailed Structural Pattern: Progenitor Domains in the Embryonic Mammalian CNS

Plate 95

In Plate 95, the primary FMUs shown in Plate 93, Fig. C, are subdivided into smaller, molecularly defined entities. The diagram is largely based on a recent review of Puelles (2013) on the developmental basis of the genoarchitecture of the mammalian CNS. The precise rank (secondary or higher order FMUs) of some of these entities is not entirely clear at present; however, they all represent radially organized progenitor domains.

The lateral wall of the spinal cord (represented by the myelomere to the right) is divisible into eleven longitudinally arranged microzones, five basal ones, V0, V1, V2, MN and V3, and six alar ones. D1–D6.

The wall of the rhombencephalon is also divisible into a large number of distinct longitudinal progenitor domains. There are five basal domains, which are all directly comparable to those in the spinal cord. The rhombencephalic alar zones form two groups, a dorsal one and an intermediodorsal one. The dorsal group, adjacent to the (largely choroid) roof plate, contains the rhombic lip and is composed of three zones, A1, A2/3 and A4. The intermediodorsal group comprises four zones, B1–B4. The relations between the rhombencephalic and spinal alar zones are less clear than in the basal plate; however, it is well established that the rhombencephalic B4 zone corresponds with the spinal D6 zone, and that the rhombencephalic B2 zone has no spinal equivalent. In general, it may be said that the rhombic lip gives rise to the remarkable long tangential migrations to be discussed in Sect. 10, and to the special somatosensory, cochlear and vestibular nuclei (the designation 'special' is used to indicate that the rhombencephalic nuclei in question have no spinal equivalents). The B1–B3 zones produce the general somatosensory trigeminal nuclei. as well as the general and special (gustatory) viscerosensory centres.

Much of the alar domains of the isthmic and r1 segments is involved in the histogenesis of the cerebellum.

Whereas the floor plate at spinal and rhombencephalic levels develops into a glial septum, known as the raphe, the mesencepalic and diencephalic sectors of this structure show a significant neurogenetic activity, producing large numbers of dopaminergic neurons.

A detailed histogenetic and genoarchitectonic analysis of the mesencephalic and diencephalic basal plate domains is not available yet. Here, we tentatively subdivide these two domains ventrodorsally into medial (paramedian), intermediate and lateral microzones, designated in the plate as m, i, and l, respectively.

The mesencephalic and diencephalic alar plate domains are both subdivisible into three histogenetic zones, a small dorsal zone (d), a much more ample dorsolateral zone (dl), and a small ventrolateral zone (vl). The large dorsolateral zone of m1 produces, among other centres, the gray of the superior and inferior colliculi (SC, IC). The dorsal zones of p2 and p3 comprise the habenular area (Hb) and the prethalamic eminence (PThE), respectively. The diencephalic dorsolateral zone gives rise to the reticular thalamic nucleus (p3), the thalamic nuclear complex (p2) and various pretectal nuclei (p1). As extensively discussed in the previous section, the pretectal nuclei are organized into three distinct anteroposteriorly arranged histogenetic entities, called the precommissural (PcP), juxtacommissural (JcP) and commissural (CoP) domains.

For the sake of continuity, the hypothalamic region is included in the plate. The progenitor domains within this region will be discussed below, in connection with those in the other parts of the secondary prosencephalon (Plate 96, Fig. A).

9. The structural pattern, 'Bauplan' or 'blueprint' of the CNS
9.3 Progenitor domains in the embryonic mammalian CNS

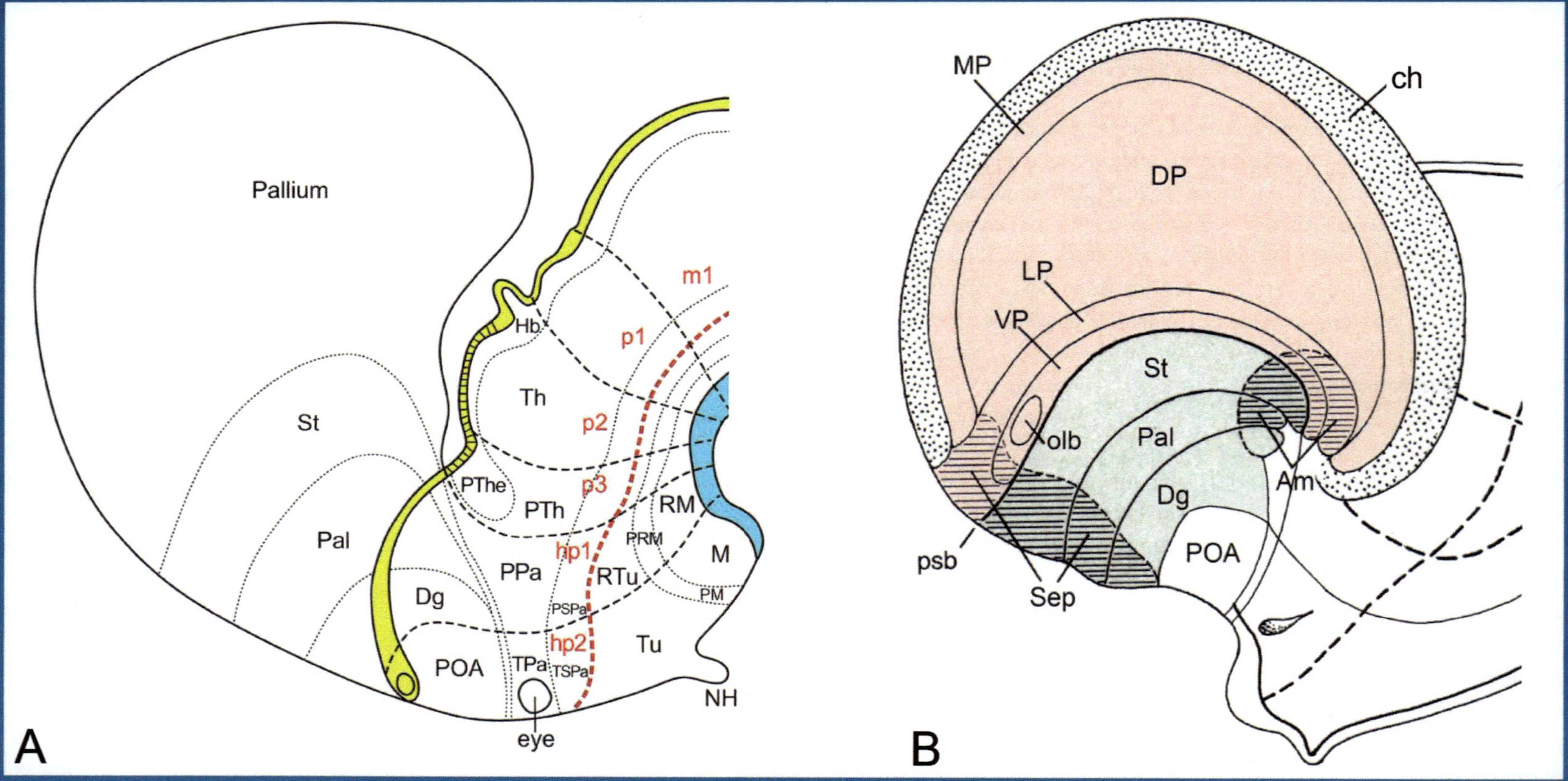

Model of progenitor domains in the CNS of a generic mammalian embryo; II intermediate and anterior parts (Puelles 2013)

Scheme of a flattened out, 'de-evaginated' cerebral hemisphere of a 19-mm human embryo to show the topology of the pallial and subpallial zones (Nieuwenhuys et al. 2008)

96

Plate 96

The general relations in the secondary prosencephalon, which forms the most rostral part of the brain, have already been discussed in Sect. 6 (Plate 63). It may be recalled that this region encompasses two neuromeres, the caudal hp1 and the rostral (and terminal) hp2, and that the alar/basal boundary, which meets its counterpart of the opposite side just ventral to the eye/chiasmatic anlage, divides these two neuromeres into basal and alar domains (Plate 96, Fig. A). Whereas the basal domains of hp1 and hp2 are confined to the hypothalamus, the alar domains of these neuromeres include hypothalamic as well as telencephalic territories.

The telencephalic territory of hp2 is confined to the small, unevaginated preoptic area (POA), but that of hp1 expands enormously and evaginates to form the bilateral cerebral hemispheres.

Before discussing the principal hypothalamic progenitor domains, it may be recalled that the hypothalamic sectors of hp1 and hp2 are designated as peduncular hypothalamus and terminal hypothalamus, respectively (Plate 63).

The medial, intermediate and lateral histogenetic zones in the basal midbrain continue through the diencephalic tegmentum into the basal hypothalamus, forming distinct separate progenitor areas in the peduncular and terminal hypothalamic sectors. Thus, in the basal plate of the terminal hypothalamus, mamillary (M), perimamillary (PM) and tuberal (Tu) areas can be distinguished, whereas the basal peduncular hypothalamus is divisible into retromamillary (RM), periretromamillary (PRM) and retrotuberal (RTu) areas. Similarly, the dorsolateral and ventrolateral zones in the alar midbrain have distinct hypothalamic counterparts, which again form separate progenitor areas in the terminal and peduncular hypothalamus. The resulting four principal progenitor areas in the alar hypothalamus are designated as the terminal paraventricular (TPa), terminal subparaventricular (TSPa), peduncular paraventricular (PPa) and peduncular subparaventricular (PSPa) areas (Fig. A).

In order to show the topological relations of the principal progenitor domains in the evaginated part of the telencephalon, we graphically flattened out a cerebral vesicle of a 19-mm human embryo by subjecting its wall to a 'de-evagination' algorithm (Nieuwenhuys et al. 2008; Plate 96, Fig. B). It will be seen that the evaginated subpallium is divisible into separate diagonal (Dg), pallidal (Pal) and striatal (St) domains, whereas the pallium differentiates into ventral (VP), lateral (LP), dorsal (DP) and medial (MP) pallial sectors. The medial pallium, which gives rise to the hippocampal and parahippocampal cortices, appears in Plate 96, Fig. B as much stretched, due to the 'de-evagination' procedure. Caudally, it is bordered by an epithelial derivative of the roof plate, which forms the telencephalic choroid plexus (ch). Posteriorly, this structure is directly continuous, via the interventricular foramen, with the diencephalic roof plate, also choroidally differentiated in large part (Plate 96, Fig. A). Rostral to the choroidal telencephalic roof area there extends the non-choroidal septal commissural plate (hippocampal and callosal commissures), which ends at the anterior commissure. The dorsal pallium, which forms an island enclosed by the medial and lateral pallial sectors, produces the six-layered neocortex. The lateral and ventral pallial sectors, finally, are involved in the formation of the olfactory prepiriform and piriform cortices and the pallial amygdala (Am). The olfactory bulb (olb) develops from the rostral part of the ventral pallial sector. It should be added that the description of the four pallial progenitor domains and their derivatives, as presented here (and during the Leipzig course), was based on the studies of Puelles et al. (2000) on the development of the telencephalon in the chick and the mouse. In the mean time, the first author of that publication has advanced the concept that in mammals the lateral pallium gives rise to the claustrum plus the insular cortex, and that the whole olfactory cortex and pallial amygdala develop from the ventral pallium (Puelles 2014).

Note that the medial and ventral pallial zones as well as the various subpallial zones, which are separated from each other by the pallial-subpallial boundary (psb), participate at their ends in the formation of the septal region (Sep) and the amygdaloid complex (Am). The primordia of both these structures are hatched in Plate 96, Fig. B.

9.4 Summary of the CNS Structural Pattern

Two salient maps of the CNS are presented in this section. The first map (Plate 93, Fig. C) represents our current views on the molecularly based overall structural pattern of the vertebrate CNS. The second map (Plate 95 and Plate 96 A) shows the positional relations of the most important progenitor domains in a generic mammalian embryo. This map will serve as a scaffold in our discussion of the development of grisea, to be presented in the next section.

10 The development of grisea

10.1 Introduction

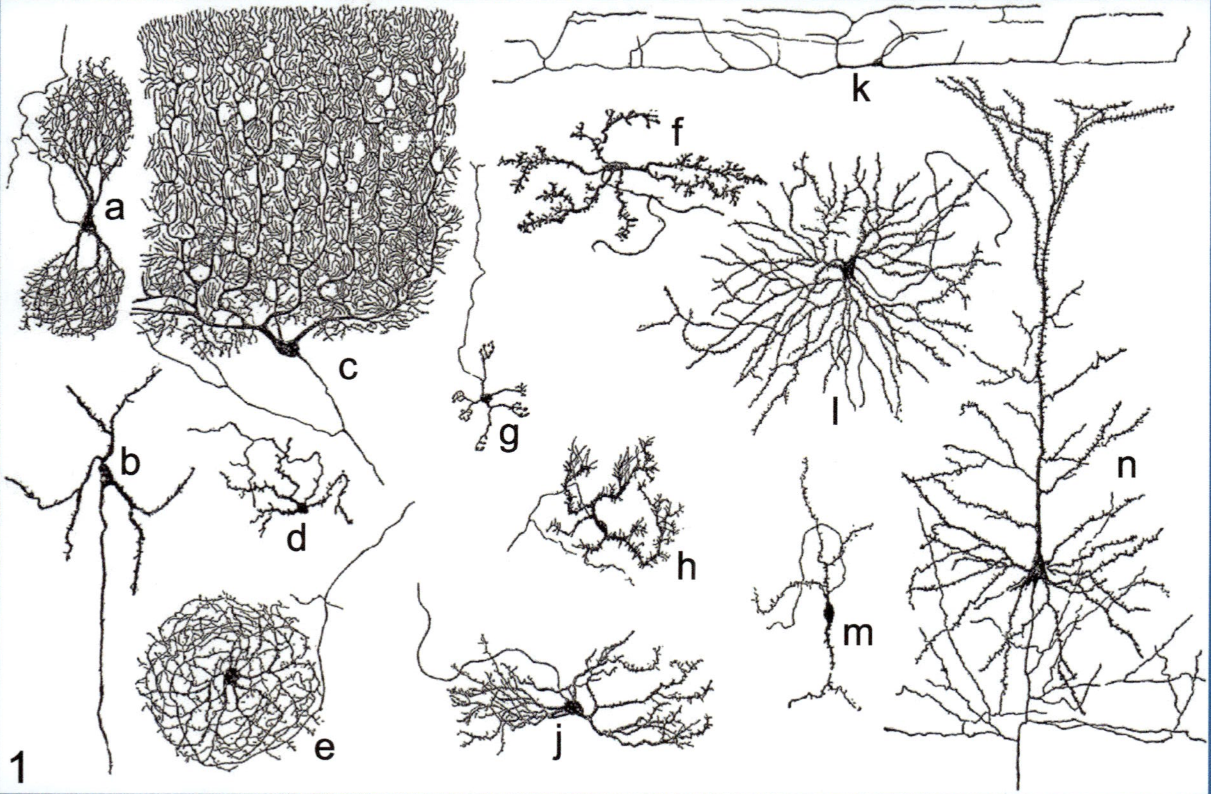

1

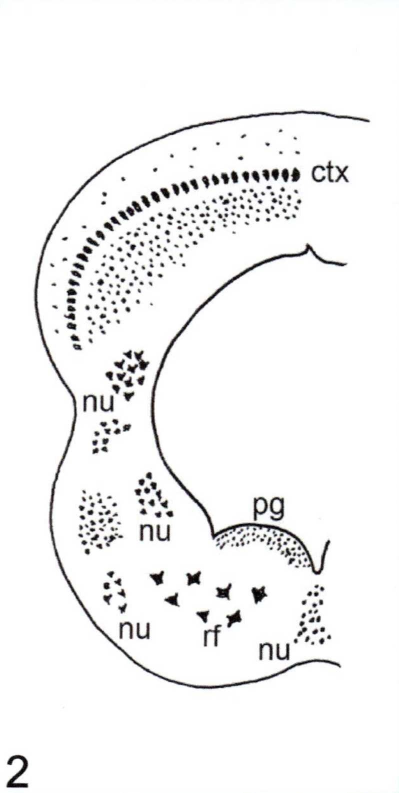

2

1. A sample of neurons from different parts of the brain. a, medial cuneate nucleus, human; b, ambiguus nucleus, cat; c, cerebellar Purkinje cell, human; d, dorsal cochlear nucleus, cat; e, inferior olive, human; f, inferior colliculus, cat; g, cerebellar ganule cell, rabbit; h, interpeduncular nucleus, rabbit; j, pontine nucleus, human; k, cortical Cajal-Retzius cell, rabbit; l, medial geniculate body, cat; m, lateral mamillary nucleus, mouse; n, cortical pyramidal cell, rabbit, Combined from drawings of Ramón y Cajal (1909, 1911)

2. Idealized transverse section through the hindbrain, to show the various kinds of grisea.
ctx, (cerebellar) cortex; nu, nuclei; pg, periventricular gray; rf, reticular formation

10 The Development of Grisea

10.1 Introduction

Plate 97

The mature brain contains more than a hundred types of neurons. These may differ in the size and shape of their cell bodies or somata, in the size, shape and complexity of their dendritic trees, in the length and destination of their axons, in receptor repertoire, in neurotransmitter profile, in their mode of influencing other neurons (excitation versus inhibition), and so on.

Neurons have been categorized in many different ways, such as long axoned projection (Golgi type I) neurons versus short axoned local circuit (Golgi type II) neurons; unipolar, bipolar or multipolar neurons, on account of the number of dendritic trunks; spiny or aspiny neurons, on account of the presence or absence of small, spiny processes on their dendrites, and excitatory (generally glutamatergic) versus inhibitory (generallly GABAergic) neurons. These categories may show a considerable overlap. Thus, most though not all long axoned projection neurons are excitatory, and most though not all local circuit cells or interneurons are inhibitory.

Plate 97, Fig. 1 presents a sample of neurons taken from different parts of the brain. It is important to realize that each of these neurons stem from the progenitor zone of one particular FMU, and that all represent the end stage of a complex differentiation and specification process in which numerous developmental regulatory genes are involved.

Neurons are not randomly spread throughout the brain. They form aggregations or grisea which, on account of their shape and position may be categorized as (1) zones of periventricular gray, (2) cell masses or nuclei, (3) contingents of loosely arranged neuronal elements or reticular formations, and (4) zones of superficially situated, laminated gray matter or cortices (Plate 97, Fig. 2). An alternative subdivision is into simple or uniform grisea composed of neurons of a single type, or alternatively complex grisea, composed of neurons of two or more types. Grisea are nodes or centres in the 'wiring' of the brain. Their overall function is the processing of neural information in the sense of filtering, correlating, summating, gating, synchronizing or triggering signal transmission to a different brain centre or peripheral effector.

The aim of the present section is to provide some insight into the development of grisea, i.e. regarding the transformation of the embryonic progenitor areas into the complex cellular organization of the mature brain. The following three processes play a prominent role in this transformation: (1) the progressive differentiation of the mantle layer via radial migration within the various FMUs; (2) the formation of plurisegmental complexes, and (3) the export of neuroblasts from certain FMUs to other neural territories via tangential migrations. These processes will now be discussed.

R. Nieuwenhuys and L. Puelles, *Towards a New Neuromorphology*,
DOI 10.1007/978-3-319-25693-1_10

10 The development of grisea

10.2 The progressive differentiation within the FMUs

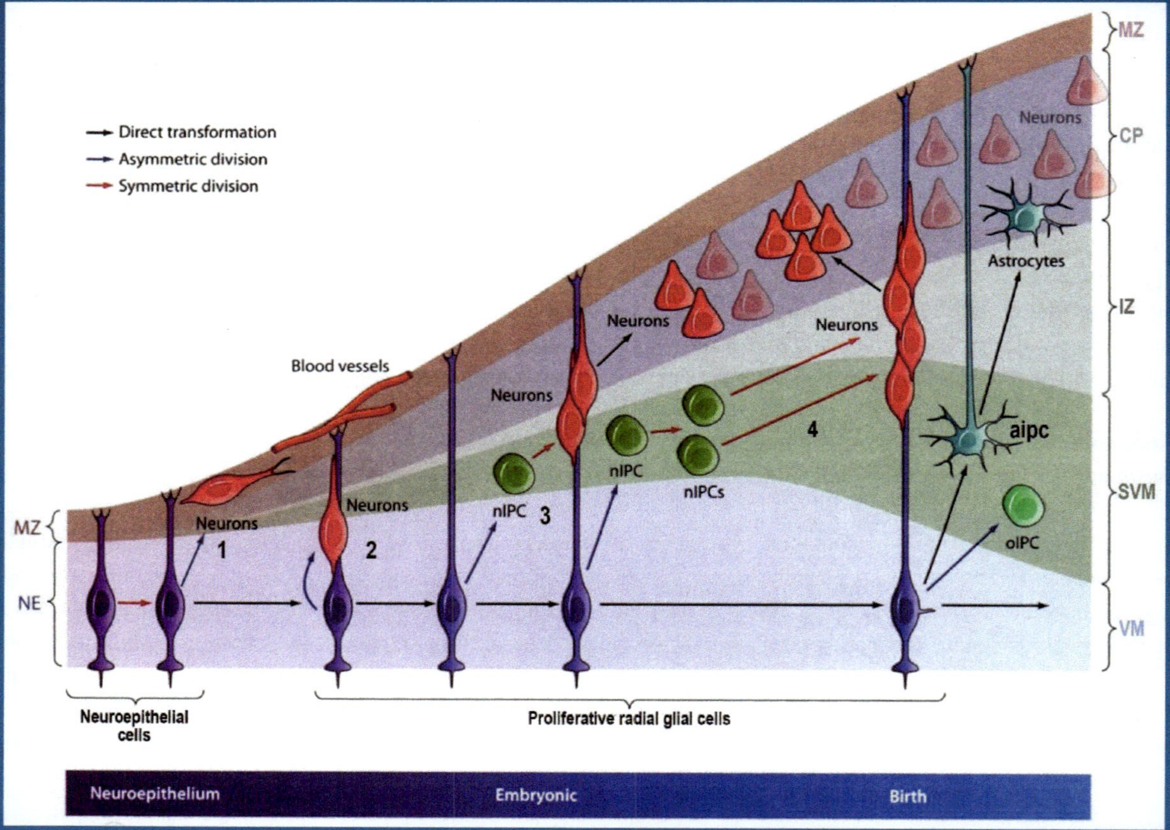

Four modes of neurogenisis during cortical development. Neuroepithelial matrix cells produce neuroblasts (1) directly through asymmetric or symmetric division; (2) indirectly through proliferative radial glial cells (PRGCs); (3) more indirectly through PRGCs and one generation of neurogenic intermediate precursor cells (nIPCs), and (4) still more indirectly through PRGCs and two or more generations of nIPCs. The picture also shows that PRGCs during later development may produce astrocytes, via astrocyte intermediate precursor cells (aIPCs), as well as oligodendrocytes, via intermediate precursor cells (oIPCs). Modified from Kriegstein and Alvarez-Buylla (2009). CP, cortical plate; IZ, intermediate zone; MZ, marginal zone; NE, neuroepithelium; SVM, subventricular matrix zone; VM, ventricular matrix zone

98

10.2 Progressive Differentition Within the FMUs

Plate 98

The principal insight to be gained from the previous sections is that the developing CNS of vertebrates typically has a modular structure; it is composed of radially oriented and differentially specified fundamental morphological units (FMUs), which occupy fixed AP/DV topological positions with respect to each other. All FMUs have a specific gen expression profile, referred to as their 'molecular identity'. Early delimitable FMUs include the basal and alar sectors of the various neuromeres (Plate 93). It has already been mentioned that during further development, these initial FMUs show a tendency towards splitting up into smaller radial entities. Thus, it has been established that the basal as well as the alar sectors of the various rhombomeres and myelomeres (spinal segments) split into a number of dorsoventrally arranged microzones, and that the pretectum, i.e. the alar sector of the first prosomere, first splits up into three rostrocaudally arranged areas, each of which subsequently differentiates dorsoventrally into three or more radial subareas (Plates 73, 89). We designate these subsequent subdivisions of the primary FMUs provisionally as 'secondary', 'tertiary', and 'final' FMUs. A problem is that in a number of mesencephalic, diencephalic and telencephalic brain regions we presently do not know how far this splitting process proceeds. In these regions the identity of the terminal or definitive FMUs has not been determined as yet.

Each of the FMUs has its own periventricularly located compartment of undifferentiated, proliferative matrix cells. The descendents of these matrix cells migrate radially to populate their FMU; hence, these units are also designated as *progenitor domains*. The offspring of the matrix cells include neuronal and glial elements, the generation and specification of which occurs according to highly ordered temporal and spatial patterns.

The genes that together define the boundaries of a given FMU are also involved in the specification of the elements, which will develop within its confines. The simplest (but still very complex!) mode of development is that neuroepithelial cells in the periventricular matrix zone proliferate during a certain period and, by dividing symmetrically, produce more undifferentiated elements. This period of 'pure proliferation' is followed by a period of asymmetric mitoses, during which each dividing matrix cell produces one matrix cell and one postmitotic neuroblast. Still later, most or all neuroepithelial matrix cells will, by symmetric mitosis, produce two postmitotic neuroblasts, thereby gradually depleting the matrix layer. During the decline of the matrix layer, and after its complete exhaustion, there follows a period of differentiation in which the neuroblasts are transformed into neurons.

Recent studies, summarized by Kriegstein and Alvarez-Buylla (2009), and Sun and Hevner (2014) have shown that at least in the pallial and subpallial districts of the telencephalon, many or most neuroepithelial cells convert into *proliferative radial glial cells* (PRGCs), and that these elements may produce neuroblasts, either directly or indirectly through one or several generations of *intermediate progenitor cells* (IPCs).

The PRGCs maintain their contacts with the ventricular and meningeal surfaces by means of central and peripheral processes and their somata are retained within the matrix zone. Their nuclei show, just as those of the matrix cells, the remarkable interkinetic migrations or 'elevator movements' discussed in Chap. 8 and depicted in Plates 75 and 76. The PRGCs divide asymmetrically at the ventricular surface to self-renew and simultaneously produce either postmitotic neuroblasts, or IPCs destined to produce neuroblasts; the latter no longer reside in the ventricular matrix layer because they lose the neuroepithelial character of matrix cells and PRGCs, and form a subventricular matrix zone (nIPCs; VM; SVM; Plate 98). At the end of development, most of the PRGCs detach from the ventricular and meningeal surfaces and convert into astroglial cells.

The nIPCs accordingly do not contact the inner and outer surfaces of the neural tube and divide in situ, either to generate daughter neuroblasts directly, or to produce one or several generations of niPCs with more restricted fates. The number of times nIPCs divide may vary in different brain regions and in different species.

The nIPCs tend to accumulate in a separate subventricular matrix zone (SVM), which, in terms of size and duration of activity, clearly exceeds the relatively early declining ventricular matrix. In the neocortex of primates, the SVM can be subdivided into an inner SVM (ISVM) and an outer SVM (OSVM). The latter is strongly developed and shows a typical radial organization in the developing visual cortex (Smart et al. 2002).

So, postmitotic neuroblasts may derive from various precursors, including neuroepithelal matrix cells, PRGCs, and nIPCs of different orders. It is plausible, though not yet proven, that specific neuronal subtypes develop from these different precursors, and the same holds, *mutatis mutandis*, for the various matrix compartments. However, it is well known that within the FMUs, the neuroblasts assemble either by stacking in layers, or by packing into nuclear configurations according to regular spatiotemporal gradients. Such gradients are called 'inside-out' when the neuroblasts that are generated at successively later times migrate past those formed earlier and take up successively more peripheral positions. This inside-out pattern of assembly is, shown *inter alia* by the inhibitory neurons in the cerebellar cortex, and the excitatory neurons in the neocortex (see below). In several other primary FMUs, including the dorsal thalamus and the pretectum, the neuroblasts detach from the matrix zone(s) to form one or several layers of migrated cells simply stacked according to time of arrival, i.e. according to an 'outside-in' pattern. These 'migration layers' may split up into two or sometimes even three sublayers. During later development, each of these layers or sublayers may be transformed via selective adhesive cell aggregation into one or more separate nuclear cell masses (Bergquist and Källén 1954; Plate 83, and Ferran et al. 2007, 2009; Plate 89).

During the last decades considerable progess has been made with unraveling the genetic mechanisms underlying neurogenesis. Investigations summarized by Guillemot (2007) have shown that combinations of members of several classes of transcription factors, including progenitor proteins, proneural proteins and neuronal differentiation proteins, determine where and when different types or classes of neurons are generated during development. It is of great practical importance that many of these transcription factors may remain detectable for a long time in developing neurons, and hence can be used as 'markers' to determine the origin, migration path, and destination of particular types or classes of elements.

10 The development of grisea
10.3 The formation of plurisegmental complexes

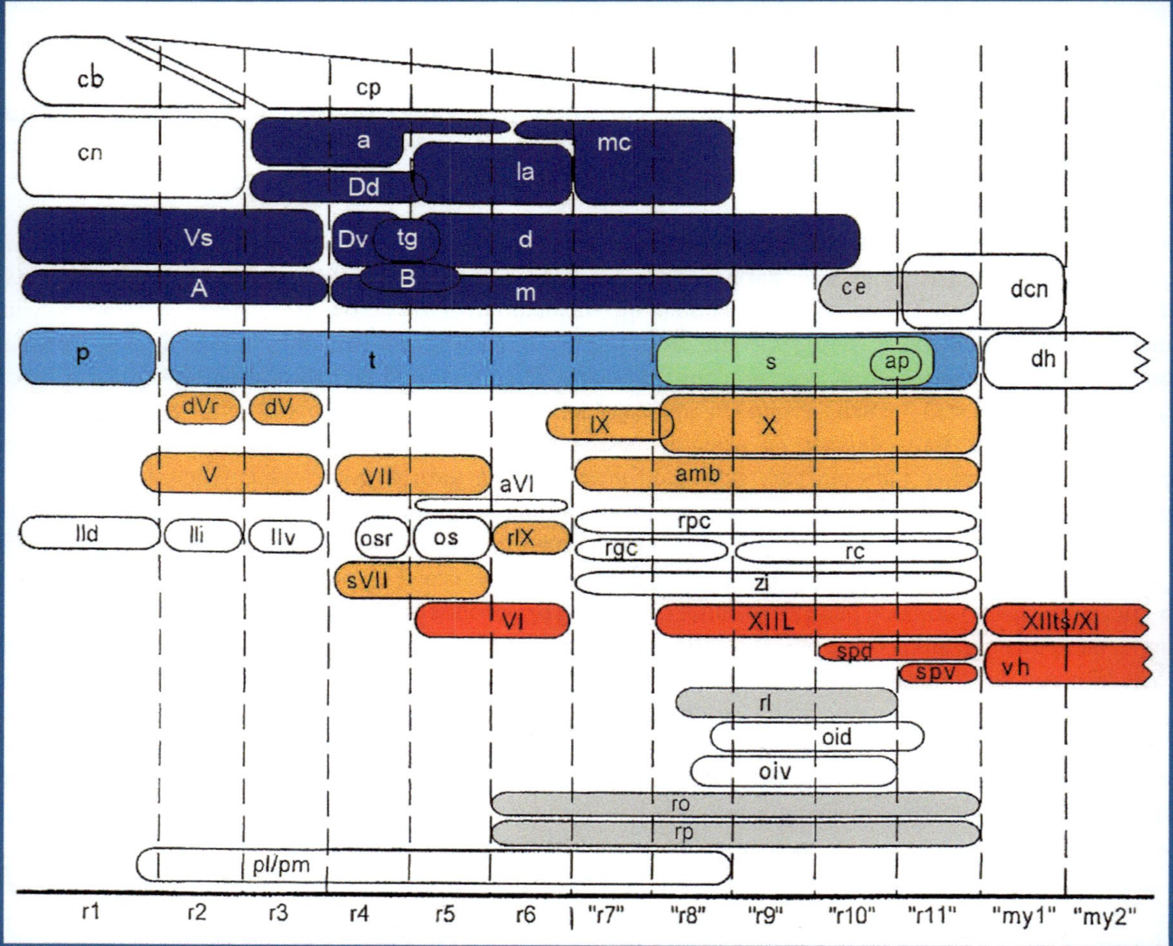

Map of the avian rhombencephalon showing the position of nuclei relative to the various segmental boundaries. (Cambronero and Puelles 2000)

99

10.3 The Formation of Plurisegmental Complexes

Plate 99

Our 'scaffold' map of progenitor domains in the mammalian (and largely also avian) embryo shows a distinct pattern of intersecting transverse (neuromeric) and longitudinal zones (Plate 95; Plate 96, Fig. A). In many places along the neural wall, shared dorsoventral patterning mechanisms produce structurally related cell contingents, which occupy mutually corresponding positions in neighbouring neuromeres, a process leading to the formation of plurisegmental nuclei or columnar formations. At first glance these are continuous, building a longitudinal columnar anatomo-functional unit, but they can be seen to be composed of individual neuromeric entities, each having some differential peculiarity due to the particular neuromeric identities (Marin et al. 2008; Tomás-Roca et al. 2014).

The phenomenon just indicated is very prominent in the rhombencephalon. The fate mapping studies with quail chick chimeras of Marin and Puelles (1995) and Cambronero and Puelles (2000), summarized in Plate 99, have shown that most sensory and motor cranial hindbrain nuclei (marked with different colours) are bi- or plurisegmental, and that the limits of most of these nuclei (and nuclear parts) coincide with interneuromeric boundaries. Similar conclusions also hold true for most "higher order" hindbrain nuclei, as exemplified by the external cuneate nucleus (ce), lateral reticular nucleus (rl), and the raphe pallidus (rp) and raphe obscurus nuclei (ro) (all shown in gray). It may be added that in teleost fishes the rhombencephalic reticulospinal neuron populations retain a distinct segmental pattern through embryonic, larval and adult periods (Gilland et al. 2014).

The colour-code used in the plate is as follows: special somatosensory (cochlear and vestibular) nuclei: dark blue; general somatosensory (trigeminal) nuclei: light blue; viscerosensory (IX and X) nuclei: green; special visceromotor or branchiomotor (V, VII, IX, X) and general visceromotor (VII, IX, X) nuclei: orange; somatomotor nuclei: red

10 The development of grisea
10.3 The formation of plurisegmental complexes

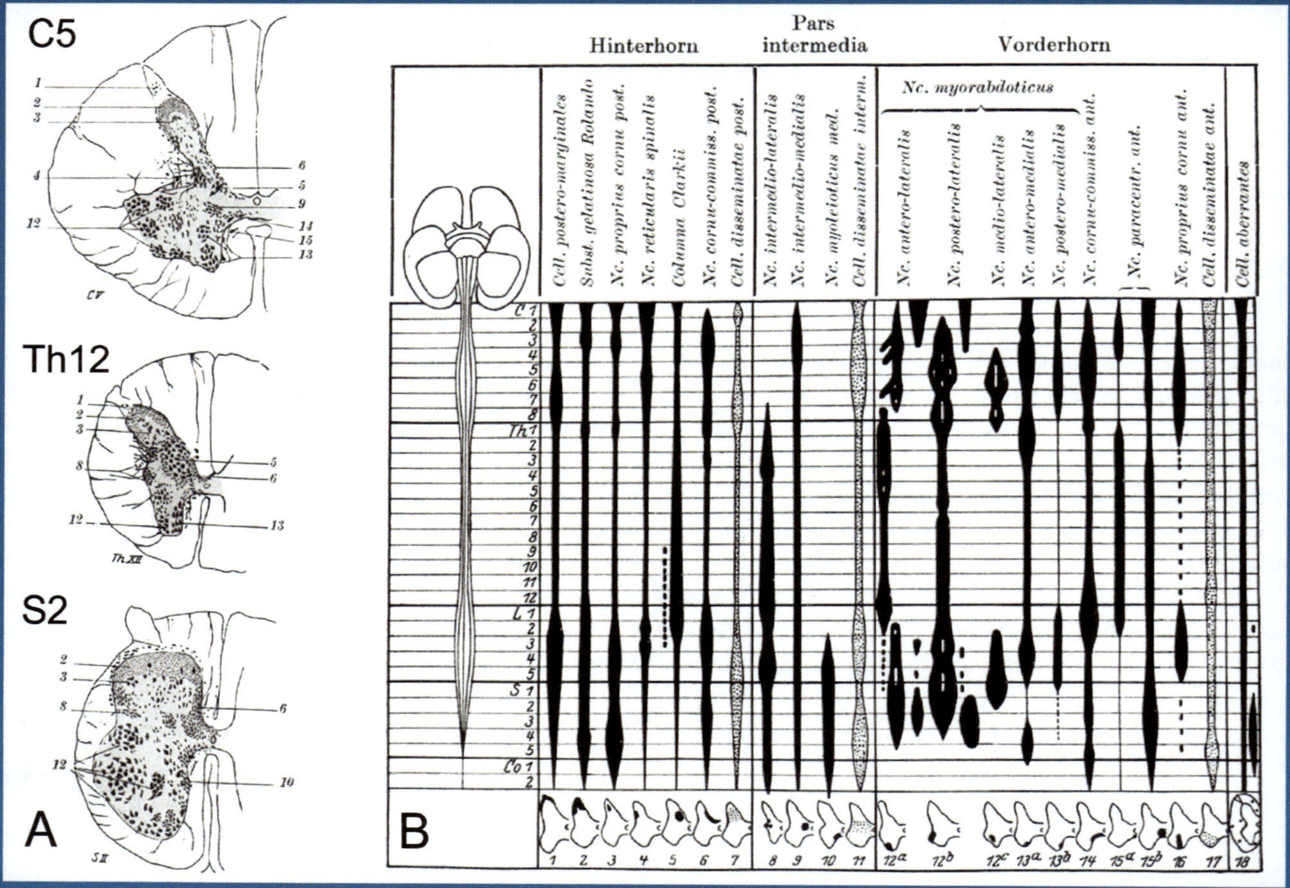

The human spinal cord. A. Transverse sections showing the cytoarchitecture at three different levels. B. Graphical reconstruction of the rostrocaudal extent and local development of the various spinal cell masses (Massaza 1924)

100

Plate 100

The formation of plurisegmental complexes is in the spinal cord even more prominent than in the hindbrain. Massaza (1922, 1923, 1924) who made a thorough study of the cytoarchitecture of the human spinal cord, found that all of the18 spinal cell groups distinguished by him form plurisegmental columnar formations. However, as shown by Plate 100, the rostrocaudal extent and local development of these formations show considerable differences.

Plurisegmental complexes also occur in more rostral parts of the CNS. The compact part of the substantia nigra represents a striking case in point (see below).

10 The development of grisea

10.4 Tangential migrations 10.4.1. Spinal cord

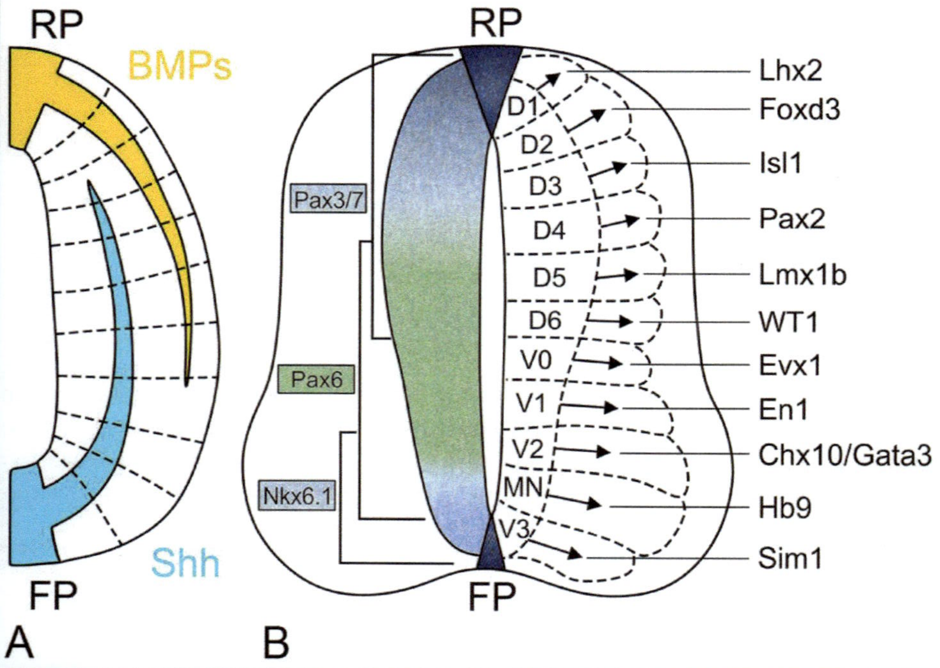

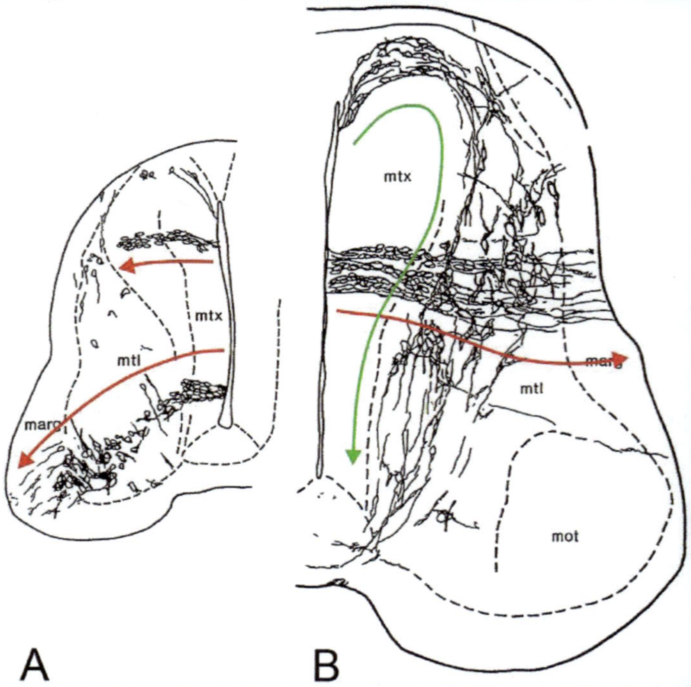

1. schematic cross sections through the developing spinal cord showing (A) the early dorsoventral patterning by gradients of morphogenetic proteins, and (B) later specification processes leading to the formation of eleven progenitor domains (D1-V3) in the ventricular matrix layer, each of wich giving rise to a distinct population of postmitotic neural elements in the mantle layer. Selected transcription factors expressed within the progenitor domains are depicted on the left and those specific to the postmitotic neuronal populations are shown to the right. Based on Goulding (2009) and Grossman et al. (2010)

2. Patterns of migration in the embryonic chick spinal cord revealed by injections of recombinant retroviruses.
A. injected at stage 8, sacrified at stage 22
B. injected at stage 9, sacrified at stage 29
(Leber and Sanes 1995)

101

10.4 Tangential Migrations

It is a very common phenomenon that streams of neuroblasts, glioblasts or even undifferentiated matrix cells leave 'their own' FMU to migrate tangentally to other, sometimes amazingly remote territories of the neuraxis. Here we present a number of representative examples of this intriguing phenomenon, taken from different parts of the CNS.

10.4.1 Spinal Cord

Plate 101

We have seen (Sect. 4.3; Plate 33) that during the early development of the neuraxis a ventral gradient of Sonic hedgehog (Shh) and a dorsal gradient of bone morphogenetic proteins (BMPs) provide positional signals to the epithelial neural wall (Plate 101, Fig. 1A). These general inductive processes lead to the restricted activation of patterning factors in discrete dorsoventral domains, including Nkx6.1 (ventral), Pax6 (intermediate), and Pax3 and Pax7 (dorsal) (Plate 101, Fig. 1B). These patterning factors function instructively in a combinatorial manner to subdivide the spinal neural wall into eleven parallel, dorsoventral microzones, D1-V3. As development proceeds, postmitotic elements migrate radially out from these progenitor zones, thus giving rise to distinct populations of neuroblasts that acquire different molecular specifications. Some of the transcription factors involved in these specification processes are indicated in Plate 101, Fig. 1B. Before discussing the (possible) involvement of (some of) the eleven populations of postmitotic spinal elements in tangential migrations, the results of two previous studies on the subject should be briefly mentioned, and a widely used subdivision of the spinal gray matter should be introduced.

Leber and Sanes (1995) injected recombinant retroviruses into the lumen of the spinal cord of chick embryos. They studied the migration of cells out of the matrix zone (mtx) by infecting at a fixed stage and varying the time of analysis. Isolated clusters of cells were interpreted as single clones. At first, most clones consisted of radial arrays of cells, suggesting that the initial migration is predominantly radial (Plate 101, Fig. 2A; red arrows). However, in later sacrificed embryos it was observed that in some clones neuroblasts turned orthogonally, to migrate ventrally or dorsally, parallel to the surface of the matrix layer (Plate 101, Fig. 2B, green arrow). The majority migrated from dorsal to ventral. Most of these migrating elements were unipolar in shape, with leading processes extending in the direction of apparent migration. Comparable observations have been made by Phelps and Vaughn (1995), who studied the development of a group of early differentiating cholinergic neurons in the cervical spinal cord of the rat. They found that at the time of its first appearance, this group, occupies the most ventral part of the matrix zone, but later it migrates dorsally to become a group of cholinergic dorsal horn cells seen in older animals. During their translocation, the immature cholinergic elements appeared to be directly apposed to arcuate fibres passing ventrally along the external surface of the matrix layer.

10 The development of grisea

10.4 Tangential migrations 10.4.1. Spinal cord

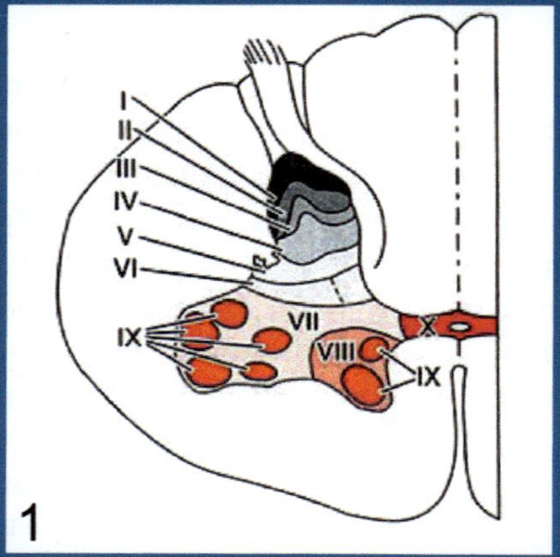

1. The gray matter of the sixth cervical segment of the human spinal cord, subdivided according to Rexed.

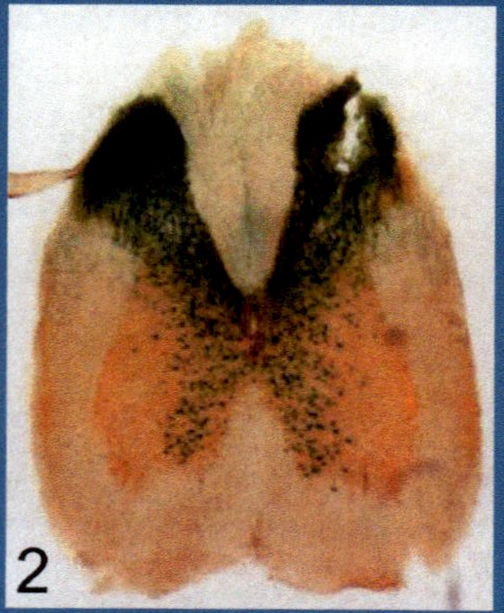

2. LacZ- labeled alar plate progeny in the cervical cord of a Pax7- LacZ transgenic adult mouse.

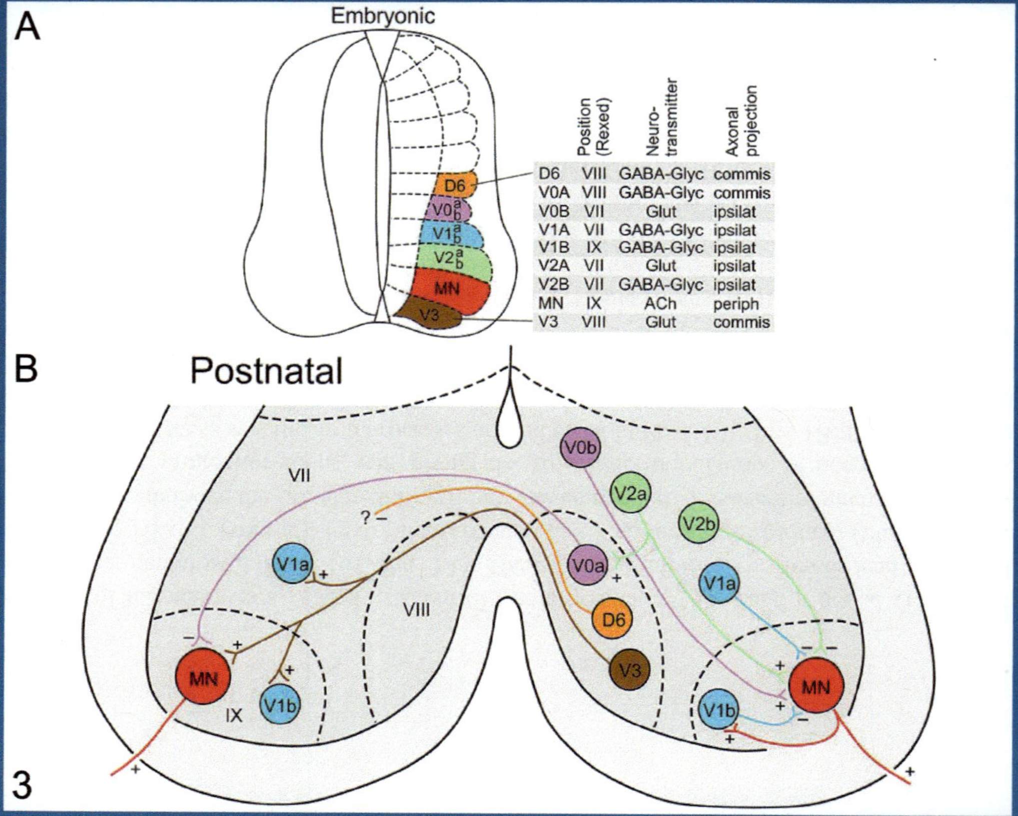

3. Genetic classification, properties and organization of spinal neurons. A. Postmitotic neuronal (sub) populations in the embryonic spinal cord, and the location, neurotransmitter profile and axonal destination of the neurons stemming from these (sub)populations. B. Circuitry in the postnatal ventral spinal cord. The colour code used is the same as in (A). Based on (Grossmann et al. 2010). Glut, glutamate; Glyc, glycine.

102

Plate 102

Rexed (1954) subdivided the gray matter of the spinal cord of the cat into ten dorsoventrally arranged zones or layers (Plate 102, Fig. 1: I–X), a subdivision, which appeared to be applicable to many other mammalian species. Laminae I–V belong to the dorsal horn, laminae VI and VII to the intermediate zone; the motoneuronal cell groups constitute lamina VIII and IX, and lamina X surrounds the central canal. (Rexed's lamina IX is not properly a lamina, since it consists of separate motor nuclei; his system works best with the dorsal horn. A different problem with Rexed's laminar model is that it does not account for differences due to the developmental dorsoventral microzones; e.g., Clarke's thoracic column seems to be a more 'dorsal' microzonal derivative than the dorsal horn in general (which is an organized postmigratory mixture of derivatives from several dorsolateral microzones, rather than representing a deeper lamina by itself).

The neurons of layers I–IV are concerned with the processing of sensory information entering the cord via the dorsal roots. Most neurons of laminae VII and VIII are interneurons giving rise to short axons terminating on motoneurons (with added migrated sympathetic preganglionic neurons). Lamina VIII contains apart from motoneurons, numerous commissural neurons the axons of which contact elements in the contralateral anterior horn.

Plate 102, Fig. 2 shows that numerous neurons that populate the medial part of the anterior horn (lamina VIII) in the cervical cord of the mouse are derived from the (Pax7-positive) alar plate, and apparently have reached their destination via a tangential migration (Puelles, unpublished observations).

Goulding (2009) and Grossmann et al. (2010) collected data concerning the position, axonal projection and neurotransmitter profile of neurons derived from the six most ventral populations of postmitotic cells in the mouse spinal cord (Fig. 3). These populations give rise to core elements of the spinal motor circuitry. Three of these populations, V0, V1 and V2, appeared to diverge into smaller subpopulations, indicated in Fig. 3 as V0a, V0b, V1a, V1b, V2a, and V2b.

The dorsal part of the V0 domain (V0a) gives rise to GABA/glycinergic commissural interneurons, whereas the ventral part (V0b) produces glutamatergic ipsilaterally projecting interneurons. Elements derived from the V1 domain differentiate into two types of ipsilaterally projecting GABA/glycinergic elements, the so-called Ia inhibitory interneurons (V1a) and also the inhibitory Renshaw cells (V1b). Remarkably, these two types of inhibitory interneurons represent less than 25 % of all V1-derived cells, indicating that there are additional, as yet uncharacterized V1-derived neuron types. The V2 domain gives rise to two molecularly distinct classes of ipsilaterally projecting interneurons, the glutamatergic V2a cells and the GABA/glycinergic V2b elements. The D6 and V3 populations both produce commissural interneurons, the former using GABA/glycine, the latter glutamate as their neurotransmitters. The MN population, finally, gives rise to peripherally projecting cholinergic somatomotor neurons. The data just presented are, as already indicated, derived from the review articles of Goulding (2009) and Grossmann et al. (2010), to which we refer for sources and details concerning the molecular specification of the various cell types.

The precise migration paths of the various groups of elements, which participate in the circuitry of the spinal anterior horn cannot be derived from the data reviewed. Comparison of the embryonic origins of these cell groups (Plate 102, Fig. 3A) with their ultimate locations (Fig. 3B) renders it likely however, that at least the populations D6, V0a, and V1b reach their respective destinations via tangential migrations.

As regards the spinal autonomic system, it is known that (cholinergic) sympathetic preganglionic neurons are produced in the ventral part of the embryonic basal plate at thoracic levels, though later their somata translocate into the 'lateral horn', i.e., a lateral extension of the (alar) D6 domain (Levi-Montalcini 1950; Yip et al. 2000). Parasympathetic preganglionic neurons in the sacral spinal cord show a similar translocation. Note that these spinal visceral elements behave analogously to the hindbrain visceromotor and branchiomotor neurons, which also translocate into the alar plate (see below).

Finally, it should be mentioned that oligodendrocytes are produced in one of the microzones of the basal plate. These elements subsequently invade the whole spinal gray and white matter, intermixing with other oligodendrocytes of alar origin (Richardson et al. 1997; Noble et al. 2004).

If we survey the data reviewed above, it may be concluded that tangential migration is a prominent phenomenon in the spinal cord, and that numerous neuronal precursors pass from the basal plate to the alar plate and vice versa, thereby following and contributing to the system of embryonic arcuate fibres (Plate 65, Fig. C). The elements emanating from the six most ventral spinal microzones do not seem to follow simple parallel paths towards their destinations. Rather, they intermix extensively in passing to their specific positions in the circuitry of the anterior horn. As regards the specification of neuronal subtypes, there seem to be no microzones (or groups of adjacent microzones) that produce specifically, and to the exclusion of all others, neuronal subtypes using a particular neurotransmitter.

10 The development of grisea

10.4 Tangential migrations 10.4.2. Visceromotor and branchiomotor nuclei

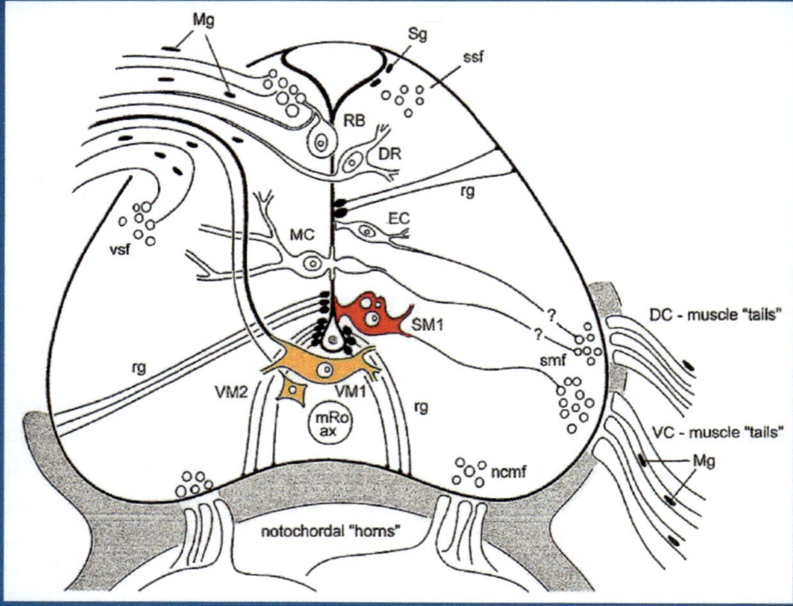

1. Schematic transverse section of the spinal cord of amphioxus.
A somatomotor neuron (SM1) is shown in red; two visceromotor neurons (VM1,VM2) are shown in orange.
(Wicht and Lacalli 2005)

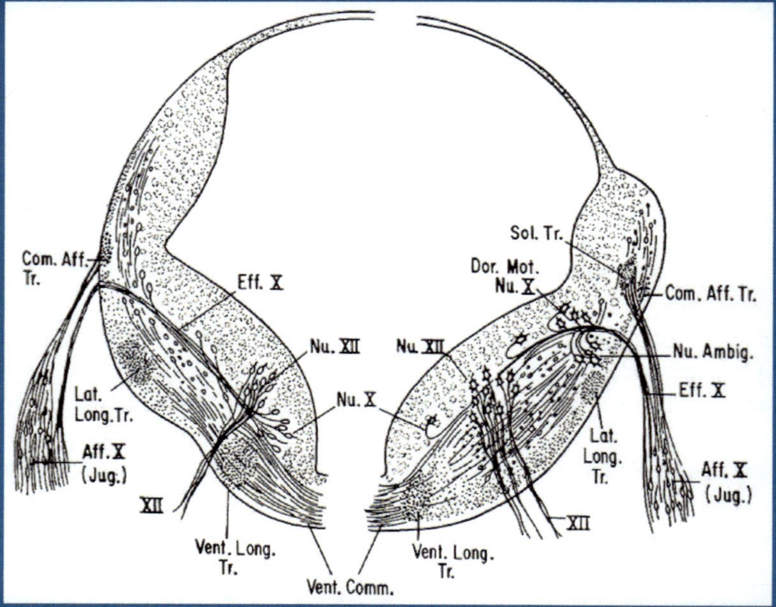

2. Diagrammatic representations of silver-impregnated sections through the rhombencephalon, at the level of the Xth and XIIth cranial nerves in human embryos of the sixth week (left) and the seventh week (right).
(Windle 1971)

10.4.2 Branchiomotor and Visceromotor Nuclei in the Rhombencephalon

Plate 103

Earlier during this course, we made an excursion to the remarkable brain of the cephalochordate amphioxus (Sect. 4, Plates 28–30). Here we mention a few features of the equally remarkable spinal cord of that form. In the spinal cord of amphioxus, which is triangular in cross section, the cell bodies of most of the neurons are located close to the slit-like ventricular cavity. The visceromotor neurons occupy the most ventral part of this narrow grey zone; the somatomotor neurons are situated slightly more dorsally (Plate 103, Fig. 1). The axons of the visceromotor elements pass dorsally and emit via the nerve roots (Bone 1960). [It is noteworthy that amphioxus lacks any counterpart of the ventral nerve roots of vertebrates. Instead, the myotomal muscle cells send long processes ("muscle tails") toward the ventrolateral surface of the cord, where they contact the axons of the somatomotor neurons; the latter elements remain entirely confined within the spinal cord (Flood 1966).]

In the embryonic midbrain and rhombencephalon of vertebrates, the special visceromotor or branchiomotor (br) and general visceromotor (vm) elements are produced in a continuous zone, which extends from the midbrain to the spinal cord. This precursor zone is situated in the ventralmost part of the developing midbrain and hindbrain, thus occupying a position corresponding to that of the visceromotor elements in the spinal cord of amphioxus. The postmitotic elements derived from this zone develop axonal processes, which generally pass tangentally dorsalward along the surface of the matrix layer and exit through the roots of the branchial cranial nerves (V, VII, IX, X, XI), though the preganglionic fibres of the accessory III nucleus in the midbrain incorporate straightforwardly into the ventrally exiting III root. During further development the bm/vm neuroblasts of each rhombomere translocate their somata tangentially within their own axonal processes across the basal plate. They pass the molecularly defined alar/basal boundary and reach the most ventral (B4) zone of the (Pax7-positive) alar plate (Ju et al. 2004). Most preganglionic vm cells simply cease migrating once they reach the B4 zone and differentate further at local periventricular levels; in contrast, the bm cells migrate further, but now radially and with a trailing axon, into more superficial B4 loci (Plate 33, Fig. e), where they aggregate in the bisegmental branchiomotor V and VII nuclei, and in the plurisegmental (IX + X + XI) ambiguus nucleus.

The tangential translocation of bm and vm elements is beautifully illustrated in Plate 103, Fig. 2, derived from Windle (1971). It will be seen that in a six weeks old human embryo (left side) the bm and vm precursor cells of the vagus nuclei (Nu. X), are still situated mediobasally and ventral to the incipient somatomotor hypoglossus nucleus (Nu. XII), but that in an embryo one week older (right side) almost all of these elements are translocated laterally to participate in the anlagen of the dorsal motor vagus (Dor. Mot. Nu. X), and ambiguus (Nu. Amb.) nuclei.

The migration path of the elements destined to form the branchiomotor facial nucleus, which are born in r4–r5, is exceptionally complex and involves, in addition to the tangential and radial migrations discussed above, a caudal shift into rhombomere 6.

10 The development of grisea

10.4 Tangential migrations 10.4.3. Rhombic lip

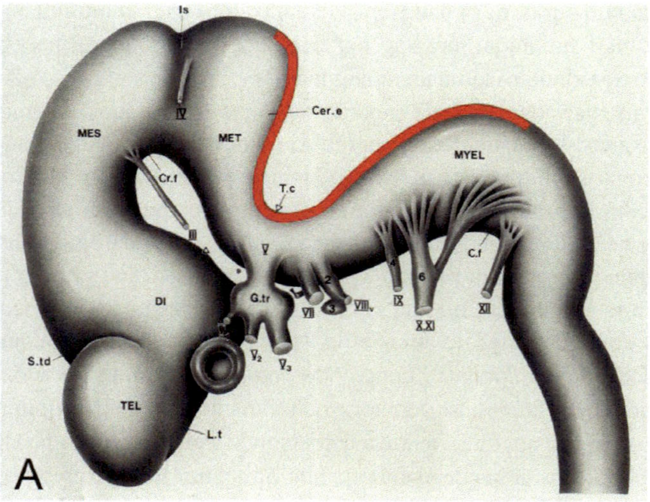

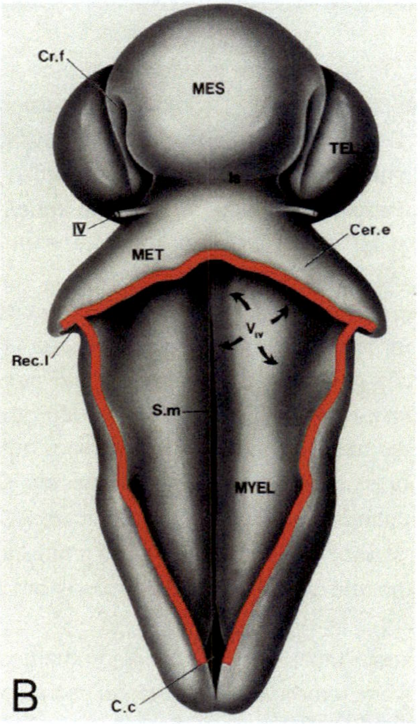

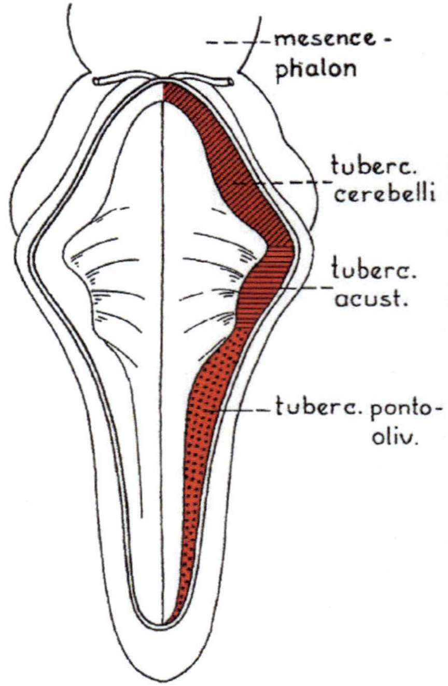

1. The brain of an 11.5-mm CRL embryo of harbour porpoise, Phocaena phocaena (Schweinswal) (Buhl and Oelschläger 1988)

A. lateral view

B. dorsal view

2. Dorsal view of the rhombencephalon of a human embryo of about one month

rhombic lip in red

104

10.4.3 Rhombic Lip

Plate 104

The hindbrain or rhombencephalon (‘Rautenhirn’) is called so because its ventricular cavity is rhomboidally-shaped. The ventral and lateral walls of this cavity are formed by the bilateral neural basal and alar plates, respectively; dorsally it is covered by the thin epithelial roof plate. The dorsal parts of the alar plates directly adjacent to the lines of attachment of the roof plate, contain during embryonic development a highly proliferative zone, designated by His (1890) as the ‘Rautenlippe’: rhombic lip (Plate 104, Figs. 1, 2). This zone, which is very prominent in mammals, is present in all gnathostome vertebrates, but it is lacking in cyclostomes (Wullimann et al. 2011; Pose-Méndez et al. 2014).

10 The development of grisea

10.4 Tangential migrations 10.4.3. Rhombic lip

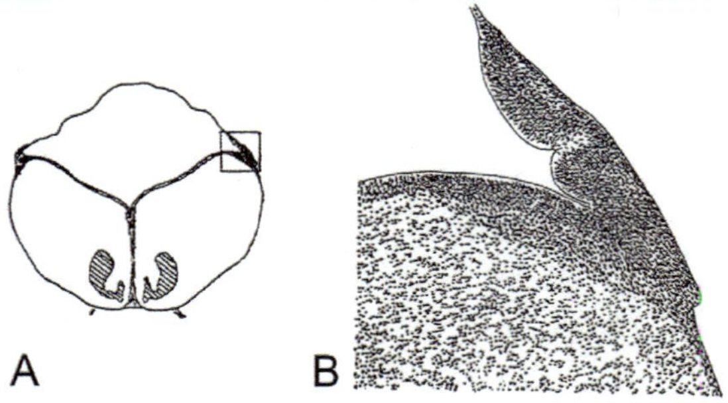

1. Transverse section through the caudal rhombencephalon of a human embryo of about 2 months, showing location (A) and structure (B) of the rhombic lip. (Essick 1912)

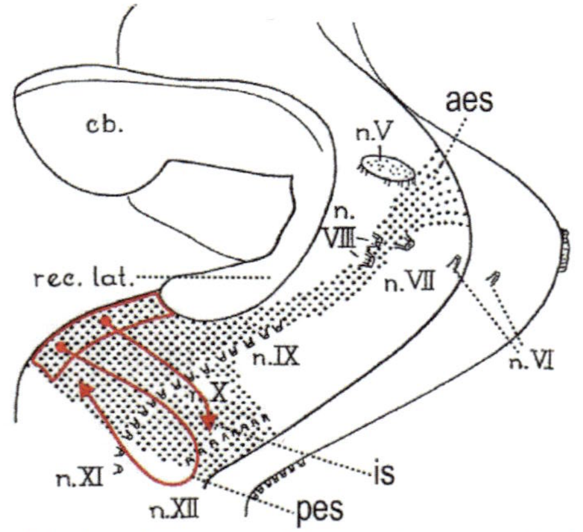

2. Sagittal scheme of long tangential migrations in a 23 mm-human embryo (Essick 1912)

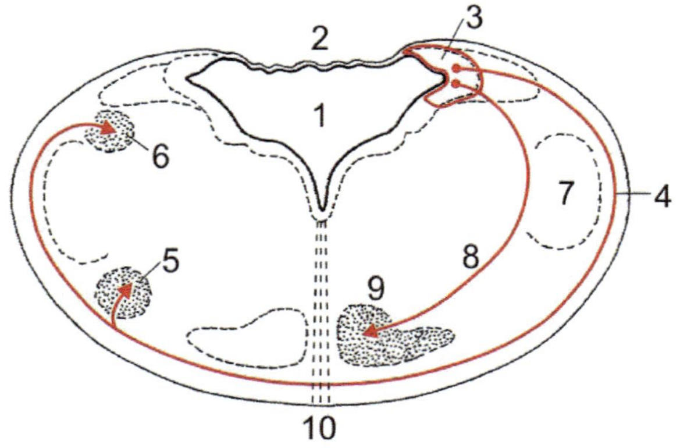

3. Long tangential migrations in the rhombencephalon of the rat. (Nieuwenhuys et al. 2008)

1. fourth ventricle
2. roof plate
3. caudal rhombic lip
4. posterior extramural migratory stream (pes)
5. lateral reticular nucleus
6. lateral cuneate nucleus
7. nucleus of the spinal trigeminal tract
8. intramural migratory stream (is)
9. inferior olivary complex
10. floor plate

105

Plate 105

The structure of the rhombic lip in a human embryo of about two months is shown in Plate 105, Fig. 1.

The caudal part of the rhombic lip (i.e. the tuberculum ponto-olivare in Plate 104, Fig. 2) is the source of some spectacular tangential cellular migrations, giving rise to a number of cell masses, including the lateral reticular nucleus, the lateral cuneate nucleus, the inferior olivary complex, and the pontine nuclei (His 1890; Essick 1912, Harkmark 1954). Because all of these cell masses project to the cerebellum, they are collectively designated as precerebellar nuclei. The axons of the inferior olivary cells terminate as climbing fibres in the molecular layer of the contralateral half of the cerebellum, where they synapse with the dendritic trees of Purkinje cells. The axons of all the remaining precerebellar nuclei terminate as mossy fibres in the granular layer of the contralateral cerebellum. Before discussing these various 'precerebellar' tangential migrations, it may be recalled that molecular studies have shown that the rhombic lip is divisible into three longitudinally arranged microzones, A1, A2/3 and A4 (Plate 95). The rhombic lip as a whole is molecularly defined by the expression of W*nt1* and O*lig3*. The A1, A2/3 and A4 microzones are characterized by the local expressions of *Math1 (Atoh1)*, *Ngn1*, and *Ptf1a*, respectively (Landsberg et al. 2005; Storm et al. 2009).

After leaving the rhombic lip, the precerebellar neuroblasts assemble in three migratory streams, the posterior and anterior extramural migratory streams and the intramural migratory stream (Altman and Bayer 1987a–d; Plate 105, Figs. 2, 3).

The elements constituting the *posterior extramural migratory stream (pes)* originate from the A1 rhombic lip microzone and are destined to form the lateral reticular (LRn) and external cuneate (ECn) nuclei. They move circumferentially and subpially (preceded by their elongating axons) through the wall of the caudal medulla oblongata (rhombomeres r8–r11) across the median plane (floor plate) and finally reach the contralateral basal (LRn) and alar (ECn) plates (Plate 105, Fig. 3). The axons enter the cerebellum through the inferior cerebellar peduncle in r1 (it remains unknown where and how they adopt an ascending course).

The cells forming the *intramural migratory stream (is)* originate from the r8–r10 sectors of the A4 rhombic lip microzone (Storm et al. 2009). Preceded by their axons just like the caudal extramural stream, they follow a strictly transverse, though deeper tangential course towards the superficial part of the medial basal plate, where they settle to form the inferior olivary complex (Plate 105, Fig. 3). Their crossed axons apparently join the *pes* ones, and likewise enter the cerebellum via the inferior cerebellar peduncle. In humans this complex differentiates into a large, highly convoluted structure, which produces a distinct elevation of the external surface of the medulla oblongata.

10 The development of grisea
10.4 Tangential migrations 10.4.3. Rhombic lip

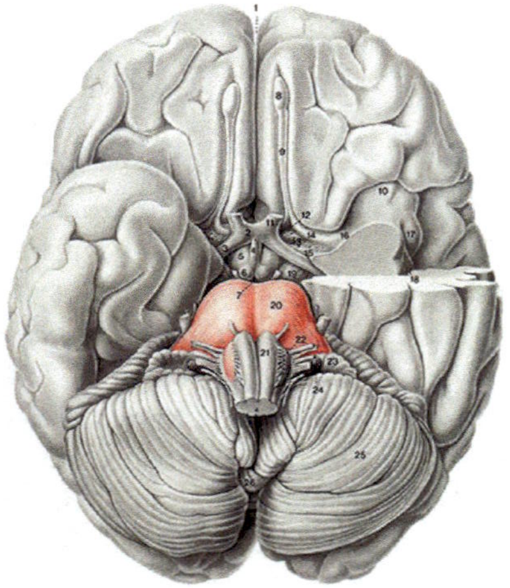

1. Human brain, basal surface

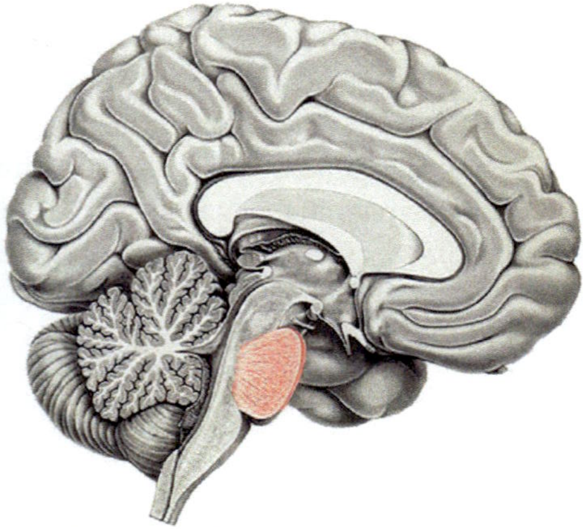

2. Human brain, medial surface

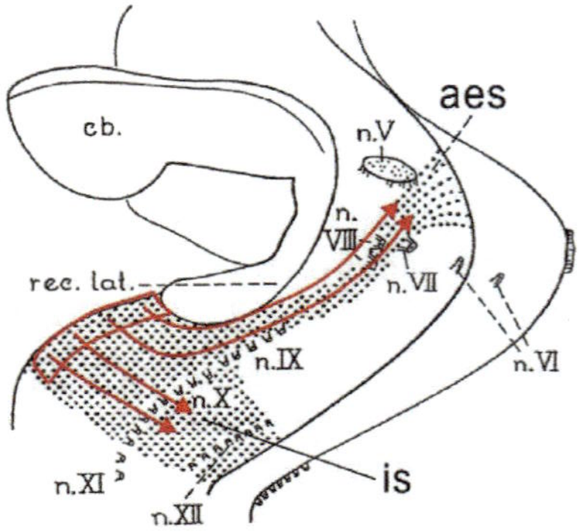

3. Sagittal scheme of long tangential migrations in a 23 mm-human embryo (Essick 1912)

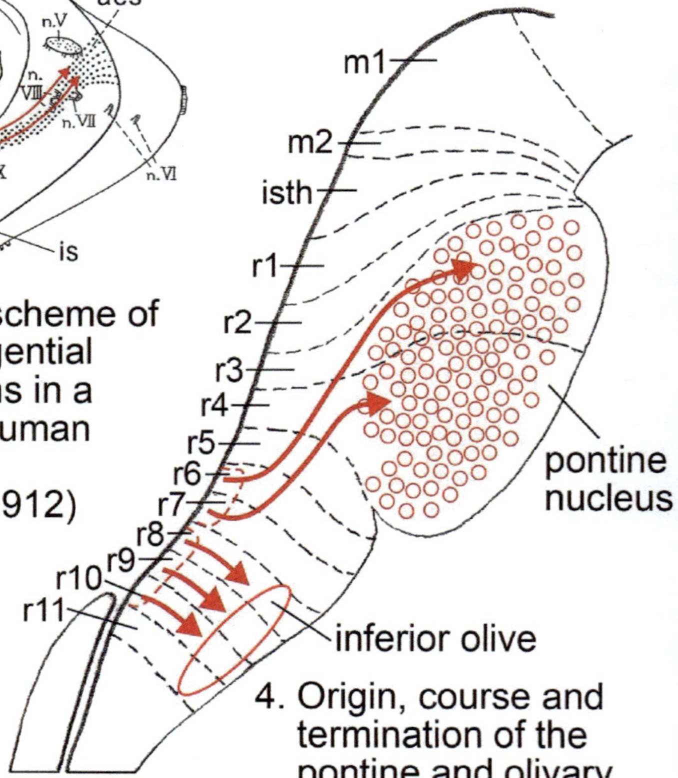

4. Origin, course and termination of the pontine and olivary migrations, projected upon the adult human brainstem.

106

Plate 106

The cells in the *anterior extramural migratory stream (aes)*, finally, originate from the A1 rhombic lip microzone, just as those of the other mossy fibre producing nuclei, in r6–r7 (Farago et al. 2006). They migrate only a short distance basally and then turn anteriorly, always subpially, passing between the VII and VIII nerve roots in r4. From there they change course again, arching ventrally and fanning out into the basal plate of r3 and r4, where they form the basilar pontine nuclei (Plate 106, Fig. 3). Their preceding crossed axons pass dorsalward either rostrally or caudally to the V root in r2, and always rostrally to the VII/VIII roots, and enter the cerebellum through the middle cerebellar peduncle in r1. These nuclei receive large projections from the cerebral cortex and target the contralateral cerebellar cortex. They attain very large dimensions in primates, elephants and whales. Plate 106, Figs. 1 and 2 show that the human pontine nuclei not only effect a large external elevation, but also protrude inwardly, 'compressing' the adjacent tegmentum.

Plate 106, Fig. 4 shows a sagittal schema of the human rhombencephalon, in which outlines of the embryonic lower rhombic lip and of the segmental organization of the adult stage are combined. It will be seen that elements produced in a small, dorsally situated germinal zone leave their FMUs to invade other, basal domains of the same or different rhombomeres.

As regards the mechanism of the tangential migrations just discussed, it has been observed that the neuroblasts generated in the rhombic lip develop leading axons which apparently pioneer the migration paths, and that later the somata move down these processes (Bourrat and Sotelo 1990).

It may be mentioned in passing that (1) cells stemming from the intermediate (r3–r5) sector of the rhombic lip migrate radially outward to form the cochlear nuclei; (2) the caudal, intermediate and rostral parts of the rhombic lip are all involved in the development of the vestibular nuclei; (3) the caudal B2/3 microzone contributes cells to the nucleus of the solitary tract (Storm et al. 2009), and (4) excitatory neurons (mostly glutamatergic) in a variety of rhombencephalic nuclei are derived from the *math1* expressing most dorsal microzone of the lower rhombic lip (Wang et al. 2005; Rose et al. 2009). These include all of the following: the somatosensory dorsal column (gracile and medial cuneate) nuclei and the spinal and principal trigeminal nuclei, the auditory related superior olive and lateral lemniscal nuclei, the respiration related Kölliker-Fuse nucleus, and the bladder-control related Barrington's nucleus.

The wholly astonishing performances of the matrix cells in the most rostral part of the rhombic lip will be highlighted in the next subsection.

10 The development of grisea

10.4 Tangential migrations 10.4.4. Cerebellum

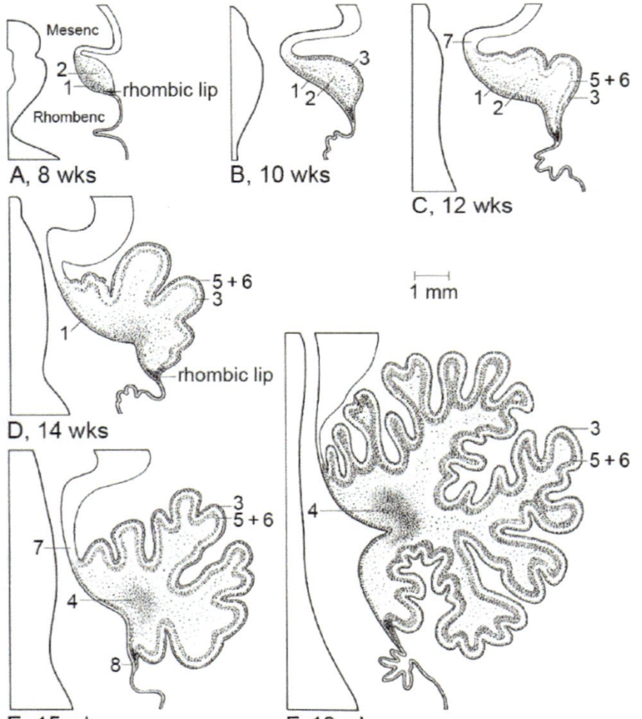

1

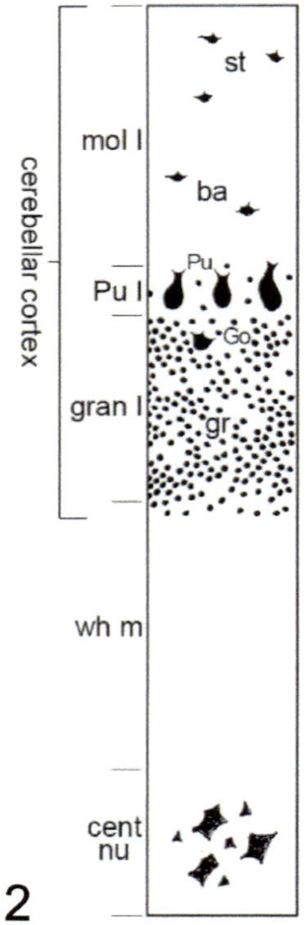

2

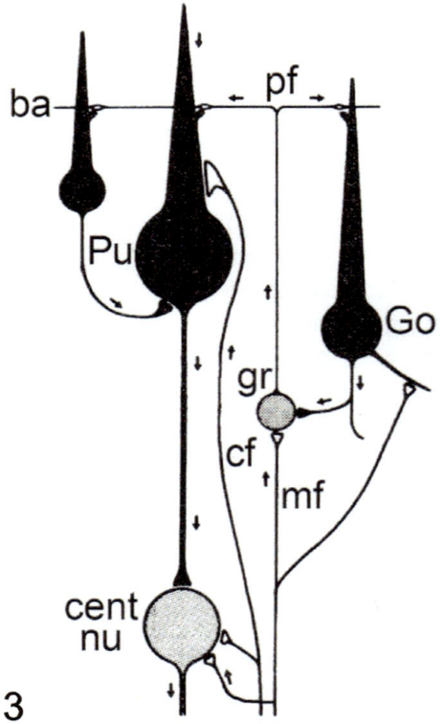

3

1. Sagittal sections of the developing human cerrebellum (Nieuwenhuys et al. 2008)

2. Cytoarchitecture of mammalian cerebellum

3. Cerebellar microcircuitry inhibitory cells are shown in black

1. Ventricular matrix
2. Mantle layer
3. External granular layer
4. Central cerebellar nuclei
5. Purkinje layer
6. Internal granular layer
7. Superior medullary velum
8. Inferior medullary velum

ba, basket cell
Go, Golgi cell
gr, granule cell
gran l, granular layer
cf, climbing fibre
mol l, molecular layer
pf, parallel fibre
Pu, Purkinje cell
Pu l, Purkinje layer
st, stellate cell
wh m, white matter

107

10.4.4 Cerebellum

Plate 107

The cerebellum develops from the most rostral part of the rhombencephalon. The alar regions of the isthmic and first rhombencephalic segments are both involved in its formation. Initially, the bilateral cerebellar anlagen form a sharp ($< 90°$) angle to each other (Plate 104, Fig. 2), but due to the influence of the pontine flexure (Chap. 5, Plate 47), this angle gradually increases. In the human embryo, the bilateral anlagen of the cerebellum form at the end of the second month a thickened transversely oriented plate, which caps the most rostral part of the fourth ventricle (Plate 107, Fig. 1A). This cerebellar plate is rostrally connected with the tectal region of the midbrain by the superior medullary velum, formed by the roof plate of the isthmic neuromere. The vermal midline of the cerebellum also contains a cryptic astroglial (raphe-like) roof plate, which likewise is isthmic in nature. In fact, the entire cerebellar vermis derives from the isthmus, whereas the cerebellar hemispheres develop within r1. Caudally, in topographic terms, the cerebellar primordium thins out into the epithelial roof of the fourth ventricle. Its caudal rim (dorsal in topological terms) is occupied by the upper part of the rhombic lip (Plate 104). During further development, the cerebellar plate greatly thickens and its external surface becomes increasingly folded, thus forming transversely oriented lobi, lobuli and folia (Plate 107, Fig. 1).

The adult cerebellum contains a superficially situated cortex, whereas its periventricular zone is occupied by a series of mediolaterally arranged central nuclei. The cortex and the central cerebellar nuclei are separated by a wide zone of white matter (Plate 107, Fig. 2).

The cerebellar cortex consists of an external, cell-poor molecular layer, an intermediate layer formed by the somata of the large Purkinje cells, and a deep layer made up of densely packed, small granule cell elements. Although the cerebellar cortex is cytoarchitectonically homogeneous, histochemical, hodological and molecular studies have shown that intersecting longitudinal zones and transverse bands subdivide this structure into numerous modular entities (Dastjerdi et al. 2012).

The inhibitory Purkinje cells, which form the only output elements of the cerebellar cortex, send their axons to the central cerebellar nuclei (Plate 107, Fig. 3). The flattened, sagittally oriented dendritic trees of the Purkinje cells extend into the molecular layer, where they synapse with transversely running parallel fibres, i.e. the branches of the bifurcating ascending axons of the excitatory granule cells. The cerebellar cortex contains three types of inhibitory interneurons, the stellate and basket cells, which are situated in the molecular layer and act upon the Purkinje cells, and the Golgi cells, which are found in the superficial zone of the granular layer and act upon the dendritic tips of the granule cells at the level of so-called cerebellar glomeruli, which also receive the excitatory axonal terminals of mossy fibres (Plate 107, Figs. 2, 3).

The cerebellar cortex receives two types of afferents, climbing fibres, which arise from the inferior olivary complex, targeting Purkinje cells, and mossy fibres, which originate from the remaining precerebellar nuclei and target the granule cells. The synaptic relations of these fibres and of the various cerebellar neuronal elements are diagrammatically represented in Plate 107, Fig. 3.

It is of note that the central nuclei, which give rise to the cerebellar efferent systems, contain, in addition to (glutamatergic) excitatory cells, also (GABAergic) inhibitory elements.

10 The development of grisea
10.4 Tangential migrations 10.4.4. Cerebellum

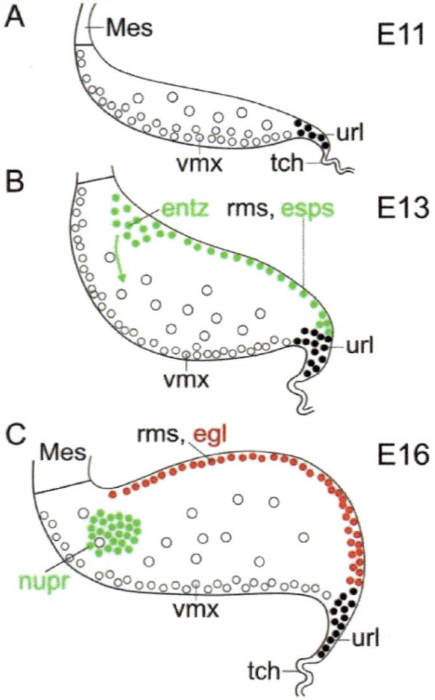

1

1. Diagrammatic sagittal sections through cerebellum of mouse embryos of 11, 13 and 16 days. egl, external granular layer; entz, early nuclear transition zone; esps, early subpial stream; nupr, nuclear primordia; rms, rostral tangential migratory stream; tch, tela chorioidea; url, upper rhombic lip; vmx, ventricular matrix. Based on Fink et al. (2006), Wullimann et al. (2011) and Yeung et al. (2014)

2

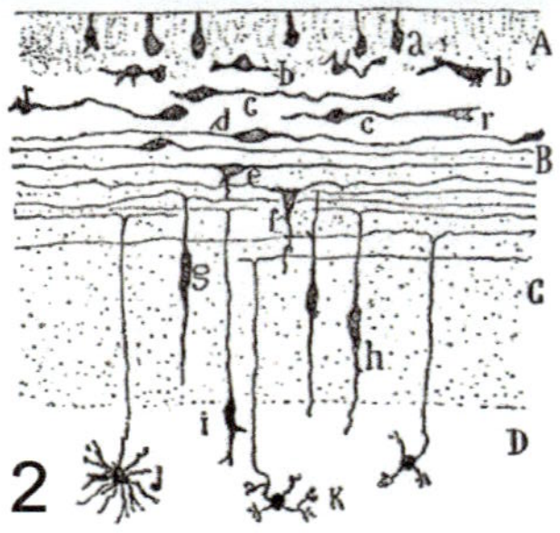

2. Various stages of granule cell development. From Ramón y Cajal (1911)

3

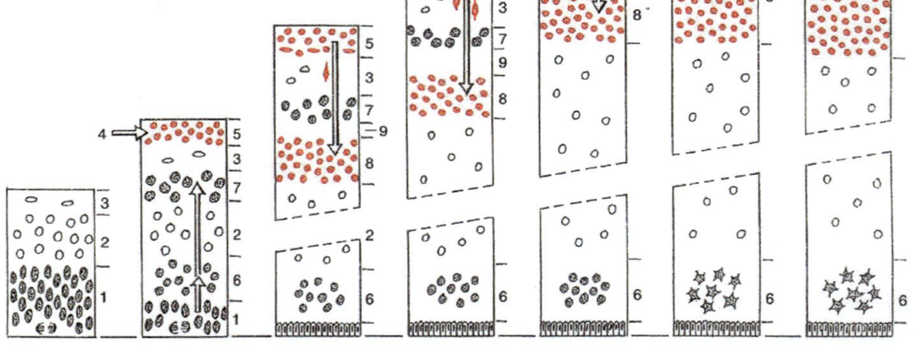

3. Semidiagrammatic summary of the histogenesis of the human cerebellum. The *arrows* indicate the main directions of cell migrations. p.n.m., postnatal months; the external granular layer and its derivatives are shown in *red*. From Nieuwenhuys et al. (2008)

1. Ventricular matrix
2. Mantle layer
3. Marginal layer
4. Rhombic lip
5. External granular layer
6. Deep cerebellar nuclei
7. Purkinje layer
8. Internal granular layer
9. Lamina dissecans
10. Molecular layer

108

Plate 108

In early developmental stages (eleven days old mouse embryos: Plate 108, Fig. 1A; eight weeks old human embryos: Plate 108, Fig. 3A) the cerebellar anlage contains an active ventricular matrix, the topologically dorsal most part of which is formed by the rostral or upper rhombic lip (URL). The histogenesis of the human cerebellum may be summarized as follows (Plate 108, Fig. 3). The cerebellar ventricular matrix is mainly active during the second and third month. It produces (1) some cells of the central cerebellar nuclei, (2) Purkinje cells and Golgi cells, and (3) stellate and basket cells in that order (Leto et al. 2006, 2009). Leto et al. (2006) emphasize that the different types of cerebellar GABAergic interneurons originate from a common pool of multipotent progenitor cells. The Purkinje cells first accumulate periventricularly and later migrate radially into the overlying emerging cortex (possibly after initiating axonal contact with the nuclei).

From the tenth week onwards, undifferentiated matrix cells migrate out from the URL to form the *rostral migratory stream (rms)*, leading to the formation of the subpially situated external granular layer. (It is of note that the term 'rostral migratory stream' indicates in fact the *topographical* direction of the stream. Because the rhombic lip represents the topologically most dorsal structure of the hindbrain, all migrations emanating from this structure are in fact directed ventrally.) The external granular layer forms a second cerebellar germinal zone. Whereas the cerebellar ventricular matrix declines at the end of the third month, the external granular layer remains mitotically active until well after birth. During this protracted proliferative period, the steadily expanding external granular layer (Plate 107, Fig. 1) produces enormous numbers of neuroblasts destined to form the internal granular layer. After their final mitosis, the precursors of the granule cells become bipolar and send out two processes oriented transversely, i.e. parallel to the long axis of the folia. Then, another process develops at the soma and grows out centrally (radially inwards). The nucleus and surrounding cytoplasm transfer into this process and the cell soma now assumes a radially oriented elongated spindle shape, and attaches to one of the radial processes of Bergmann astroglia cells. The radially oriented neuroblasts continue migrating inward following these glial guidelines to finally form the internal (or definitive) granular layer (Plate 108, Fig. 2). This mechanism was discovered by Ramon y Cajal in 1890, using the Golgi method.

Molecular studies (Wang et al. 2005; Fink et al. 2006) have shown that the matrix cells which form the rostral tangential migratory stream, originate from the *Math1*-positive A1 microzone of the URL. These studies have also shown that the formation of the external granular layer from the URL, which occurs between E13 and E16 (Plate 108, Fig. 1C), is preceded by an earlier (E10–E12) superficial stream of excitatory (glutamatergic) central cerebellar nuclear cells (Plate 108, Fig. 1B). These elements, which express Pax6, Tbr2 and Tbr1 sequentally, accumulate at the cerebellar periphery in a transitory nuclear zone, from where they descend ventrolaterally to the primordial central cerebellar nuclei (Plate 108, Fig. 1B, C) to differentiate into projection neurons. According to Fink et al. (2006), these findings imply that cerebellar neurogenesis, like that of the cerebral cortex, is compartmentalized. All GABAergic neurons are produced in the cerebellar ventricular matrix, whereas all glutamatergic elements appear to come from the rhombic lip.

Machold and Fishell (2005) and Wang et al. (2005) reported that a set of elements developing early (prior to E13 in the mouse) from the *Math1*-expressing microzone of the r1 segment of the rhombic lip form an extracerebellar branch of the rostral migratory stream, by which several prepontine and isthmic cell masses are populated. These cell masses include the lateral parabrachial, laterodorsal tegmental, pedunculopontine tegmental and parabigeminal nuclei, all lying within r1. It is of note that the laterodorsal tegmental and pedunculopontine tegmental nuclei are cholinergic (they form part of the so-called Ch5-Ch6 tegmental complex), whereas all other elements produced by the *Math1*-expressing microzone are (as far as we know) glutamatergic. Green et al. (2014) recently and remarkably reported that the elements forming the various extracerebellar nuclei mentioned above originate from a separate FGF-dependent isthmic extension of the *Atoh1* (*Math1*)-expressing microzone of the rhombic lip. [It is to be regretted that Green et al. (2014) *torture* the concept of homology in the discussion of their otherwise very interesting results.]

10 The development of grisea
10.4 Tangential migrations 10.4.5. Pretectum and Hypothalamus

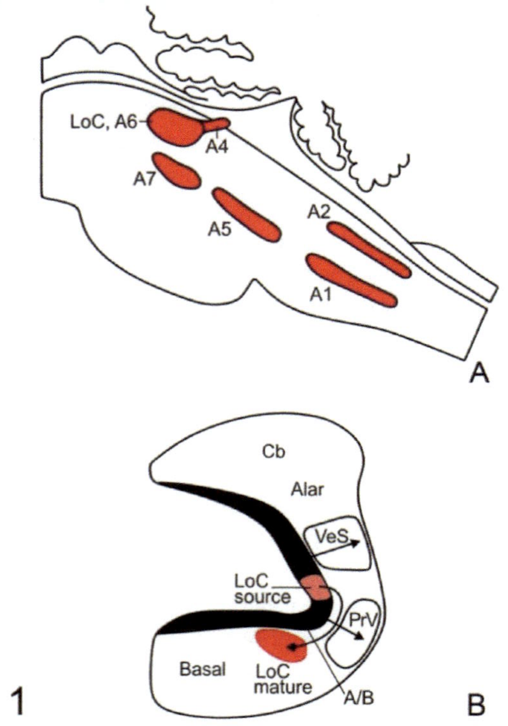

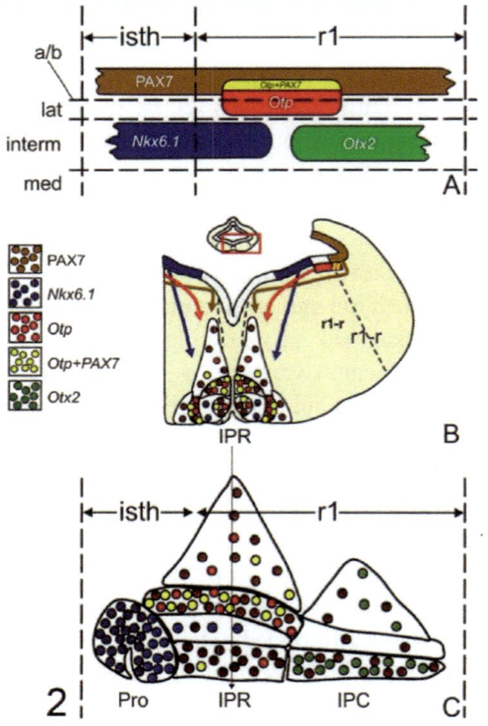

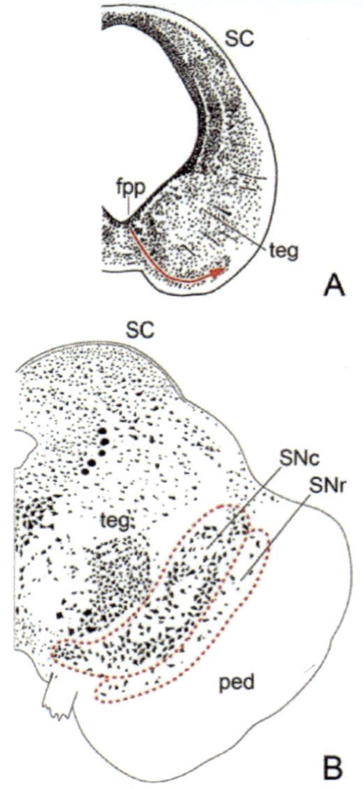

1. The locus coeruleus (LoC). A: Sagittal scheme of the human brainstem, showing the position of the LoC and other noradrenergic centres. B: Origin, tangential migration (thick arrow) and ultimate position of LoC cells, projected on a transverse hemisection through r1 of an older embryo. Thin arrows indicate direction for radial migration of Ves and PrV derivatives A/B, alar-basal boundary; Alar, alar plate; Basal, basal plate; Cb, cerebellum; Ves and PrV indicate respectively the vestibular and trigeminal sensory column portions present in r1.

2. Origin (A), migration routes (B), and definitive position (C) of neuronal elements forming the nucleus interpeduncularis. B and C reproduced from Lorente-Cánovas et al. (2012). For further explanation see text.

3. The substantia nigra. A: Transverse section through the midbrain of an eight-week old human embryo. The red arrow indicates the direction of the migration stream, leading to formation of the compact part of the subtantia nigra (SNc). B: Similar section through the adult human midbrain. The position and cellular structure of the compact and reticular parts of the substantie nigra (SNc, SNr) are indicated, fpp, floor plate proliferation zone; sc, superior colliculus; teg, tegmentum; ped, cerebral peduncle.

109

10.4.5 Locus Coeruleus, Interpeduncular Nucleus and Substantia Nigra

Plate 109

The *locus coeruleus* (LoC), or A6 cell group (see below), is a prominent noradrenergic formation in the brain of vertebrates. It occupies a periventricular position in the rostral part of the rhombencephalic tegmentum (Plate 109, Fig. 1A).

Aroca et al. (2006) investigated the origin and development of the LoC in chick and mouse embryos, using several markers among which the key noradrenergic marker gene *Dbh*. They found that LoC neurons originate from a longitudinal, band-like matrix zone in the alar plate, extending along the whole length of r1. This band-like Loc origin is wedged dorsoventrally between the sources of the prospective vestibular and trigeminal sensory columns. It was demonstrated that the LoC neurons migrate tangentially from the alar plate to a final periventricular position in the lateral basal plate (Plate 109, Fig. 1B).

The LoC (A6) is the most rostral of a series of noradrenergic cell groups, A1-A7, which are all located in the rhombencephalon (Plate 109, Fig. 1A). It is noteworthy that the A1, A5 and A7 groups are produced at the Phox2b-expressing B2 alar microzone (Ju et al. 2004; Plate 95), and apparently invade the basal domain via a tangential migration just as has been demonstrated for the LoC.

The *interpeduncular nucleus* (IP) is a highly conserved structure in all vertebrates, occupying a superficial position across the median floor plate in the isthmic and first rhombencephalic (r1) segments. It is a complex structure consisting of several cyto- and chemoarchitectonically distinct cell groups (Hamill et al. 1984). The avian IP can be divided into three main parts, the isthmic prodromal nucleus (Pro), and the rostral and caudal interpeduncular nuclei (IPR, IPC), situated in the corresponding halves of r1. IPR and IPC are both composed of several subnuclei (Plate 109, Fig. 2C).

Lorente-Cánovas et al. (2012) investigated the origins, migration paths and molecular profiles of cells populating the IP of the chick. They detected four distinct progenitor domains for IP neuronal subtypes characterized by the expression of PAX7, *Otp, Nbx6.1, and Otx2,* and a fifth one co-expressing PAX7 and *Otp*. These progenitor domains are situated in the liminal rim of the alar zone, and in the lateral and intermediate basal zones of the isthmus and r1. Their position is diagrammatically indicated in Plate 109, Fig. 2A. These domains give rise to distinct subpopulations that target the IP with variable subnuclear specificity. The final fate of these subpopulations in the IP complex is shown in Plate 109, Fig. 2C. Each of the five subpopulations follows a specific route of migration. As indicated in Plate 109, Fig. 2B, the postmitotic Pax7-positive and *Otp*-positive elements, stemming from the liminal alar and lateral basal zones, respectively, proceed tangentially medialwards (topologically ventralwards) within the basal plate, later switching to a more radial course at the medial basal zone. [The fact that in this analysis several IP subnuclei remained only sparsely populated suggests, according to the authors, the existence of additional, as yet unknown populations.]

The *substantia nigra* (SN) (Plate 109, Fig. 3 B) is a large, flattened mesodiencephalic complex, lying between the tegmentum (teg) and the cerebral peduncle (ped). Its most rostral part extends into the diencephalon. The SN can be divided into a deeper, cell-rich pars compacta (SNc), consisting of large dopaminergic neurons, and a superficial, less cellular pars reticulata (SNr), composed of somewhat smaller, GABAergic elements.

The mesencephalic and diencephalic floor plate is expanded into a wide and distinct progenitor zone (Plate 109, Fig. 3A: fpp). Neuroblasts originating from the lateral part of this zone first migrate radially ventralwards almost to the pial surface of the floor plate, and then migrate tangentially lateralwards through the marginal zone of the mesencephalic basal plate, possibly reaching partially the alar plate. The crescent of cells that develops from this remarkable tangential migration forms the SNc (Cooper 1946; Verney et al. 2001). The GABAergic cells destined to form the SNr stem from quite another plurisegmental source. These elements are apparently generated in the lateral zone of the isthmic, mesencephalic and diencephalic basal plate (Plate 95), migrate radially through the developing SNc, and form a more superficial primordium, later covered by the descending fibres of the cerebral peduncle. Some dopaminergic SN compacta cells become intermixed with the reticular elements.

10 The development of grisea
10.4 Tangential migrations 10.4.6. Pretectum and Hypothalamus

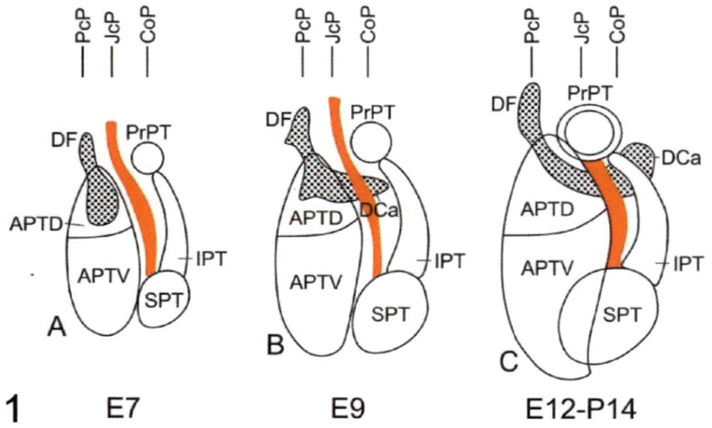

1. Lateral schematic representation of the pretectum of the chick, showing three stages of development (7, 9 and 12 days of incubation). Rostral is to the left. The levels of the three principal pretectal regions, PcP, JcP and CoP, are indicated. The position of JcP is shown in orange. The APTD/APTV cell masses symbolize the main PcP derivatives; PrPT, IPT and SPT represent characteristic CoP elements. The figure illustrates the development of a caudally directed, tangential migratory stream, leading to the formation of a dorsocaudal nucleus (Dca). Reproduced from Ferran et al. (2009).

2. Schematic colour-coded maps of the hypothalamus showing the derivatives of the various progenitor areas at periventricular (A), medial (B), lateral (C), and superficial levels (D).
Reproduced from Puelles et al. (2012)

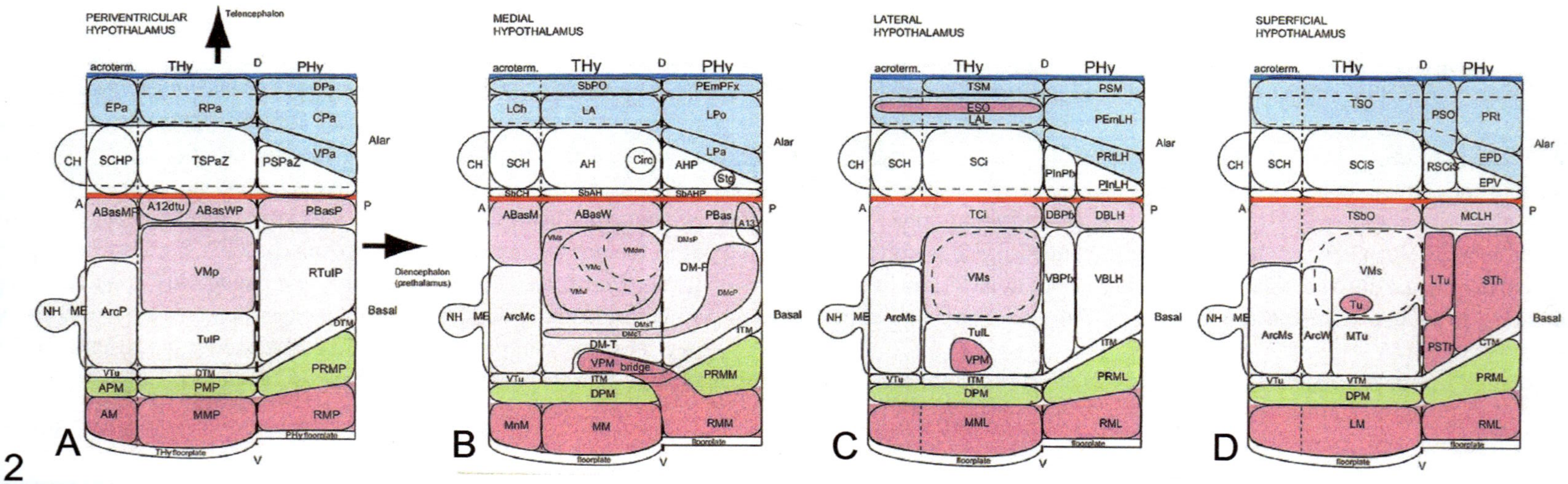

110

10.4.6 Pretectal Region and Hypothalamus

Plate 110

The *pretectal region* represents a primary FMU occupying the alar region of the first prosomere (p1). As extensively discussed in Sect. 8, this region splits up during early development into three distinct anteroposterior domains: CoP, JcP, and PcP, each of which subsequently subdivides into a number of dorsoventrally arranged subdomains (Plates 69, 73, 89). The histogenesis occurring within these subdomains involves the radial migration of neuroblasts and the formation of periventricular, intermediate and superficial cell layers, from which the various pretectal nuclei develop. During later development, some local shifts of cell masses occur, but in general it may be said that throughout the pretectum radial migration is the dominant mode of cell displacement. However, there is one notable exception to this rule in birds. As shown by Plate 110, Fig. 1, a primordial dorsofrontal nucleus (DF), which develops in the intermediate layer of the rostrally situated PcP domain, produces a ventral extension. This ventral extension is the source of a caudally directed tangential migration, eventually leading to the formation of a dorsocaudal nucleus (DCa). This cell mass extends through the intermediate layers of JcP and CoP, to end at the pretecto-mesencephalic border (Ferran et al. 2009). Mammals apparently lack this migration (Ferran and Puelles, unpublished observations).

Plate 110, Fig. 2 presents four orthogonalized maps of the *hypothalamus*, showing the derivatives of the various hypothalamic progenitor domains (see Plates 93, Fig. C, and 95, Fig. A), at periventricular (A), medial (B), lateral (C), and superficial levels (D), taken from the recent detailed genoarchitectonic analysis of that brain part by Puelles et al. (2012). The close correspondence of the outlines of the cell masses delineated at these four levels suggests that is in the hypothalamus, just as in the pretectum, radial migration is the prevailing mode of cell displacement. However, the analysis just mentioned revealed that signs of tangential migration are not entirely lacking in this brain part. Thus, it appeared that several nuclei are products of this type of migration, including the episupraoptic nucleus (ESO: Plate 110, Fig. 2C), ventral premamillary nucleus (VPM: Fig. 2B), subthalamic nucleus (STh: Fig. 2D), the ventromedial core nucleus (VMc), and the continuum formed by the terminal and peduncular dorsomedal core nuclei (DMcT, DMcP: Fig. 2B). All of these nuclei consist of glutamatergic neurons. The elements forming the ESO, VPM, and STH originate from the mamillary/retromamillary areas (M/RM: Plate 94, Fig. C), whereas those of the VMc, and DMcT-DMcP continuum stem from the dorsal tuberal and dorsal retrotuberal zones (TuD, RTuD: Plate 94, Fig. C). The remarkably long, dorsally directed tangential migration of the cells destined to form the ESO can be followed with the aid of the mamillary-selective marker *Foxb1*.

10 The development of grisea
10.4 Tangential migrations 10.4.7. Neocortex

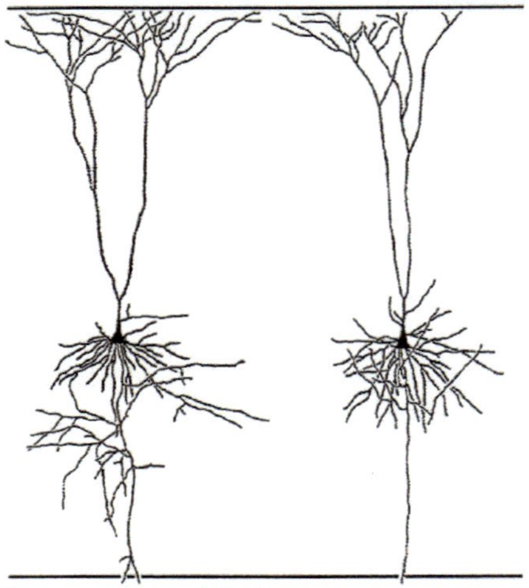

1. Neocortex, pyramidal neurons, excitatory (Glu)

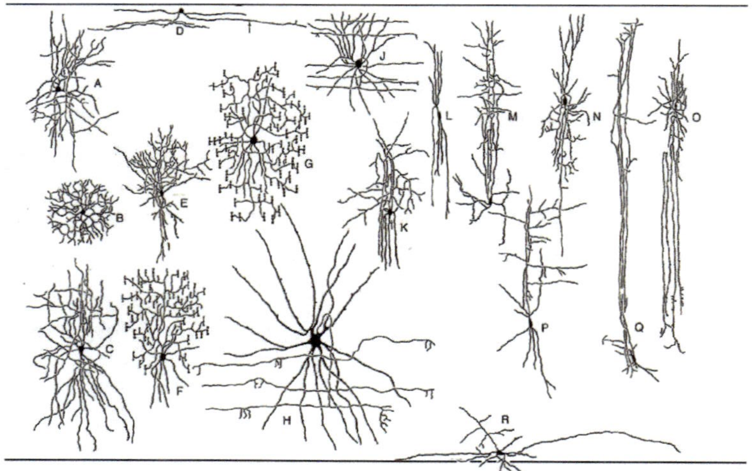

2. Neocortex, local circuit neurons, inhibitory (GABA)

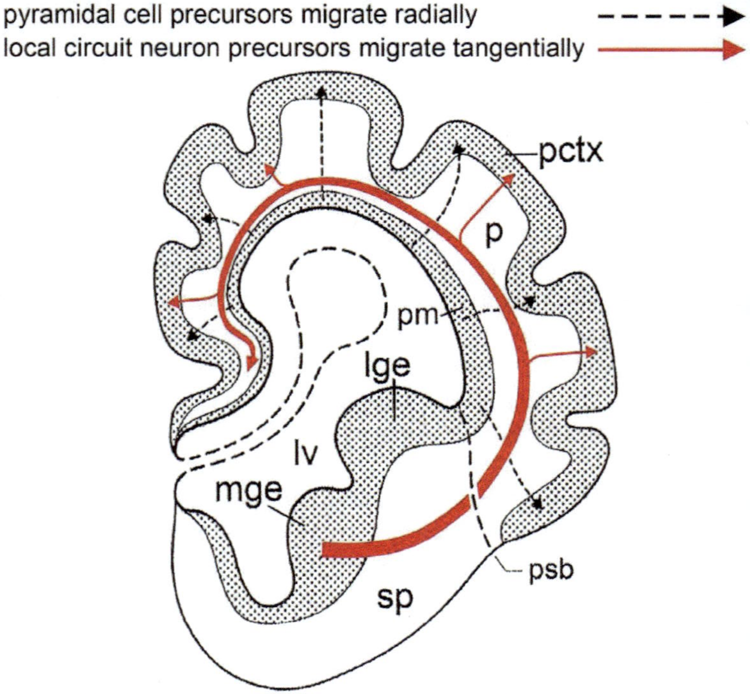

3. Diagrammatic transverse section through the fetal telencephalon of a generic mammal

lge, lateral ganglionic eminence;
lv, lateral ventricle;
mge, medial ganglionic eminence;
p, pallium; pctx, pallial cortex;
pm, pallial matrix;
psb, pallial-subpallial boundary;
sp, subpallium

111

10.4.7 Neocortex

Plate 111

The neocortex is an ultracomplex, six-layered structure that develops from the dorsal pallial sector of the telencephalic hemispheres (Plate 96, Fig. B). All mammals have a neocortex, but its size varies greatly among the various mammalian groups. In insectivores, such as the hedgehog, its size does not exceed that of the 'older' parts of the cortex, but in primates and cetaceans it attains amazing proportions, becoming by far the largest centre of the brain. Local differences in size, shape and arrangement of the cortical neuronal cell bodies render it possible to parcellate the neocortex into a number of separate cytoarchitectonic areas (Plate 84). The number of these varies among mammals and seems largest in the human brain. It is widely though by no means generally accepted that these cortical areas are in turn composed of large numbers of small, column-like, vertically arranged structures that span the cortical layers. According to Rakic (1990, 2009) these columnar structures are the product of radial progenitor units that consist of proliferative matrix and radial glial cells, which, via various intermediate stages, and after having become glutamatergic postmitotic neuroblasts, migrate outward along radial glial fibres, to form cylindrical cell assemblies with a diameter of about 30 μm.

The neocortex contains a significant population (up to 25 %) of extrinsic inhibitory GABAergic interneurons as well as the intrinsic excitatory glutamatergic neurons.

The excitatory glutamatergic population consist mainly of pyramidal projection neurons, i.e. of elements with pyramid-shaped cell bodies, the axons of which pass either to other cortical areas or to subcortical targets (Plate 111, Fig. 1).

The inhibitory GABAergic interneurons form a morphologically very heterogeneous population (Plate 111, Fig. 2). They are classified by their dendritic and axonal arborizations, synaptic targets, neurochemical characteristics, and firing properties. As for neurochemistry, it has been established that some subpopulations of inhibitory cortical interneurons express the cytoplasmatic calcium binding proteins parvalbumin (PV) or calretinin (CR), and that other subpopulations contain, in addition to GABA, a neuropeptide, as for instance vasoactive intestinal polypeptide (VIP), cholecystokinin (CCK), neuropeptide Y (NPY) or somatostatin (SST); still other subpopulations selectively express the serotonin receptor 3 AR (5-HT3AR).

With regard to the histogenesis of the neocortex, experimental studies have conclusively shown that the neocortical, excitatory, glutamatergic neurons are generated by the pallial ventricular and subventricular matrix zones, whereas the subpallium, including the ganglionic eminences and the preoptic area, is the source of the neocortical inhibitory GABAergic interneurons. These immature inhibitory elements cross the palliosubpallial boundary and reach their final position in the developing neocortex via a tangential subventricular or subpial migration, followed by radial penetration of the cortical plate (Plate 111, Fig. 3). This highly remarkable long tangential migration of neocortical inhibitory elements was first described for rodents (Anderson et al. 1997; Marin and Rubenstein 2001), but has also been observed recently in primates (Hansen et al. 2013; Ma et al. 2013).

10 The development of grisea
10.4 Tangential migrations 10.4.7. Neocortex

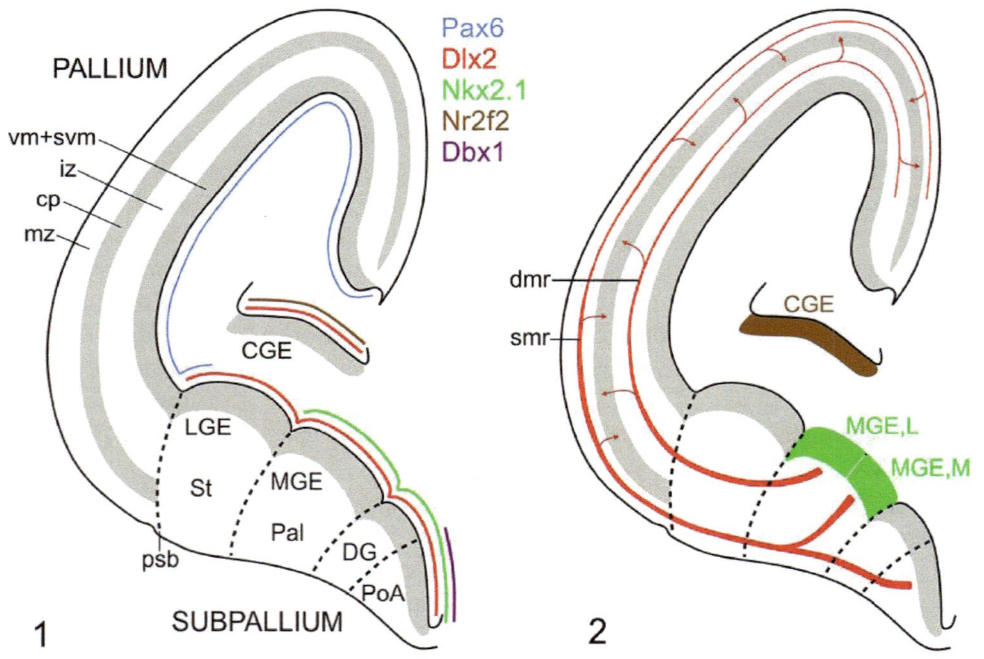

1 2

1. Diagrammatic transverse section through the embryonic mammalian telencephalon, showing location of principal structures / compartments and expression domains of some transcription factors. The caudal ganglionic eminence (CGE), which is located at a more caudal level, is included. cp, cortical plate; Dg, diagonal area; iz, intermediate zone; LGE, lateral ganglionic eminence; MGE, medial ganglionic eminence; Pal, pallidal area; PoA, preoptic area; psb, pallio-subpallial boundary; St, striatal area; svm, subventricular matrix; vm, ventricular matrix

2. Precursor domains of cortical inhibitiory interneurons, and principal migration routes of these elements. dmr, deep migration route; MGE,L and MGM,M lateral and medial sectors of MGE; smr, superficial migration route

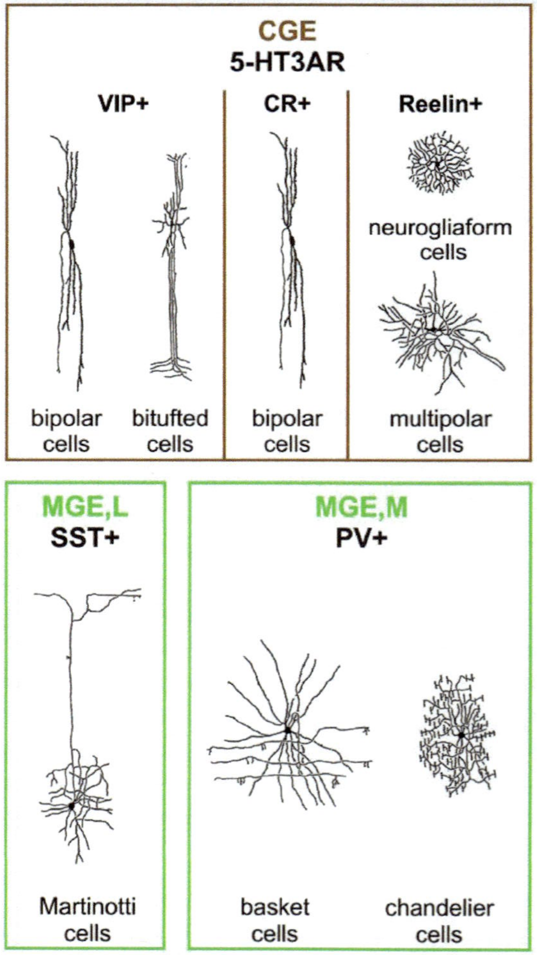

3

3. Subtypes of cortical inhibitory interneurons produced in the various subpallial precursor domains indicated in Fig. 2

112

Plate 112

The subpallium includes the preoptic area (PoA), which is a product of the second telencephalo-hypothalamic neuromere (hp2), and the diagonal (Dg), pallidal (Pal) and striatal (St) domains, which are derivatives of the first hypothalamo-telencephalic neuromere (hp1) (Plate 96, Fig. A). During ontogeny, the precursor areas of the striatum and globus pallidus protrude into the telencephalic lateral ventricle, forming the lateral and medial ganglionic eminences (LGE, MGE), respectively. Posteriorly, these two eminences merge into a single caudal ganglionic eminence (CGE), associated with the subpallial parts of the amygdala (Plate 112, Fig. 1). It has been shown that the MGE, CGE, and PoA are all involved in the production of neocortical inhibitory interneurons, and that distinct interneuron subtypes are produced in spatially distinct subpallial progenitor zones (Flames et al. 2007; Wonders et al. 2008; Miyoshi et al. 2010). The various subpallial precursor domains are specified by transcription factors, the combinatorial activities of which determine the subtype of neurons to be produced. The expression domains of some of these transcription factors are indicated in Plate 112, Fig. 1. The data included in this figure are derived from the review articles of Sultan et al. (2013) and Kepecs and Fishell (2014), supplemented with data from Gelman et al. (2013) and Ma et al. (2013).

It is noteworthy that in primates the subventricular matrix (SVM) in the ganglionic eminences expands massively during the period immediately preceding the production of cortical inhibitory interneurons. During that period, the SVM becomes densely populated with neural stem cells and intermediate progenitor cells. In contrast with the neocortex, the stem cells in the ganglionic eminences do not maintain radial processes or radial orientation (Hansen et al. 2013).

The MGE is responsible for the production of about 60 % of neocortical inhibitory interneurons (Wonders and Anderson 2006; Gelman and Marín 2010). Within the MGE, different interneuronal subtypes are generated in spatially separated pools of progenitors. Thus, it has been established that the medial part of the MGE gives rise to a subgroup of parvalbumin-expressing interneurons, including the conspicuous basket and chandelier cells, while the lateral part of the MGE produces a somatotostatin-expressing population, of which the Martinotti cells with ascending axons form the largest subset (Flames et al. 2007; Wonders et al. 2008; Plate 112, Figs. 2, 3).

The CGE is the origin of about 30 % of the total cortical inhibitory population (Miyoshi et al. 2010). The elements produced here belong to diverse subtypes, which are largely marked by the expression of the serotonin 3A receptor (5-HT 3AR) and include CR and/or VIP expressing bipolar and double bouquet cells, as well as a group of Reelin-expressing neurogliaform and multipolar cells {Lee et al. 2010; Miyoshi et al. 2010; Ma et al. 2013; Plate 112, Fig. 2).

The remaining 10 % of neocortical inhibitory interneurons are produced in the PoA. This region appears to give rise to a surprisingly diverse cohort of neocortical interneurons, which all stem from *Dbx1*-expressing progenitor cells. Indeed, the PoA appears to contribute to almost all of the major groups of cortical GABAergic interneurons, including PV + basket cells, SST + cells, Reelin + neurogliaform and bipolar cells, and VIP/CR + bipolar/bitufted cells (Gelman et al. 2011). These remarkable findings suggest that the mechanisms controlling the specification of different subtypes of cortical interneurons are more complex than previously expected. Gelman and Martin (2010) indicate that it remains to be determined to what extent MGE- and PoA-derived cortical interneurons with similar characteristics indeed belong to the same class of interneurons, because there might be distinguishing features not yet investigated. As regards the wide variety of interneurons that emerge from Dbx1-expressing progenitors, they consider it conceivable that this pool of cells indeed consists of several classes of progenitors, each one of them expressing additional, as yet unidentified transcription factors, or that, alternatively, the same progenitor elements give rise to very different subtypes of interneurons.

Additionally, the PoA also forms a source for large numbers of oligodendrocyte precursors, which also migrate tangentially into other telencephalic regions including the pallium (Tekki-Kessaris et al. 2001). The densely spiny GABAergic projection neurons of the caudate-putamen (CP) are provided by the lateral ganglionic eminence (LGE) (Yun et al. 2003).

Neocortical inhibitory interneurons generated in the subpallium enter the pallium through two different routes, a superficial trajectory within the marginal zone, and a deep path within the intermediate zone (Plate 112, Fig. 2). It has been reported that these elements use corticofugal axons as a scaffold during their tangential migration along these routes (Parnavelas et al. 2000). After having completed their tangential displacement, they turn radial to reach their ultimate location in the developing neocortex.

10 The development of grisea

10.4 Tangential migrations 10.4.8. Subpallium

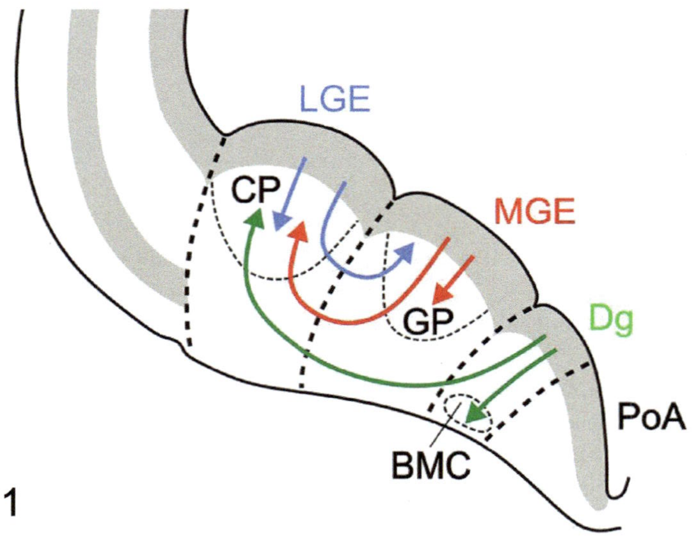

1

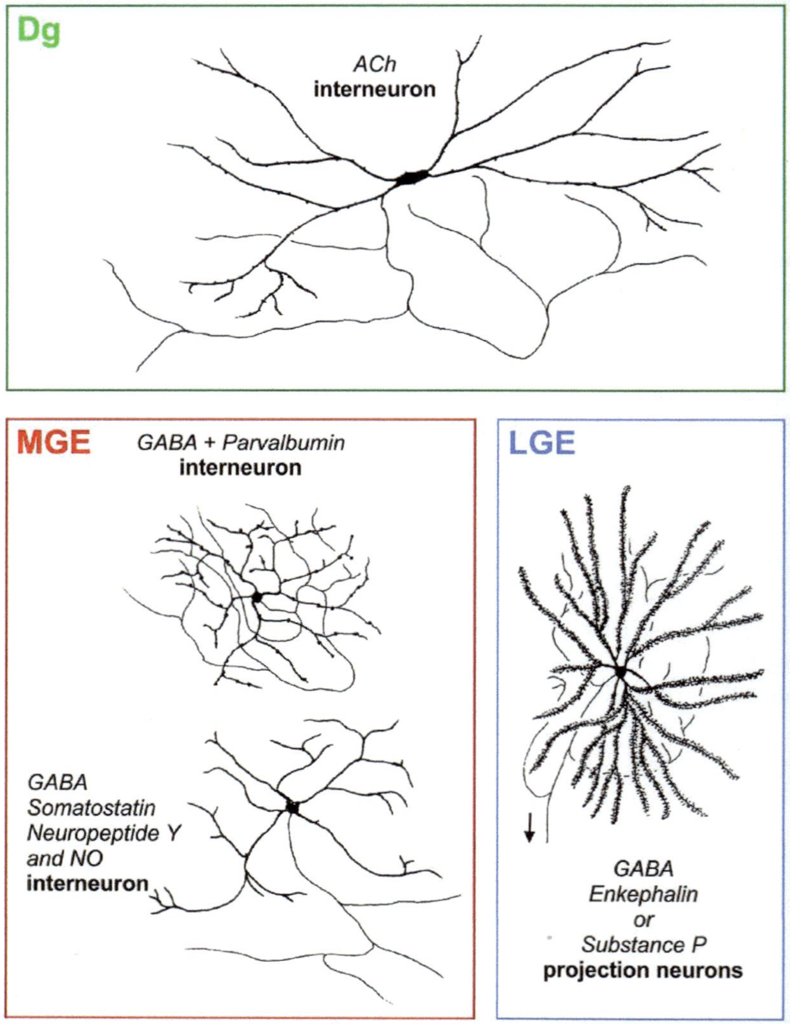

2

1. Migratory routes, their sites of origin and destination in the embryonic mammalian subpallium. BMC, basal magnocellular complex; CP, caudate-putamen complex; GP, globus pallidus; LGE, lateral ganglionic eminence; MGE, medial ganglionic eminence; PoA, preoptic area

2. Principal neuronal (sub)types in the caudate-putamen complex. The sites of origin of the elements depicted are indicated in the upper left hand corners of the panels.

113

10.4.8 Subpallium

Plate 113

The progenitor zones of the medial and caudal ganglionic eminenences (MGE, CGE) and the preoptic area (PoA) produce projection—as well as interneurons that populate the adult subpallial centres, apart from inhibitory interneurons that migrate to the cortex. Thus a radial migration from the diagonal area (Dg), a progenitor zone intercalated between the MGE and the PoA, generates the cholinergic projection neurons of the basal magnocellular complex (BMC), whereas tangential migrations from the same area produce the large cholinergic interneurons of the caudate-putamen/striatal complex (CP) (Marín et al. 2000: Garcia-Lopéz et al. 2008). Similarly, a radial projection from the MGE produces the GABAergic projection neurons of the globus pallidus (GP), whereas tangential migrations from the MGE give rise to the GABAergic interneurons of the CP (Marín et al. 2000; Pauly et al. 2014). The latter interneurons can be divided into several subtypes. Suffice it to mention here that the elements of one subtype express the calcium binding protein parvalbumin (PV), whereas those of another subtype contain the neuropeptides somatostatin (SST) and neuropeptide Y (NPY), as well as the diffusable secondary messenger nitric oxide (NO). The densely spiny GABAergic projection neurons of the CP are provided by the lateral ganglionic eminence (LGE) (Yun et al. 2003). It is noteworthy that in the MGE, the production of GP projection neurons precedes that of the cortical and striatal interneurons (Pauly et al. 2014). It is also of note that a subpopulation of the GABA-ergic projection neurons in the globus pallidus originates from the LGE (Nóbrega-Pereira et al. 2010; Bupesh et al. 2011).

Finally, it may be mentioned that the LGE, MGE, CGE, Dg, PoA, ventral pallium (PV), and alar hypothalamus, all contribute cells to the amygdaloid complex (Garcia-Lopéz et al. 2008; Bupesh et al. 2011; Morales-Delgado et al. 2011; Medina et al. 2014). It has been established that some of these cell contingents, including those from the PoA, VP, and alar hypothalamus reach their final positions within the amygdaloid complex by way of tangential migration (Bupesh et al. 2011). Hirata et al. (2009) found that a ventral pallial *Dbx1*-positive progenitor pool generates the excitatory output neurons in the amygdalar cortical nuclei and in the basolateral complex, and that additional *Dbx1*- positive cells derived from the PoA migrate by way of a tangential migratory stream to the medial amygdala where they form a unique subtype of inhibitory neurons.

10.4.9 Tangential Migrations; Summary and Concluding Remarks

The phenomenon of tangential migration includes sets of cells that leave the FMUs where they are produced, and migrate to other territories orthogonally to radially oriented structures. The data collected above on this phenomenon will play a prominent role in the overview of the formation of grisea, to be presented below. Here we confine ourselves to some concluding remarks.

1. Whereas most FMUs that are involved in tangential migrations export only part of the cells produced, some seem to exhaust themselves completely in the formation of their migratory stream. The origins of the rhombencephalic intramural and extramural migratory streams, and the mesodiencephalic stream leading to the formation of the compact part of the substantia nigra, are striking cases in point.
2. The lengths of the migration trajectories vary considerably. At one end of the range are the minimal local intrusions into neighbouring areas, e.g. by the dorsocaudal pretectal nucleus (DCa) and the hypothalamic ventral premamillary nucleus (VPM), or the local exchanges of cells between the various striatal primordia, while at the other extreme are the long journeys undertaken by the cells destined to form the pontine nuclei, or those assembling in the subpallio-pallial migrations.
3. It is remarkable that the tangential migrations of certain groups of functionally related cell masses seem to be orchestrated in the sense that they move from mutually strictly corresponding sites of origin to mutually strictly corresponding ultimate locations at different levels or the neuraxis. The branchiomotor nuclei and general visceromotor nuclei in the hindbrain show such orchestrated tangential migrations, in that they all migrate to the B4 zone of the alar plate. The traditional explanation offered for this phenomenon is the teleological platitude that these developing cell masses apparently move to positions that are favourable for their integration into the wiring of the CNS. The observation that some noradrenergic neurons (A7, A5, A3, A1) converge with the migrated branchiomotor nuclei or with the migrated parasympathetic preganglionic vagal complex at the hindbrain B4 alar zone suggests the existence of some common molecular signalling mechanism.

10 The development of grisea
10.5 Overview of the development of grisea

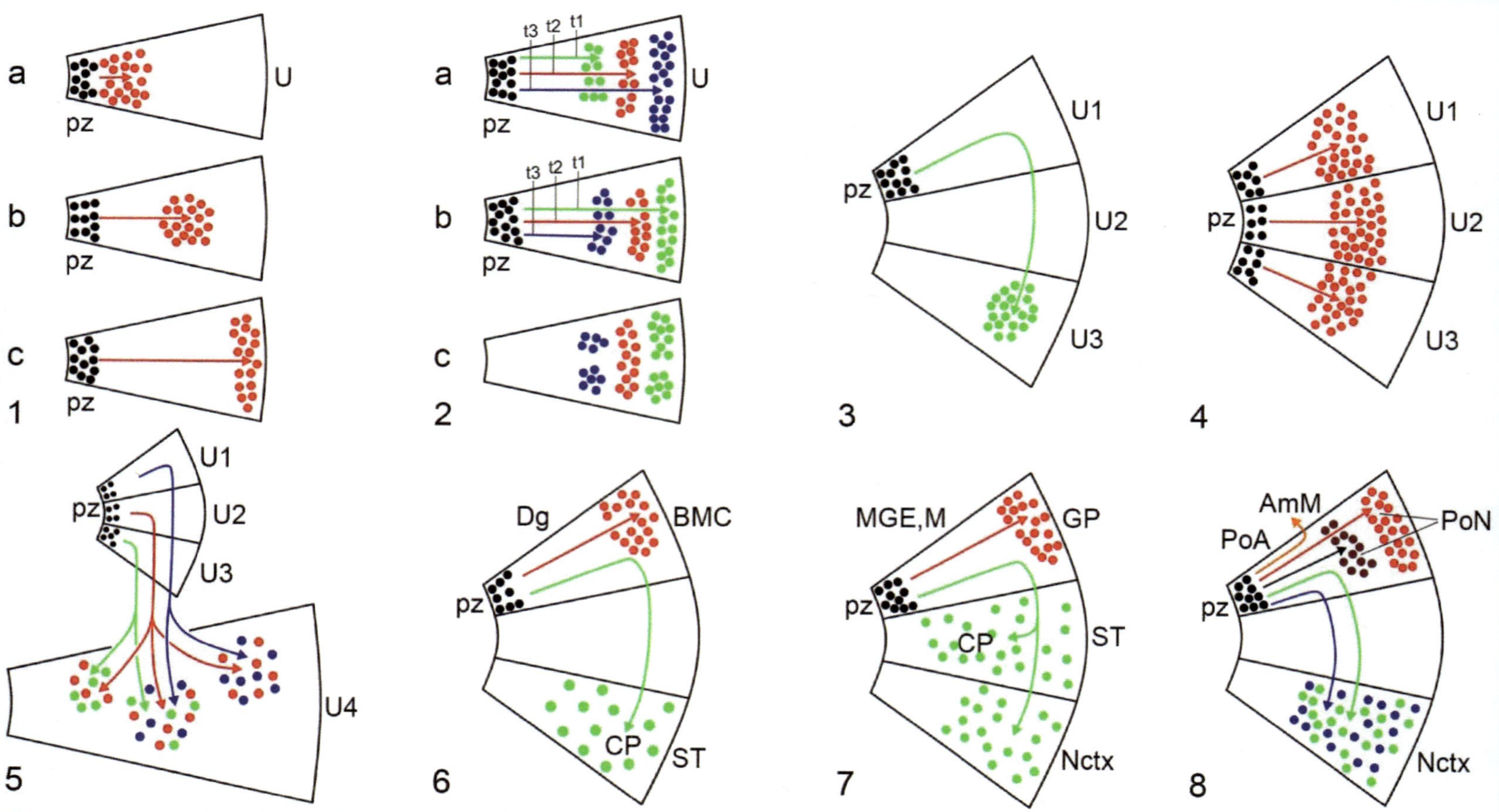

Diagrams illustrating the various modes of development of grisea discussed in the text.
AmM, medial nucleus of amygdala; BMC, basal magnocellullar complex; CP, caudate-putamen complex; Dg, diagonal area; MGE,M, medial part of medial ganglionic eminence; Nctx, neocortex; PoA, preoptic area; PoN, preoptic nuclei; pz, progenitor zone; St, striatal area; U, fundamental morphological unit

114

10.5 Overview of the Development of Grisea

Plate 114

The data assembled in the previous subsections enable us to present the following *tentative* overview of the ontogeny of grisea.

1. The entire ventricular surface of the developing CNS is occupied by a mosaic of progenitor zones (Plate 95). These form part of the molecularly defined fundamental morphological units (FMUs), which can also be designated as progenitor domains.
2. Progenitor zones generally consist of a single ventricular proliferative matrix. However, in most telencephalic regions a second, subventricular matrix zone is present, and in the dorsal pallial domain of primates, separate inner and outer subventricular matrix zones can be distinguished.
3. The progenitor zones produce neurons of one or several (sub)types. The nature of the elements produced by a given progenitor zone depends on specific gene expression patterns within the FMU to which this progenitor zone belongs. The activated genes produce different classes of transcription factors that control the various steps in the differentiation of progenitor cells, and determine the identity of the neuronal (sub)type(s) produced (Guillemot 2007).
4. Progenitor zones which produce different (sub)types of neurons commonly do so in a fixed temporal sequence, correlated with progressive changes in the genomic read-out state of the FMU involved.
5. Sets of neurons produced by a particular progenitor zone commonly share one or more characteristics, such as projection pattern, neurotransmitter profile, presence of certain neurochemical markers, such as particular calcium binding proteins, firing patterns, and so on.
6. Many progenitor zones produce neuroblasts which migrate radially outward, to settle within the confines of ‘their own’ FMU. In simple cases, a single periventricular, intermediate, or superficially situated cell mass is formed in this way (Plate 114, Fig. 1a–c). In more complex cases, a particular progenitor zone produces successively several laminar zones of cells, according to ‘inside-out’ (Fig. 2a) or ‘outside-in’ patterns (Fig. 2b). During later development, each of these primordial layers may form one or several separate cell masses (Fig. 2c). The latter developmental mode prevails in the diencephalon proper (prosomeres p1–p3) and in the hypothalamus.
7. Several progenitor zones produce neuroblasts which migrate tangentially, to settle within the domain of one or more ‘foreign’ FMU(s) (Fig. 3).
8. It frequently occurs that cell masses which are the products of either the radial or the tangential migrations indicated above under 6 and 7, associate to form larger cellular complexes (e.g. the spinal dorsal horn). As discussed in subsection 10.2, shared dorsoventral patterning mechanisms in adjacent neuromeres may lead to the formation of structurally related cell contingents which occupy mutually corresponding positions, and which during further development unite to form bi-, tri- or plurisegmental cell masses or columnar formations. Similar ‘internuclear associations’ may also develop in dorsoventrally adjacent FMUs (Fig. 4). This phenomenon manifests itself prominently in the pretectum. As illustrated by Plates 89 and 90, three out of the five concentric zones of this structure are partly or wholly occupied by such ‘internuclear associations’.

9. The so-called 'rhombic lip', composed of the A1–A4 microzones of rhombomeres 1–11, and molecularly defined by the expression of *Wnt1* and *Olig3*, forms a highly remarkable formation, which has evolutionarily developed during the transition from agnathans to gnathostomes (Plate 104). Progenitor zones within its confines give rise to an astonishing variety of cell masses, including the cochlear and vestibular nuclei, the various mossy fibre- producing precerebellar nuclei*, the climbing fibre-producing inferior olivary complex, the superior olivary complex, the lateral lemniscal nuclei, the deep cerebellar nuclei*, the parabrachial nuclei, the Kölliker-Fuse nucleus, the parabigeminal nucleus*, the laterodorsal tegmental nucleus*, the pedunculopontine tegmental nucleus*, and last but not least the transitory external granular layer, from which the definitive cerebellar internal granular layer* will develop (Plate 108). Moreover, it is known that the viscerosensory and general somatosensory centres, which are the products of radial migrations from the rhombencephalic B1–B4 microzones, also receive tangentially migrating elements, from some of the rhombic lip microzones (Storm et al. 2009).

The cell masses marked with an asterisk in the previous paragraph are all known to stem from specific segmental sectors of the *Math1* (*Atoh1*)-expressing A1 microzone of the rhombic lip. It is of note that this microzone produces glutamatergic neurons thoughout its entire extent. However, a rostral part of the rhombencephalic A1 microzone is involved in the production of cholinergic neurons destined for the pedunculopontine and laterodorsal tegmental nuclei within r1.

10. Neurons derived from various molecularly distinct progenitor zones may migrate to a given nuclear complex, to intermix differentially within the constituent nuclei of that complex. This may be supposed to be always the case whenever a nucleus contains several different types of neurons. This phenomenon is illustrated in Plate 114, Fig. 5, which is based on the remarkable development of the r1 portion of the interpeduncular nuclear complex (cf. Plate 109, Fig. B). A similar intermixing of neurons from different sources has been observed in, among other structures, the spinal dorsal horn, the bed nucleus of the stria terminalis and the amygdaloid complex (Garcia-López et al. 2008; Bupesh et al. 2011) An interesting question related to the formation of such composite nuclei is whether, and, if so, on which grounds, may the cellular elements derived from one of the contributing progenitor zones of a given cell mass, be designated as its *principal neurons*.
11. Some progenitor zones produce two or more entirely different sets of neurons, usually at different time periods viz. (1) neurons which migrate radially outward to settle within the confines of 'their own' FMU, and (2) elements which migrate tangentially to contribute to grisea formed in other FMUs. We illustrate this mode of development with outlines of the differentiation and migration patterns of three subpallial FMUs, viz, (1) the diagonal area; (2) the medial sector of the medial ganglionic eminence, and (3) the preoptic area.

Within the *diagonal area* (Dg; Plate 114, Fig. 6) cells migrate radially outward to form the large cholinergic projection neurons, which populate the nucleus basalis magnocellularis (BMC), whereas a laterally directed stream of tangentialy migrating elements provides the caudate-putamen complex (CP) with large cholinergic interneurons.

The *medial sector of the medial ganglionic eminence* (MGE, M; Plate 114, Fig. 7) participates in the production of the GABAergic projection neurons, which populate the globus pallidus (GP). These elements attain their destination via a radial migration (Flandin et al. 2010). The MGE, M progenitor zone also gives rise to a group of parvalbumin-expressing GABAergic interneurons, which includes both basket and chandelier cells. These elements migrate tangentially toward the neocortex (Nctx), but the caudate-putamen complex (CP), the hippocampus and the amygdaloid complex are also supplied with parvalbumin + basket cells (Xu et al. 2008; Hirata et al. 2009).

The *preoptic area* (PoA; Plate 114, Fig. 8), finally, produces by means of radial migrations its own preoptic nuclei (PoN), but its progenitor zone also gives rise to a widely diverse group of *Dbx1*-expressing, GABAergic neocortical interneurons (Gelman et al. 2011). Hirata et al. (2009) reported that a distinct stream of these elements does not target the neocortex but rather the amygdala, to generate a unique type of inhibitory interneuron in the medial nucleus of that complex (AmM). Thus, it seems that some progenitor zones, such as e.g. the MGE, M, produce particular types of interneurons destined for different and disparate grisea, whereas others, as e.g. the PoA, may in addition, generate a specific interneuron subtype, committed to populate a single specific brain structure.

10 The development of grisea
10.5 Overview of the development of grisea

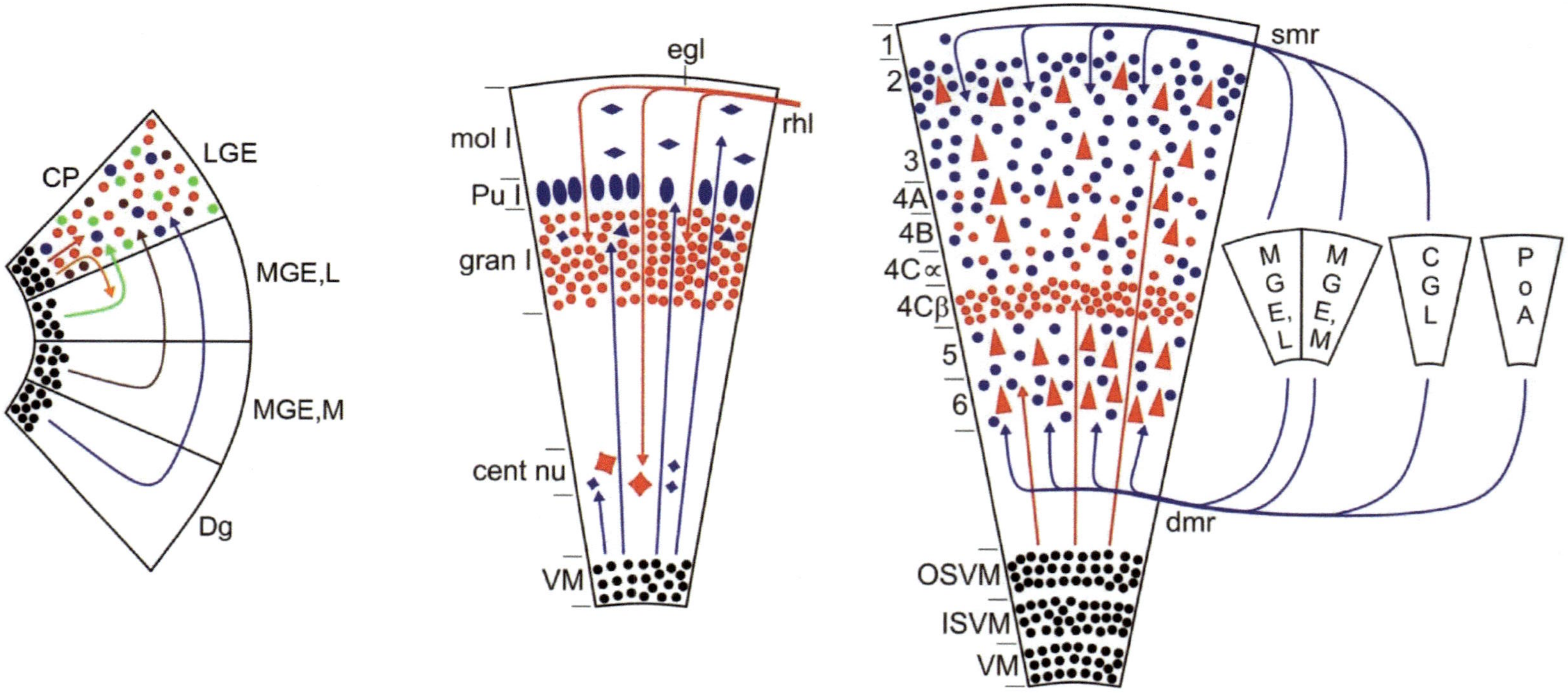

1. Caudate-putamen complex 2. Cerebellum 3. Neocortex

Diagrams illustrating the various modes of development of grisea discussed in the text. In figs. 2 and 3 excitatory neurons are in red, and inhibitory neurons are in blue. Abbreviations: cent nu, central nuclei; CGL, caudal ganglionic eminence; CP, caudate-putamen complex; Dg, diagonal area; dmr, deep migration route; egl, external granular layer; gran l, granular layer; LGE, lateral ganglionic eminence; MGE,L, medial ganglionic eminence, lateral part; MGE,M, medial ganglionic eminence, medial part; mol l, molecular layer; PoA, preoptic area; Pu l, layer of Purkinje cells; pz, progenitor zone; rhl, rhombic lip; smr, superficial migration route

115

Plate 115

12. Some grisea, as for instance, the caudate-putamen complex, the cerebellum and the neocortex, are composed of two entirely different sets of neurons, viz. radially migrating 'intrinsic' neurons, and tangentially migrating 'extrinsic neurons.

In the *caudate-putamen complex* (CP; Plate 115, Fig. 1), GABAergic spiny projection neurons comprise 90 % of the cells. These elements originate from the progenitor zone of the lateral ganglionic eminence (LGE), and migrate radially into the adjacent striatal primordium (Marín and Rubenstein 2001). The remaining 10 % of striatal neurons consist of interneurons, which are diverse in their morphological and neurochemical characteristics. They include large cholinergic elements and several subtypes of medium-sized and small GABAergic interneurons, among which parvalbumin (PV)-, neuropeptide Y (NPY)/somatostatin (SST)/nitric oxide (NO)-, and calretinin-containing elements. All of these interneurons are of subpallial origin and enter the caudate-putamen complex via tangential migrations. It is known that the cholinergic elements are produced in the diagonal area (Dg), and that the PV-containing neurons and the NPY/SST/NO-containing neurons are derived from the medial (MGE, M) and lateral parts of the medial ganglionic eminence (MGE, L) respectively (Wonders et al. 2008). Interestingly, a tangentially migrating subpopulation of GABAergic neurons that originates in the lateral ganglionic eminence and expresses the transcription factor Islet1, contributes to the formation of the globus pallidus (Nóbrega-Pereira et al. 2010). The lateral ganglionic eminence represents therewith, to our knowledge, the only progenitor domain which *exchanges* neuronal elements with another progenitor domain.

As regards the *cerebellum* (Plate115, Fig. 2), molecular studies have shown that all cerebellar GABAergic neurons derive from the cerebellar ventricular matrix and migrate radially outward (Leto et al. 2006, 2009), whereas the glutamatergic cerebellar nuclear projection neurons and the also glutamatergic granule cells emerge from the r1 segment of the *Math1*-expressing microzone of the rhombic lip (rhl). Proliferation-capable precursor elements spread as the proliferative external granular layer over the surface of the developing cerebellum, after which their neuronal daughter cells first differentiate bipolarly a parallel fibre and then translocate their somata inward to attain their ultimate locations (Machold and Fishell 2005; Fink et al. 2006; Yeung et al. 2014). Once within the internal granular layer there is a phase of excess dendrite formation, finally reducing to 3–4 participating in synaptc glomeruli.

The *neocortex* (Plate 115, Fig. 3) represents by far the most complex structure of the mammalian brain. It is composed of two major classes of neurons, glutamatergic excitatory neurons and GABAergic inhibitory elements (shown in red and blue, respectively, in Fig. 3). The glutamatergic neurons include the pyramidal cells, which constitute the long-axoned efferent elements of the cortex, and the short-axoned spiny stellate cells. The latter are to be considered as strongly modified pyramidal neurons, which have lost their long axons and their pyramidal shape (Lund 1984). They are particularly numerous in primary sensory cortices, in which they play a crucial role in the radial propagation of the activity fed in by the thalamocortical afferents. The GABAergic inhibitory cortical elements, which are all short-axoned interneurons, form a very diverse group.

The neocortex is generally subdivided into six layers, 1–6. In the primate primary visual cortex, the structure of which is diagrammatically indicated in Fig. 3, the fourth layer is strongly developed and can be divided into four sublayers, 4A, 4B, 4Cα and 4Cß (Lund 1988). All of these sublayers contain spiny stellate cells. As shown in Fig. 3, these elements are scattered in sublayers 4A, 4B and 4Cα, but densely packed in sublayer 4Cß.

The developing primate dorsal pallial region contains three matrix zones, the ventricular matrix (VM), and the inner and outer subventricular matrices (ISVM, OSVM). Although these progenitor zones are of course lacking in the mature brain, they are included in Fig. 3. The glutamatergic neurons produced in the various matrices migrate radially outward and settle in the developing cortex according to an 'inside-out pattern'.

All neocortical GABAergic interneurons arise from three subpallial progenitor zones, the medial ganglionic eminence, which can be subdivided into separate medial (MGE, M) and lateral (MGE, L) subzones, the caudal ganglionic eminence (CGE), and the preoptic area (PoA). As detailed in Sect. 10.4.7, distinct interneuronal subtypes are produced in subdomains within these various zones. These elements migrate tangentially over long distances via superficial (smr) and deep migration routes (dmr) before being radially positioned to their final location in the cortex. Within the developing neocortex, the various sets of interneurons do not randomly disperse, but rather form spatially organized horizontal or vertical clusters (Ciceri et al. 2013).

It is remarkable that in the cerebellar (Plate 115, Fig. 2) as well as in the cerebral cortex (Plate 115, Fig. 3), the glutamatergic excitatory neurons and the GABAergic inhibitory elements originate from strictly separate progenitor zones.

10 The development of grisea
10.6 The homologization of grisea

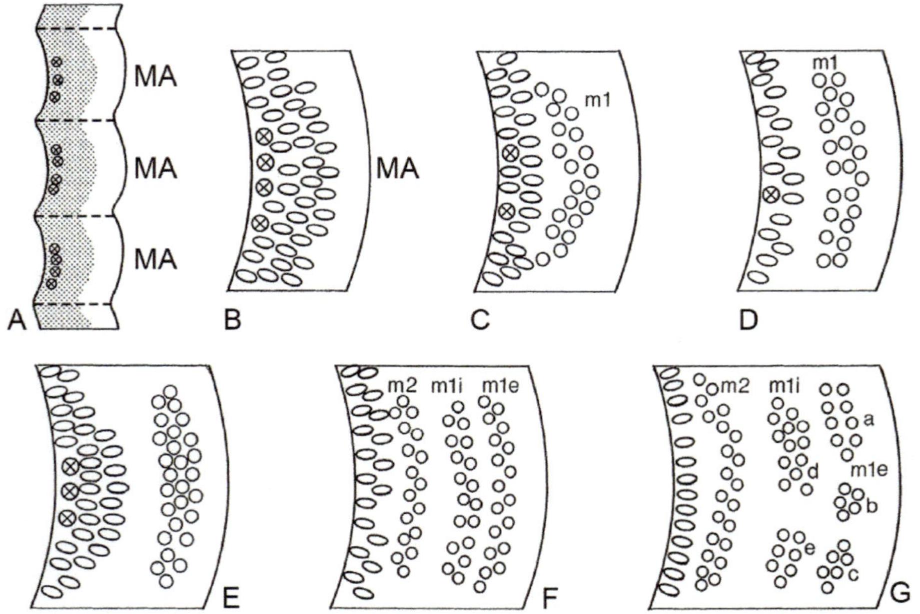

Fig. 1 Development of brain nuclei according to Bergquist and Källén (1954)

Questions pertaining to the determination of the morphological identity of nuclei (Bergquist and Källén 1954, p. 641):

(1) from which migration area does the nucleus develop,
(2) within which migration layer is it formed, and
(3) within which part of the migration layer does it develop?

Morphological and molecular interpretation of neural structures: homology[1]

Principal criterion[2]
  Similarity in primary[3] topological[4] position and associated molecular profile
Auxiliary criteria
  1. Similarity in secondary or definitive topological[4] position
  2. Similarity in fibre connections[5]
  3. Similarity in special quality[6]

[1] Homology is a purely morphological concept.
It has nothing to do with function (de Beer, 1971).
Similarity in function is denoted with the term: ANALOGY.

[2] Principal criterion, because of its *sine qua non* character: Neural structures that do not occupy corresponding primary topological positions are not homologous.

[3] Primary topological position, *i.e.* the position occupied by the progenitor zone from which the pertinent nucleus originated.

[4] Not: *topographical* position.

[5] This is an auxiliary criterion, because a cell mass may change one or more of its connections, without losing its *morphological* identity (Holmgren, 1925; Striedter, 1992).

[6] Particular cytological or histochemical properties.

116

10.6 The Homologization of Grisea

10.6.1 Introductory Note

The concept of homology expresses the existence of typical and specific correspondences between anatomic parts of members of natural groups of living organisms that share a Bauplan and a common ancestor. The methodical search for these homologous relationships stands central in morphology and requires the formulation of one or more criteria, with the aid of which homologous structures can be recognized.

10.6.2 The Swedish School

Plate 116

In the preceding sections of this course it has been repeatedly mentioned that the new neuromorphology introduced here has conceptually evolved from the work of the Swedish school of comparative neuroembryology. The members of this school, among whom stand out Palmgren (1921), Holmgren (1922, 1925), Rendahl (1924), Bergquist (1932, 1953, 1954a, b) and Källén (1951a, b, c, d, 1962, 1965), had very clear ideas about the methodology of the homologization of brain nuclei, ably summarized in Bergquist and Källén's landmark paper of 1954 (Plate 116, left panel). Building on previous studies (Nieuwenhuys 1998f, 2009a), we have carefully selected a number of criteria, which hopefuly will be useful for the institution of homologies of neural grisea within the context of our new genoarchitectonic neuromorphology (Plate 116, right panel). In what follows the two methodologies mentioned above will be compared and commented upon.

A major result of the investigations of Bergquist and Källén was that the neural walls of the embryonic vertebrate brain can be divided systematically into a number of basic developmental units or fields, formed by the intersection of longitudinally arranged zones and transversely oriented neuromeres (Plate 93, Fig. B). These units manifest themselves initially as centres of proliferation, and later on as cushion-like thickenings of the differentiating mantle layer. Because these thickenings of the mantle layer are the product of local intensifications of radial neuroblast migration, the units within the confines of which these intensifications occur were designated as *migration areas* (Plate 116, Fig. 1A, B: MA). Berquist and Källén indicated that the following developmental events may occur within these migration areas: (1) Outward migration of neuroblasts may be limited, so that the resulting cell mass remains in direct contact with the matrix zone, or later on with the ependymal layer (Fig. 1B). Such formations of *periventricular gray* prevail in the brains of anamniotes. (2) Alternatively, neuroblasts may migrate further outward, to form one or more successive layers of migrated cells (Fig. 1 C–E). (3) One or more of the layers of migrated cells may secondarily split up into two or several sublayers (Fig. 1F). (4) Each of the layers or sublayers may fractionate into two or more separate aggregations of cells, representing the direct precursors of nuclei in the mature brain (Fig. 1G).

It is of note that the morphological identification of brain nuclei is, according to Bergquist and Källén (1954), a hierarchical process involving the consecutive answering of the three questions included in Plate 116, querying ascription to migration area, layer and sublayer. Hence, as regards the homologization of nuclei, Källén (1951c, p. 6) stated: "If it can be proven that two nuclei in different species develop from the same anlage (i.e. the same migration area), and in a similar way, they must be looked upon as homologous. The longer the developments of two nuclei are similar, the stricter is the homology between the nuclei." This implies that the derivatives of the matrix zones of one particular migration area in different species can only be compared *as whole fields,* insofar as divergence may occur in layer and nuclear differentiation. (The precise meaning of the term 'field' will be elucidated below.)

If we imagine that the configurations shown in Plate 116, Fig. 1 do not only represent ontogenetic stages, leading to the configuration shown in G, but, in addition, pertain to diverse final configurations as encountered in different adult species, then the cell mass m1e in species F would be homologous *as a field* to the cell masses a, b and c in species G, the cell layer m1 in species D would be homologous *as a field* to layers m1i and m1e in species F and, finally, the single periventricular zone in species B would be homologous *as a field* to the sum of all layers and cell masses observed in each of the species C–G. It will be clear that in the context of the morphological procedure for the establishment of homologies just sketched, the availability of microscopic material of closely graded series of embryos of all species involved is an absolute requisite: "The only basis for the homologization of brain nuclei ought to be the ontogenetic development of the nuclei from the regions where the first morphologically visible cell migration takes place, the so-called migration areas." (Källén 1951c, p. 30). Note that judgements concerning similarities in development were based at that time largely on cyroarchitectonic or reduced silver data.

Most members of the Swedish school, including Palmgren (1921), Rendahl (1924) and Bergquist (1932) concurred with this 'pan-embryological' point of view. Holmgren (1922. 1925), on the other hand, accepted similarity in fibre connections as an *additional* or *auxiliary* criteron for the establishment of the homology of nuclei in adult, related species. We will comment below on the rationale and caveats for using this auxiliary criterion.

10.6.3 The Principal Homology Criterion

How does all this relate to the tentative conceptual framework for our new neuromorphology, synoptically presented in Plate 116, right panel? Let us begin by emphasizing that in both methodologies the central idea for the homologization of neural grisea is the presence of a set of topological structural invariants or *Bauplan,* common to the CNS of all species to be compared. The general plan of the vertebrate brain presented by Bergquist and Källén (1954; Plate 93, Fig. B) was based, as we have seen, on the location and distribution of centres of proliferation. Its 'new-neuromorphological' equivalent, which is a derivative of the modified prosomere model of Puelles et al. (2012), is complemented/enriched with expression patterns of characteristic sets of developmental regulatory genes at each FMU (Plate 93, Fig. C). Both maps show a pattern of intersecting longitudinal zones and transverse neural segments, and in both maps the delineated planar fields represent *progenitor zones,* though normally the molecularly delimited FMUs are smaller, being subdivisions of the classic migration areas. These progenitor zones represent, in their turn, the developmental antecedents of radially organized, three-dimensional entities, i.e. the *migration areas* of Bergquist and Källén, which feature in our new neuromorphology as *fundamental morphological units* (FMUs). It follows that the entire CNS of vertebrates is essentially composed of radial units extending throughout the thickness of the neural wall and occupying invariant topological anteroposterior and dorsoventral positions with respect to each other. All vertebrates are held to have the same set of FMUs, with few exceptions in Agnatha, so that, by definition, corresponding FMUs are homologous entities.

As detailed before (Chap. 8), the FMUs represent relatively autonomous histogenetic domains, which can be characterized as modules of neural proliferation, radial migration and differentiation. Nevertheless, it may well be that corresponding fully developed FMUs of different species show considerable differences with respect to the number, position and configuration of the respective derived cellular formations. Yet, all of these evolutionarily divergent cellular formations have one salient feature in common: they stem from, *i.e.* they can be radially projected back to the same progenitor area with a conserved molecular profile in the common Bauplan. As already mentioned above, this general correspondence is denoted as *field homology* (Puelles and Medina 2002; Medina 2007). *The composite sets of cellular formations located within the confines of corresponding FMUs of different species represent (in so far as they stem from radial migrations, see below) field homologues.* The first step in the morphological interpretation of a cellular formation concerns the determination of its field-homological relationships. This is encapsulated in Bergquist and Källén's first question, as well as in our principal homology criterion (Plate 116, right panel), but attains now greater precision by referring to smaller, molecularly defined FMUs.

Before continuing our discourse on the determination of homologies within the realm of our new neuromorphology, three general remarks should be made.

1. In the foregoing we have separated conceptually the identification of FMUs and their unification in a Bauplan from the analysis of the histogenetic events occurring within these units. Note that the addition of a characteristic molecular profile to the topologic positional definition of the FMU serves to achieve proper selection of the structural units of the brain wall that should be compared. Bergquist and Källén (1954) and their contemporaries were not aware of the complexities of patterning and detailed genoarchitectonic regionalization, which are now known to us. This led to classic tentative comparison of excessively complex primary FMUs, which caused difficulties when searching for homology, due to the high diversity of differentiative phenomena among their constituent tertiary or final FMUs. Nowadays, we purposefully attend to positional topology of derivatives within the context of final FMUs whenever this is possible. As a result, the homologization procedure is presently less arbitrary, and can be guided by the relevant genoarchitectonic variables. Advances in genoarchitectonic knowledge therefore may allow the resolution of mystifying homology problems, insofar as the problems can be restated within a new, more precise FMU context. To give an example: In the 'detailed' structural pattern of the mammalian forebrain, shown in Plate 96, Fig. A, the subpallium was subdivided into four FMUs: POA, Dg, Pal, and St. However, Flames et al. (2007) reported that the mammalian subpallium is divisible into at least 18 domains that are uniquely defined by the combinatorial expression of several transcription factors, and there is no compelling evidence indicating that these 18 domains represent 'definitive' or 'final' FMUs. That is to say, studies of the expression patterns of 'new' genes may yield another, even more detailed parcellation of the subpallial FMUs.
2. Conceptually, the new molecular neuromorphology clearly transcends classic neuromorphology, because the 'markers' consistently used in this subdiscipline are directly related to causal developmental mechanisms and form part of the regulatory genomic network involved in the ontogenetic unfolding of the CNS. This implies that the domains and histological entities studied may be designated as *natural units,* conferring a *causal dimension* to the new neuromorphology.
3. In our conceptual synopsis (Plate 116, right panel), we mentioned twice that in the process of determining homologies, the *topological* rather than the *topographical* relations of structures should be taken into consideration. This is to reemphasize that in the study of homologies, and in the characterization of the positions and relations of neural structures in general, the intrinsic, natural coordinate system of the CNS, rather than an artificial, orthogonal Cartesian system, is normative (Chap. 7). This implies that in the new neuromorphology, the descriptors 'rostral' and 'caudal' relate to the course of the natural axis of the brain, which is approximated by the molecularly defined alar/basal boundary, as well as by the axial floor and roof plates (Chap. 5; Plate 93C). The neural segments or neuromeres are oriented perpendicular to this axis. In this neuromeric context, the descriptors 'ventral' and 'dorsal' mean: 'toward the floor plate' and 'toward the roof plate', respectively.

Back to the homologization of neural grisea!

After having made an inventory of the cellular formations present within corresponding FMUs of different species, it may be investigated whether certain members of these sets can be compared in a more specific way (one-to-one comparisons of layers and nuclei). For this procedure, we recommend the application of the three auxiliary criteria indicated in our synopsis.

10.6.4 The First Auxiliary Homology Criterion

Consideration of the secondary or definitive radial topological position of cellular formations (our auxiliary criterion 1), involved in Bergquist's and Källén's procedure the answering of questions two and three of their protocol (Plate 116, left panel), based on the analysis of routinely stained (hematoxilin, cresyl violet) histological material. It is at this point that the advantages of our new genoarchitectonic neuromorphology manifest themselves emphatically: gene expression patterns can not only be used to determine the boundaries of the FMUs, but also to 'dissect' the variant histological layering and aggregating events taking place within the confines of these units. Transcription factors and other relevant gene products involved in local differentiative developmental processes, can be used as 'markers' for tracing particular cell populations from their origin in the matrix zone to their ultimate topological position within their FMU. That this so-called *genoarchitectonic approach* (Puelles and Ferran 2012) may lead to spectacular results has been conclusively shown by the detailed studies of Ferran et al. (2007, 2008, 2009) and Morona et al. (2011) on the development of the pretectum in tetrapods. The results of these studies have featured as a model or paradigm throughout this course (see Plates 69, 70, 73, 88, 89 and 90).

While comparative studies performed so far have corroborated the large-scale conservation across vertebrates of the pattern of FMU-defining molecular boundaries, the analysis of specific neuronal types that can be clearly identified molecularly has shown that radial distribution of the FMU-derived cell types may differ considerably even between close vertebrate species, so that the old criteria of layer and sublayer localization, unsupported by evidence of molecular correspondence, can easily lead to errors. A typical case known already from pre-molecular times is represented by characteristic tectal neuron types of the midbrain, which can be recognized across species, but place their cell bodies in quite different layers (implying that the layers themselves are not homologous, whereas individual neuronal populations therein can be homologous). Moreover, we now know that in some species a FMU may become stratified in an outside-in pattern (early-born cells are found superficially), whereas in other species it may show an inside-out layering sequence (early born cells stay deep), like in the case of the telencephalic pallium across tetrapods. Comparison of radial histogenesis therefore strongly requires a genoarchitectonic approach to identify cell types illuminated by cell birthdate data (because sets of early-produced, intermediate and late-produced cells need to be analysed separately with regard to the relevant stratification patterns). As a result, variable radial positions of homologous cell populations can be expected (though obviously such a finding is not obligatory). Incidentally, when a molecularly comparable cell population of a given FMU occupies in two species the same layer or sublayer (e.g. Purkinje cells in the cerebellar cortex), this does not make them 'more homologous'; they are just more similar.

10 The development of grisea
10.6 The homologization of grisea

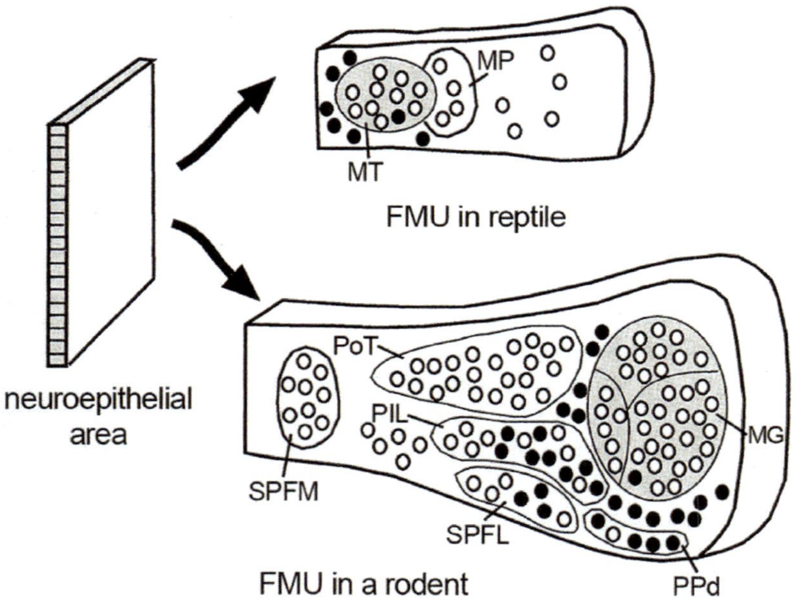

Fig. 1 Schematics showing cellular differentiation in corresponding FMUs of a reptile and a rodent.
Both FMUs are derived from the ventrolateral zone of the alar domain of prosomere 2 (see Plate 96).
Relevant abbreviations: MG, medial geniculate nucleus; MT, medial thalamic nucleus.

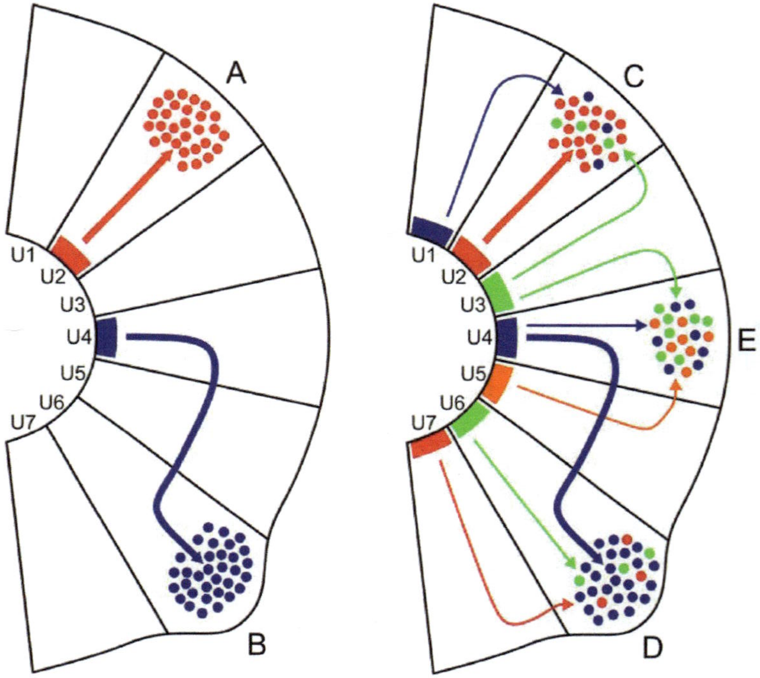

Fig. 2 Sets of seven FMUs (U1-U7) used as a framework for indicating the origin and cellular composition of various types of nuclei (A-E).
The procedure for the homologization of nuclei of these types, is discussed in the text.
The arrows indicate migration paths.

117

10.6.5 The Second and Third Auxiliary Homology Criteria

Plate 117

We believe that in the new neuromorphology there is also room for two additional auxiliary homology criteria, derived from classic neuromorphology, viz, similarity in fibre connections, and similarity in special quality. This may be demonstrated by the following example. Puelles and Medina (2002) compared the cellular differentiation of two corresponding FMUs of an adult reptile and an adult rodent (Plate 117, Fig. 1). The ventricular domains of these FMUs corresponded to the ventral zone of the alar domain of prosomere 2 (the thalamus) (Plate 95). Reference to Plate 117, Fig. 1 shows that this specific progenitor area in reptiles as well as in mammals gives rise to cell populations that migrate outward, where they are either grouped into distinct nuclei or remain dispersed. The number of cells produced in this FMU and the migration that these cells undergo are of larger magnitude in rodents than in reptiles. Experimental hodological and chemoarchitectonic studies revealed that the cell formations within these two equivalent FMUs share more than just their origin from ‘the same’ progenitor zone, indicating their overall or field homology. Thus, it appeared that the (periventricular) medial thalamic nucleus (MT) of reptiles, and the (superficial) medial geniculate nucleus (MG) in rodents receive auditory input from the reptilian torus semicircularis and its mammalian homologue, the inferior colliculus, respectively. It was also established that the neurons in the reptilan MT as well as those in the mammalian MG are glutamatergic. This means that although the reptilian MT and the mammalian MG differ in their definitive radial positions (auxiliary criterion 1), they may yet be presumed to be specifically homologous in light of auxiliary criteria 2 and 3. However, this judgment would not be valid if one of these nuclei did not develop within the same FMU, i.e. if the principal criterion would not be fullfilled.

In the same FMU of reptiles and rodents there is another example of a specific homology between cell populations formed there concerning neurons that express the neuropeptide calcitonin gene related peptide (CGRP), indicated with filled circles in Plate 117, Fig. 1. It will be seen that in both vertebrate groups, CGRP-positive cells are partly surrounding the MT/MG nuclei. In rodents, the CGRP-positive neurons are more numerous and more displaced superficially compared to the equivalent cells in reptiles, as is also true for the MG cells as compared to those in the MT. Other cell populations within these two equivalent FMUs may or may not be specifically comparable, and each case needs to be evaluated separately with relevant gene or differentiation markers. The existence side by side of three or more distinct populations within the same FMU is is probably explained by an as yet unknown subdivision of the region into smaller FMUs, each of which would produce only one cell type, with subsequent intercalation of these elements in the mantle layer.

It is important to explain why neuronal connections and chemoarchitectonic characters can be related to homology, though there are caveats when they are used on their own, leaving aside the principal criterion. This is because these properties result only partially from the molecular profile of the FMU, which controls the molecular identity of the derived cell types. As the neurons differentiate, part of their molecular profile is reflected in the adhesion proteins and other determinants (receptors, ion channels, ectoenzymes) that decorate the membrane of their axonal growth cones (see Chap. 11). These distal organs obviously interact variously with membrane-attached, or diffusible (attractant and repellent) molecules characteristically found in the diverse other FMUs that the axon crosses, until it finds where it can synapse (a thalamic axon targeting the cerebral cortex crosses at least 10 FMU boundaries along its full course). This implies that, as long as the genoarchitectonic Bauplan remains constant across all the relevant FMUs, the connections are cell-type characteristic and relatively constant as well. The caveat is that evolution introduces variations here and there, and at some point the navigational conditions may be altered, so that an old connection becomes eliminated, or is changed for a new one. The connectivity criterion is therefore essentially fallible in the long run, especially when it shows lack of similarity (the expected connection is not found), since homologous structures may have developed alternative connections. Nevertheless, in many cases hodological conservativeness still leads to corresponding connectivity of cell populations occupying similar primary topological positions, confirming their homology.

Special quality characters (neurotransmitters, neuropeptides, receptor channels, adhesivity proteins, etc.) have a similar rationale (when positive, they confirm homology, but when negative, they do not detract from a homology relationship based on the principal criterion).

10.6.6 The Problem of Tangential Migration of Grisea

So far, we have only discussed the homologization of cell formations located within equivalent FMUs (Plate 117, Fig. 2A). The preceding parts of this section have conclusively shown, however, that in many parts of the CNS cell contingents leave their 'parent' FMUs by tangential migration to colonize territories within foreign FMUs (Plate 117, Fig. 2B). This mode of development presents a problem with regard to the homologization of the resultant cell masses. A practical solution could be that our principal criterion: similarity in primary topological position and associated molecular profile, is fully maintained, meaning for instance in the case shown in Plate 117, Fig. 2B a clearcut origin from the progenitor zone of U4, and the three auxiliary criteria are declared applicable to the 'new' environment of the tangentially migrated cell mass. This would mean that the pontine nuclei of mammals can be designated homologous if it can be proven that they all stem from the r6–r7 sectors of the *Math1*-expressing A1 microzone of the rhombic lip, and that this general homology can be underpinned and specified on account of (1) their definitive postmigratory location in the basal parts of r3–r4; (2) the presence of afferents from the ipsilateral neocortex, and production of mossy fibre-efferents for the contralateral cerebellar cortex, and (3) the glutamatergic nature of their cells, apart from other features determined by the expression of additional gene markers.

10.6.7 The Problem of Multiple Origins of Grisea

Another complication, related to the establishment of the homology of cell masses is that in many final formations the constituent cells stem from several FMUs. However, we have seen that in the majority of these cases, one of the contributing cell contingents obviously features as the principal one. This holds true for, *inter alia,* the pyramidal cells in the neocortex (Plate 115, Fig. 3), the Purkinje cells in the cerebellar cortex (Plate 115, Fig. 2), the densely spiny projection neurons of the caudate-putamen complex (Plate 115, Fig. 1), as well as the somatomotor neurons produced by the spinal MN microzone (Plates 101, 102). We recommend that in all of these cases the homology of such composite cellular formations be related to the progenitor zone of its principal contribution (Plate 117, Fig. 2C), or be detailed separately for each component. This also holds true for tangentially migrated cell masses, which have received additional radial and/or tangential contributions from other sources (Plate 117, Fig. 2D).

In the foregoing text, we have presented a set of criteria for the homologization of neural grisea, and a commentary on the application of these criteria. We expect that the majority of cell formations in the CNS can be adequately homologized with the aid of these tools.

A small problematic group remains, formed by those grisea which are derived from several FMUs, without the possibility of designating one of its cellular contributions as the principal one (Plate 117, Fig. 2E). Those composite grisea may be characterized on the basis of a separate homology judgement on each cellular population with regard to *the set of progenitor zones* involved in their formation.

11 The "wiring" of the brain

11.1 Fibre systems

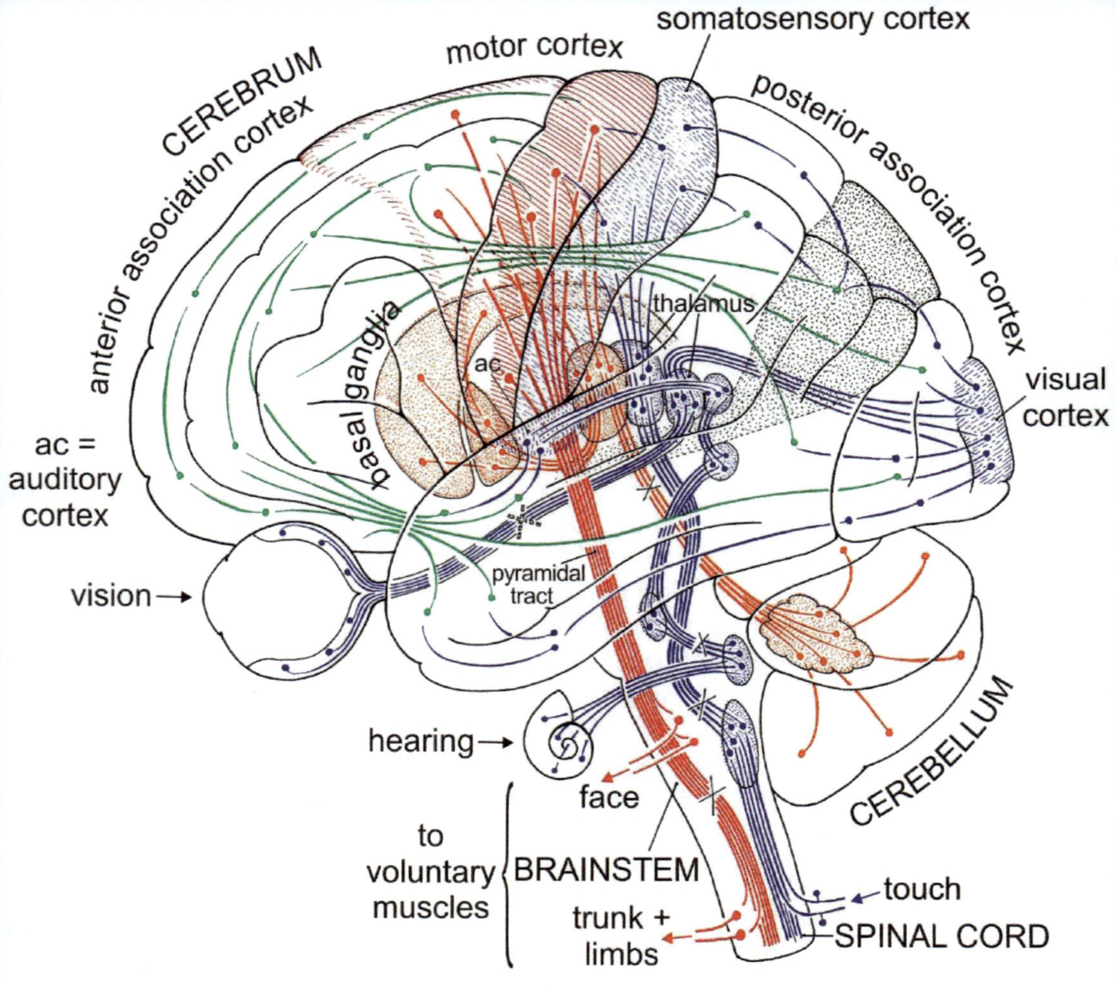

The "wiring" of the human brain (Nieuwenhuys 1992)

11 The 'Wiring' of the Brain

11.1 Fibre Systems

Plate 118

The centres or grisea in the brain discussed in the previous section are interconnected by fibres, which commonly assemble in fibre systems, bundles or tracts. These tracts constitute together the 'wiring' or 'macrocircuitry' of the CNS. (The connections between individual neurons *within* the various centres are denoted as 'microcircuitry'.) In his farewell lecture from the University of Nijmegen, now 24 years ago (!), R.N. made an attempt at summarizing the essence of the wiring of the human brain in a single picture (Plate 118). This picture shows, among other things, (1) that fibre systems conveying somatosensory (touch), auditory and visual information (all shown in blue) reach circumscribed areas of the cerebral cortex via one or several synaptic relays; (2) that a conspicuous fibre system known as the pyramidal tract (in red), which is concerned with the steering of voluntary movements, originates from the cerebral cortex, passes through the diencephalon and brainstem and descends, after having decussated largely to the contralateral side, into the spinal cord, and (3) that various cortical regions are interconnected by long association systems (in green).

R. Nieuwenhuys and L. Puelles, *Towards a New Neuromorphology*,
DOI 10.1007/978-3-319-25693-1_11

11 The "wiring" of the brain
11.1 Fibre systems

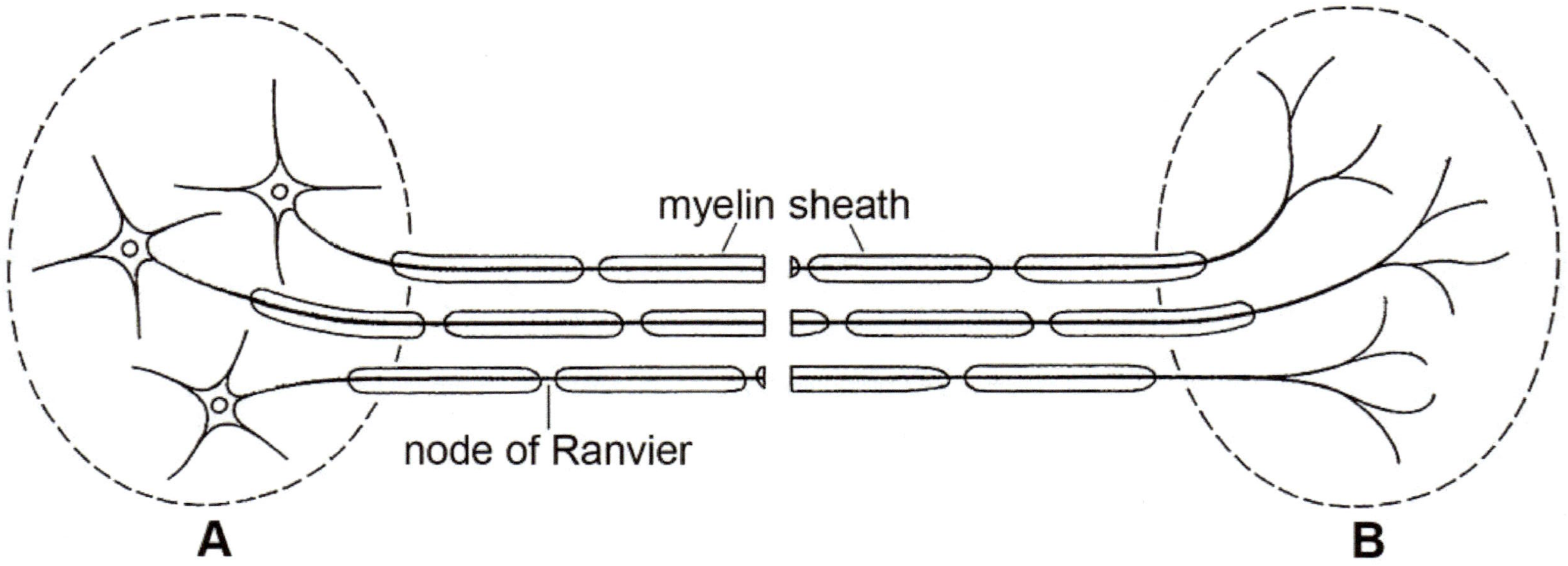

Simple fibre tract projecting from nucleus **A** to nucleus **B**

119

Plate 119

Before passing on, it should be noted that, although during the last 50 years an enormous body of experimentally obtained knowledge concerning the wiring of the CNS in almost all vertebrate groups has accumulated, the study of the structure of fibre systems *as such* is one of the most neglected fields in current neuroananatomy (see: Nieuwenhuys 1998 g).

In plate 119 a simple fibre tract connecting a given centre ‘A’ with another centre ‘B’ is depicted semidiagrammatically. The axons, which constitute this tract are surrounded by myelin sheaths. These are the products of oligodendrocytes, glia cells, which spiral their flattened processes around the axon. Adjacent glia processes leave small stretches of the axons free of myelin. These interruptions in the myelin sheaths, including the small naked axon segments, are known as the nodes of Ranvier. Myelinated fibres are confined to the nervous system of jawed vertebrates (gnathostomes). They are entirely lacking in invertebrates and cyclostomes (Werner 2013). Emergence of myelination and node formation represented important steps forward in evolutionary brain development, since they enable the axons to conduct impulses rapidly from node to node in a jumping or ‘saltatory’ fashion. This occurs quicker the larger the internodal length of the myelin sheath is (which in its turn is proportional to the fibre thickness. Unmyelinated fibres conduct impulses much more slowly and in a continuous manner.

Simple fibre systems, such as the one depicted in plate 119, do exist but are extremely rare. The most important ‘complications’ shown by fibre systems are the following:

‘Complication’ 1: Many fibre systems are composed of axons originating from more than one cell type and projecting to more than one area of termination. Thus, the primate pyramidal or corticospinal tract originates from the giant pyramidal neurons of Betz in area 4 of Brodmann, and from a much larger number of smaller pyramidal neurons in the motor, premotor and somatosensory cortical areas. The terminal ramifications of this tracts’ fibres distribute to the spinal laminae V, VI and VII of Rexed (see plate 102), ending mostly on interneurons. However, in animals that are able to execute independent finger movements, the corticospinal tract also projects directly to somatomotor neurons situated in Rexed’s lamina IX. In primates these direct cortico-motoneuronal connections are made up of the coarse axons of the giant Betz cells; hence with regard to origin, termination and function these axons constitute a recognizable subset (Kuypers 1987).

‘Complication’ 2: Many fibre systems contain fibres of opposite polarity. For example, the optic tract, which is composed mainly of centripetal fibres from different types of retinal ganglion cells, usually contains a number of centrifugal fibres targeting various retinal cell types.

‘Complication’ 3: The CNS of all vertebrates has some complex fibre assemblies composed of partially intermixed axons with very different origins and terminations. The reason that these heterogeneous fibre aggregations are treated as hodological units with a single name is simply that they appear, at least over a certain trajectory, as anatomically delimitable entities. The medial longitudinal fascicle and the medial forebrain bundle are examples of this case (Plate 120, Figs. 1, 2).

11 The "wiring" of the brain
11.1 Fibre systems

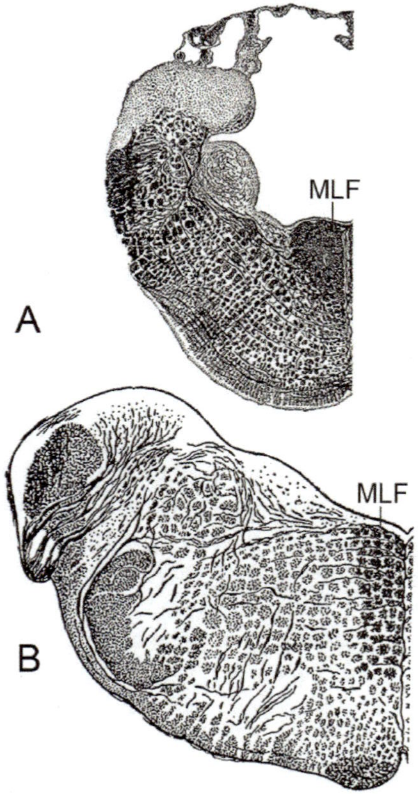

1. Transversal sections through the hindbrain of (A) the shark *Scyliorhinus caniculus*, and (B) the opossum *Didelphis virginiana*, showing the position of the medial longitudinal fasciculus (MLF)

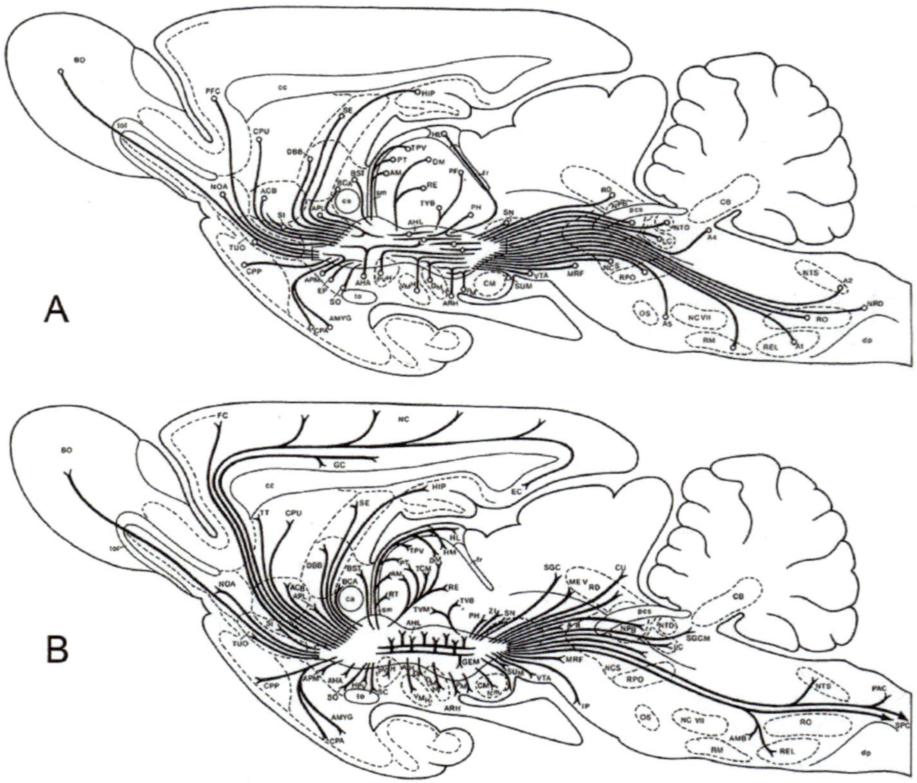

2. Diagrammatic representation of the input - (A) and output systems (B) of the medial forebrain bundle of the rat. (Modified from Nieuwenhuys et al. 1982)

120

Plate 120

The *medial longitudinal fascicle* (MLF) is a conspicuous bundle of coarse fibres which occupies a periventricular and paramedian position throughout the brainstem in all vertebrates (Plate 120, Fig. 1). Caudally, the bundle continues into the anterior funiculus of the spinal cord. Within the MLF, fibres originating from many different sources, including the interstitial nucleus of Cajal, the mesencephalic tectum, the medial reticular formation and the vestibular nuclear complex, descend to the spinal cord. However, the bundle also contains ascending fibres which connect the vestibular complex with the oculomotor nuclei. It is of note that each of the various components of the MLF occupies its own characteristic position within the bundle (Busch 1961). It thus essentially represents a conglomerate of tracts, rather than a single tract.

The *medial forebrain bundle* (MFB) is, just like the MLF, present in all vertebrates, but unlike the MLF it is mainly composed of thin, loosely arranged fibres. The MFB extends as a recognizable fibre system from the preoptc region through the peduncular hypothalamus to the rostral mesencephalic tegmentum. In mammals, the MFB presumably represents the most complex fibre system of the brain. In the rat, no less than 50 different ascending and descending components have been identified within its confines (Nieuwenhuys et al. 1982; Plate 120, Fig. 2).

11 The "wiring" of the brain

11.1 Fibre systems

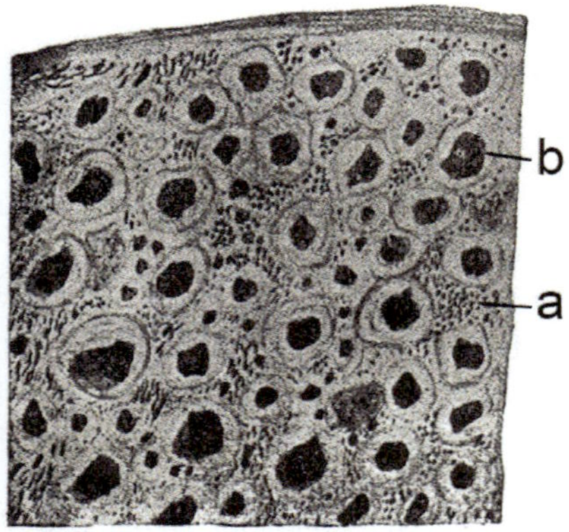

1. Transversal section through part of the lateral funiculus of the spinal cord of the rabbit, showing the presence of unmyelinated fibres (a), and myelinated fibres of different diameters (b). (From Ramón y Cajal 1909)

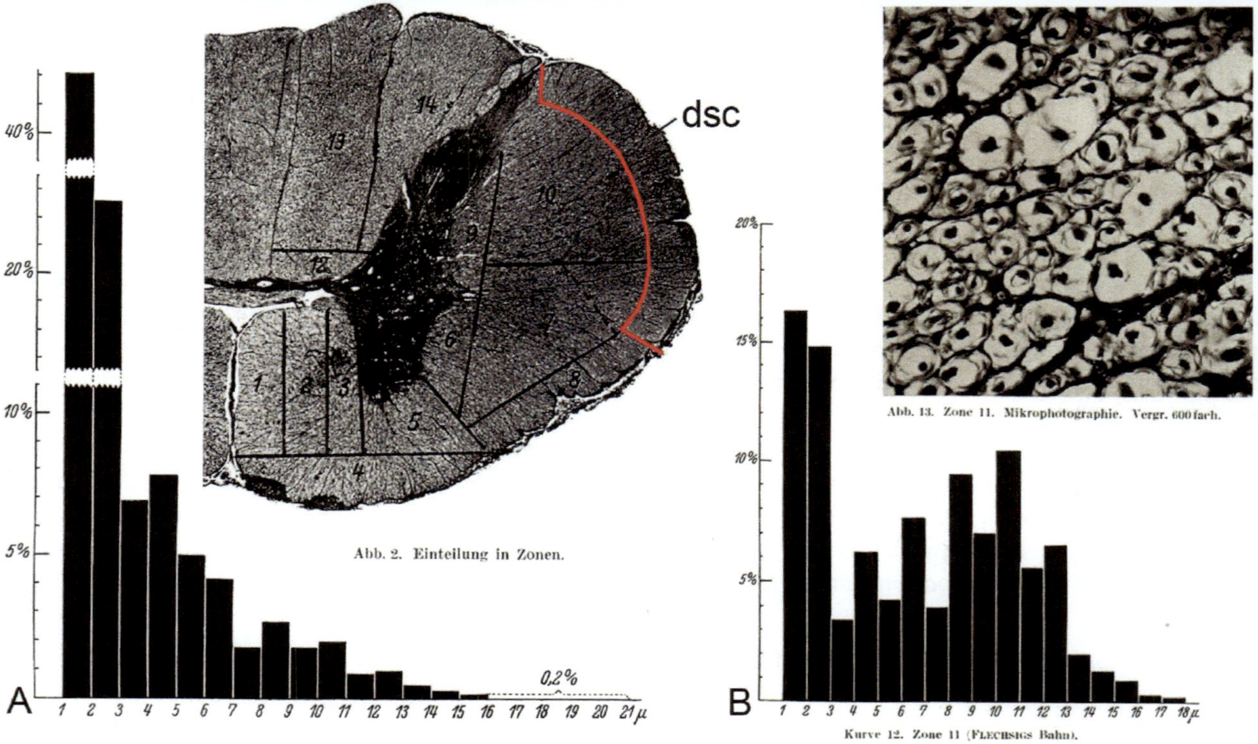

2. Fibre analysis of the third thoracic segment of the human spinal cord.
   A. Subdivision of white matter into 14 zones, and fibre spectrum of the entire section
   B. Fibre pattern and fibre spectrum of zone 11, i.e. the dorsal spinocerebellar tract, marked as dsc in A. (From Häggquist 1936)

121

Plate 121

'Complication' 4: Most fibre systems are composed of fibres with different calibres. It has long been known that the white matter of the vertebrate brain and spinal cord is composed of fibres of many diameters. Flechsig (1876) subdivided the myelinated fibres in the human spinal cord into four categories: coarse fibres, 10–15 μm in diameter, medium-sized fibres with a diameter of 7–9 μm, thin fibres measuring 5–6 μm in diameter, and very thin fibres with a diameter of 2–4 μm. Ramón y Cajal (1909) also emphasized the considerable variability in size of the fibres in the white matter. He noted that the spinal white matter contains large numbers of thin, unmyelinated axons among the myelinated fibres (Plate 121, Fig. 1).

The first investigator who subjected the fibres in the white matter to a systematic quantitative analysis was the Swedish histologist Häggquist (1936). He studied the third thoracic segment of the spinal cord of a 13-year-old girl, using a modification of Alzheimer-Mann's methyl blue-eosin staining technique, which was later named after him (Häggquist stain). The white matter of this segment was subdivided into 14 numbered areas, some of which roughly correspond to areas known to be occupied by specific fibre pathways, such as for instance area 10: lateral pyramidal tract, area 11: dorsal spinocerebellar tract, and area14: cuneate fasciculus. In each of these 14 areas, the diameters of 1200 fibres were measured; hence the total numbers of fibres measured amounted to about 17,000 (Plate 121, Fig. 2). Häggquist's (1936) main results were:

(a) Fibres of very different diameters are intermingled in all parts of the spinal white matter.
(b) The coarsest myelinated fibres measure 21 μm, the thinnest less than 1 μm in diameter.
(c) Each of the 14 areas of spinal white matter has its own characteristic fibre spectrum.
(d) Very fine myelinated fibres with a diameter of less than 3 μm constitute by far the largest fibre contingent in all parts of the spinal white matter.

Analyses of fibre systems like the one just discussed are extremely laborious and time consuming; hence most later investigators followed a more global approach, focusing on fibre patterns rather than on fibre spectra. The fibre pattern of a pathway is the aspect shown by it in silver-impregnated or Häggquist material resulting from its fibre composition. The fibre spectrum may be considered as a quantitative specification of the fibre pattern (*cf.* the upper and lower parts of Plate 121, Fig. 2).

The fibre patterns of numerous pathways in the brainstem and spinal cord of a variety of mammalian species, as observed in Häggquist material, have been documented by Verhaart and his collaborators (summarized in Verhaart 1970). These studies have shown that most pathways in the regions mentioned can be recognized by their characteristic fibre patterns, but their boundaries are not sharp, because the fibres of neighbouring pathways intermingle to a considerable extent.

The velocity of signal conduction by a myelinated fibre increases directly with its diameter, so that a doubling of the diameter gives a doubling of the transmission rate. Because of this close relation between fibre diameter and conduction velocity, the diversity of fibre diameters adds a temporal pattern to the spatial dimensions of the influences exerted by a given fibre system on its centre(s) of termination. Thus, the very coarse fibres present in the primate pyramidal tract constitute a fast subsystem which impinges directly on the motoneurons innervating the distal extremity muscles, while accompanying thinner fibres will transmit slower signals to the surrounding interneurons.

11 The "wiring" of the brain

11.1 Fibre systems

Transversal section through the medulla oblongata (A) and the midbrain (B) of the rat, to show the positions of the pyramidal tract (py) and the dorsal longitudinal fasciculus (dlf). C, D. Ultrastructure of py and dlf.
Note that in py groups of very thin, unmyelinated fibres are present (arrows), and that the dlf consists mainly of unmyelinated axons.

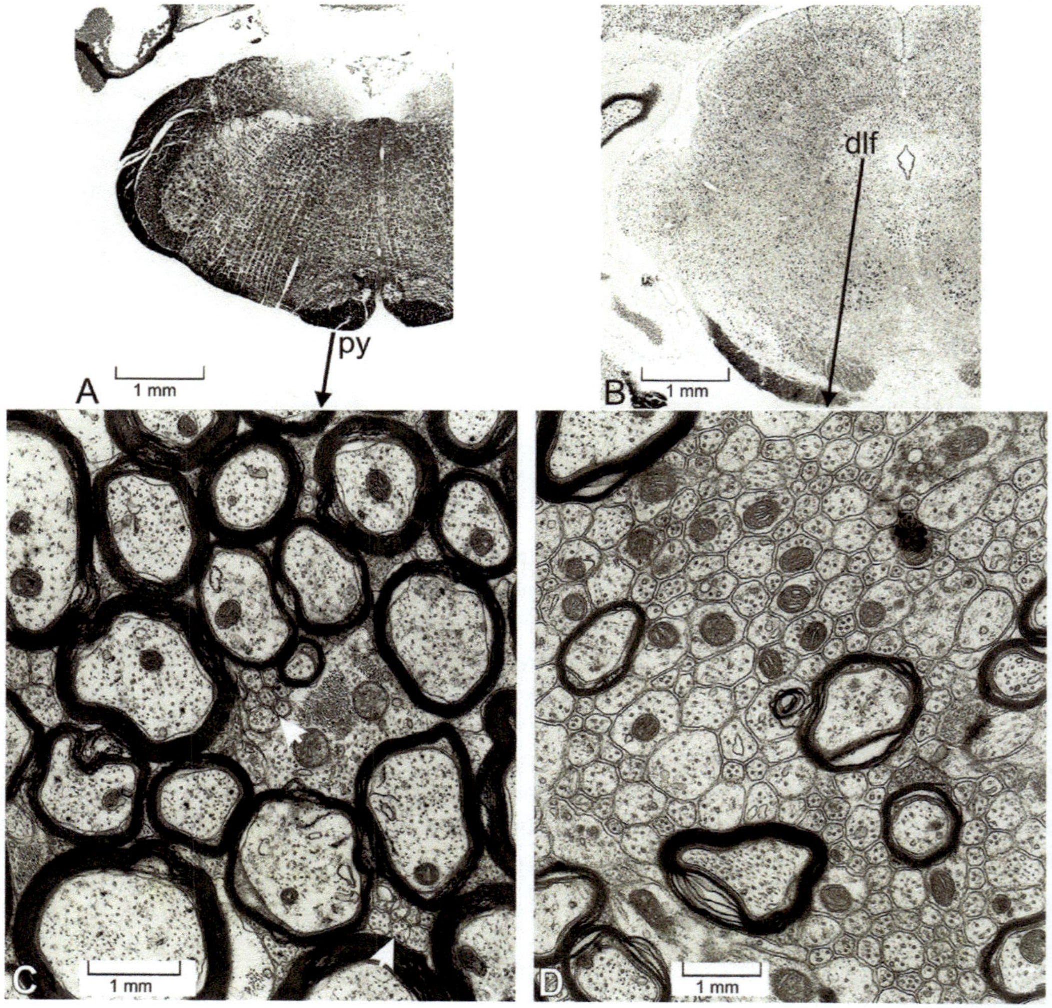

122

Plate 122

Ulirastructural studies have shown that many fibre systems in the mammalian CNS contain thin, unmyelinated fibres. Thus, Leenen et al. (1985) demonstrated that the pyramidal tract of the rat contains at lower rhombencephalic levels, no less than 140,000 ± 7000 unmyelinated fibres in addition to 103,000 ± 6000 myelinated ones. The diameter of these unmyelinated fibres ranged from 0.05 to 1.21 μm, with a mean of 0.18 ± 0.03 μm (Plate 122, Figs. A, C).

The medial forebrain bundle (MFB) and the dorsal longitudinal fasciculus of Schütz (DLF), two major conduits of the limbic system, consist largely of unmyelinated fibres. The DLF extends from the posterior part of the hypothalamus to the caudal rhombencephalon and occupies a periventricular position over its entire length. It is a composite fibre system which contains, in addition to long ascending and descending fibres directly connecting the hypothalamus with the solitary nucleus and related autonomic centres, numerous shorter links. The DLF consists of densely packed unmyelinated fibres of varying diameters, among which thin myelinated fibres are dispersed (Buma et al. 1992; Plate 122, Figs. B, D). The thin unmyelinated fibres in the MFB and the DLF (and in other limbic pathways as well) often have small varicosities. These varicosities may occur either over certain trajectories or throughout their course. Histochemical and immunohistohemical studies have shown that monoaminergic and peptidergic cell groups form prominent sources of these varicose fibres. The varicosities of the monoaminergic fibres contain small, clear vesicles, whereas those of the peptidergic fibres are characterized by the presence of large dense-core vesicles. The varicosities in both types of fibres are often involved in non-synaptic or volume neurotransmission (see below). It will be clear that whereas in the 'closed-line' axons of the fibre system depicted in Plate 119, the transfer of information is confined to their terminals, the 'open-line' monoaminergic and peptidergic fibres just discussed are potentially able to influence neurons or neuron groups throughout their extent by extrasynaptic leakage of transmitter into the extracellular space.

11 The "wiring" of the brain
11.1 Fibre systems

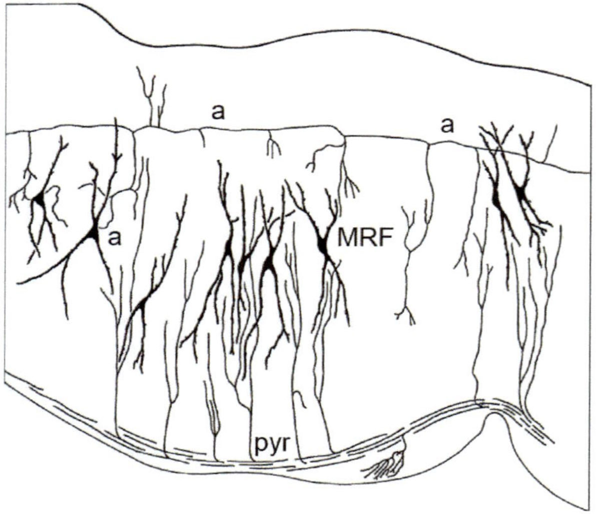

1. Sagittal section through the rhombencephalon of a 10-day-old rat, showing collaterals from the pyramidal tract (pyr) and from a single reticular cell axon targeting the medial reticular formation (MRF).
   Golgi technique.
   (Redrawn from Scheibel and Scheibel 1958).

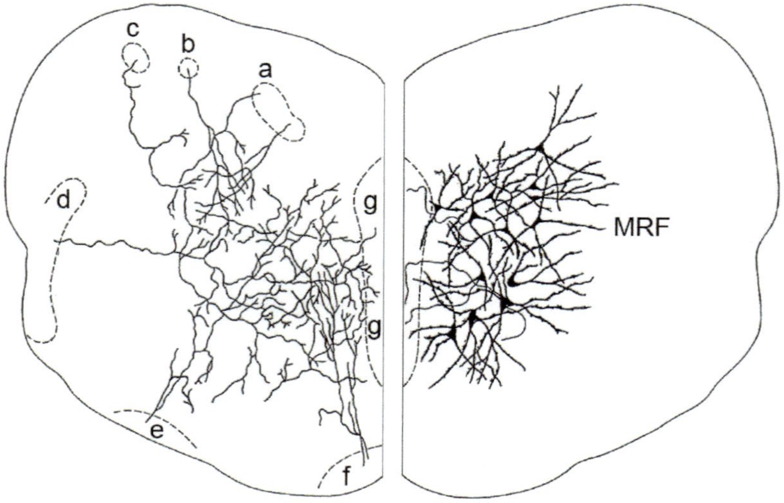

2. Transversal section through the medulla oblongata of a 0-day-old kitten, which illustrates overlapping afferents from various sources in the neuropil of the medial reticular formation (MRF). The afferents include collaterals from:
   (a) dorsal longitudinal fasciculus,
   (b) lateral vestibulospinal tract,
   (c) fasciculus uncinatus,
   (d) tractus descendens of V,
   (e) anterolateral fasciculus,
   (f) pyramidal tract, and
   (g) medial reticulospinal tract.
   Golgi technique.
   (Redrawn from Scheibel and Scheibel 1958)

123

Plate 123

'Complication' 5: The constituent fibres of many tracts emit collateral branches. This phenomenon may be exemplified with some results of Golgi-studies on early postnatal mammals by Scheibel and Scheibel (1958). These authors observed that the fibres of the pyramidal tract emit numerous collateral branches during their course through the rhombencephalon. These collaterals, which are oriented perpendicularly to the long axis of the brain, pass dorsally and enter into contact with the similarly oriented dendrites of the large neurons of the medial reticular formation (Plate 123, Fig. 1). Note that such very common periodic issuing of collaterals from a longitudinal fibre is reminiscent of a segmental pattern (each rhombomere receiving its own branch). Transverse sections revealed that, in addition to pyramidal branches, collaterals from many other fibre systems similarly converge upon the medial reticular formation (Plate 123, Fig. 2). Analogous interaction of transversely oriented dendritic fields and axon collaterals from passing fibres have been observed in the hypothalamus and in the spinal intermediate grey (Ramón y Cajal 1909; Scheibel and Scheibel 1968; Millhouse 1979; Szentagothai 1983). It will be clear that collateralized fibres, just like the varicose 'open-line' axons discussed above, are able to make contacts (synaptic ones in this case) with several grisea, rather than with a single one at their termination.

We will conclude this section on fibre systems with a comparison of the organization of two hodological extremes, viz. the simple, closed system with which we commenced our discussion, and a very complex and open system, the medial forebrain bundle. This comparison will be prefaced by a brief excursion to neurotransmission.

Excursion 7: Chemical neurotransmission

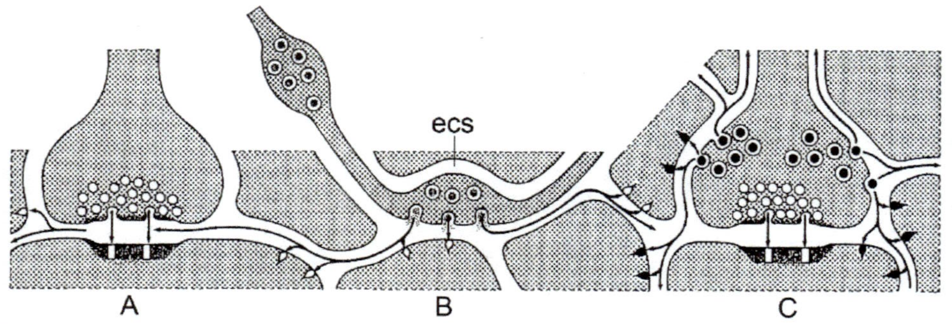

1. Synaptic and non-synaptic neurotransmission. A. Classical synapse in which a presynaptic element influences a single postsynaptic element. B. An axonal varicosity which, by means of non-synaptic secretion of a neuromediator, influences several other neuronal elements. C. A synaptic terminal containing small, clear vesicles as well as large dense-core vesicles. The small vesicles release their content in the synaptic cleft, whereas the content of the dense-core vesicles is released at non-synaptic sites of the terminal membrane in the extracellular space (ecs). The *arrows,* which indicate the flow of the neuromediator molecules, point to synaptic as well as non-synaptic membrane receptors (After Nieuwenhuys 2000).

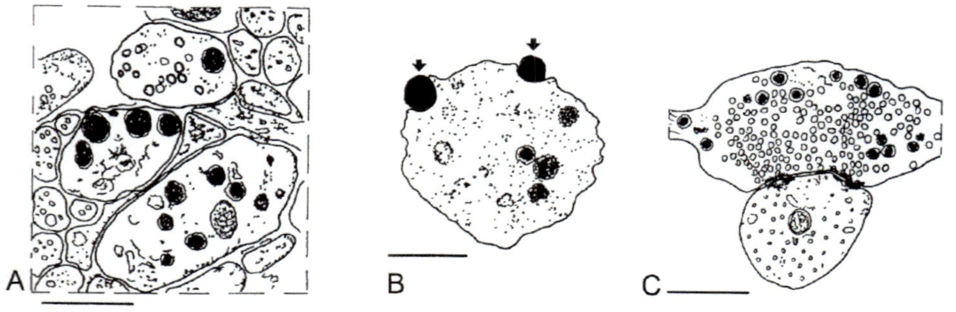

2. Ultrastructural evidence for non-synaptic neurotransmission in the CNS of the rat. A. Axonal varicosities with large dense-core vesicles (LDCVs) in the periaqueductal grey. B. Exocytosis of the content of LDCVs from an axonal varicosity in the median eminence; TARI technique. C. Axonal varicosity in the periaqueductal grey, making synaptic contacts with a dendrite and containing LDCVs at a distance from the active sites. Scale bars: 0.5 um. (After Nieuwenhuys 2000).

124

Excursion 7: Neurotransmission

Plate 124

Axons and their collaterals usually form telodendra, the branches of which show numerous terminal swellings, commonly designated as *boutons terminaux (plural for bouton terminal; French)*. These boutons form functional contacts or *synapses* with the dendrites or somata (eventually also axonal terminals) of other neurons. The term synapse was coined by Sherrington in a textbook of physiology (Foster and Sherrington 1897). It was long thought that the transfer of information in the nervous system is accomplished by the direct flow of electrical current from one neuronal element to another. However, it is now well established that this electric charge transfer is rare (electric synapses), whereas in most cases the crucial signalling is mediated by chemical substances called *neurotransmitters*, and that two different types of chemical neurotransmission, *synaptic* and *non-synaptic*, can be distinguished.

A classical chemical synapse (Plate 124, Fig. 1 A) is ultrastructurally characterized by (1) a presynaptically located cluster of synaptic vesicles, which contain the neurotransmitter, (2) electron-dense (proteic) appositions to the cytoplasmic side of the pre- and postsynaptic membranes, and (3) a widened synaptic cleft. During synaptic transmission the vesicles fuse with the plasma membrane and release their content of transmitter molecules into the synaptic cleft. These molecules diffuse to the postsynaptic membrane where they bind to and activate specific receptor complexes. The release of transmitter is brought about by depolarization of the presynaptic terminal. The consequences of post-synaptic receptor activation by the transmitter may be either depolarization (leading to excitation of the target cell) or hyperpolarization (leading to inhibition of the target).

In non-synaptic chemical neurotransmission, vesicles containing a neurotransmitter or another neuroactive substance (such as a neuropeptide) fuse with non-specalized parts of the neurolemma (cell membrane) and release their content by exocytosis in the extracellular space. The latter constitutes a continuous system of narrow aqueous voids, which extends throughout the CNS, separating the cellular elements and occupying about 20 % of the volume of the brain (Nicholson 1999). Following the non-synaptic release of a neuroactive substance, the molecules diffuse through the extracellular space and reach specific receptors in a much wider volume than release into a specialized synaptic cleft would allow (Plate 124, Fig. 1 B). It has already been mentioned that (1) monoaminergic and pertidergic neurons are generally provided with thin varicose axons; (2) the axonal varicosities of monoaminergic fibres contain accumulations of small, clear vesicles, whereas those of peptidergic fibres are characterized by the presence of large dense-core vesicles, and that (3) both types of varicosities have been demonstrated to be involved in non-synaptic neurotransmission (Buma 1989; Nieuwenhuys 2000). [This is not to say that monoaminergic and peptidergic neurons are only and exclusively involved in non-synaptic neurotransmission. The axons of both types of neurons may show fully differentiated synaptic varicosities and terminals.]

Plate 124, Fig. 2 A shows some typical peptidergic varicosities in the periaqueductal grey of the rat, and Fig. 2 B depicts the non-synaptic release of the neuropeptide vasopressin (arrows) from a varicosity in the median eminence, caught by the tannic acid Ringer incubation (TARI) technique. By this method exteriorized peptides from secretory vesicles are fixed by weak tannic acid and thus prevented from diffusing away in the extracellular space (Buma et al. 1984). Electrophysiological studies have shown that the influence of neuropeptides on their target neurons is generally slower in onset and of longer duration than the synaptic effects of classical neurotransmitters such as acetylcholine, glutamate, glycine or GABA.

The mammalian CNS contains more than 30 different neuropeptides all of which having their own specific receptor complexes (these started to evolve in evertebrates).

In the mammalian CNS a complex of centres and pathways, which may be characterized as a caudally extended limbic system, shows an extraordinary richness and density of neuroactive substances, particularly neuropeptides (Nieuwenhuys 1985, 1996, 2000; Nieuwenhuys et al. 1989). Given the fact that the pathways interconnecting and passing through the various centres of this complex consist largely of thin, unmyelinated fibres, we consider it likely that non-synaptic interneuronal communication plays an important role in this system.

Many neurons contain more than one neuroactive substance (Hökfelt et al. 1986; Lundberg and Hökfelt 1983). Two or more neuropeptides may coexist within the same neuron, and the combination of a neuropeptide with one of the monoamines is also frequently encountered. Accordingly, synaptic terminals containing small, clear as well as large dense-core vesicles have been observed (Zhu et al. 1986; Buma 1989). It is a constant finding that in terminals of this type the small, clear vesicles are concentrated in the vicinity of the synaptic contact zone whereas the dense-core vesicles are located at some distance from this zone. This suggests that these terminals release their classical neurotransmitter synaptically, but their neuropeptide(s) non-synaptically via non-specialized parts of their axolemma (Plate 124, Fig. 1 C). An axonal varicosity making synaptic contacts with a dendrite, containing dense-core vesicles in typical 'stand-back' position is shown in Plate 124, Fig. 2 C.

11 The "wiring" of the brain
11.1 Fibre systems

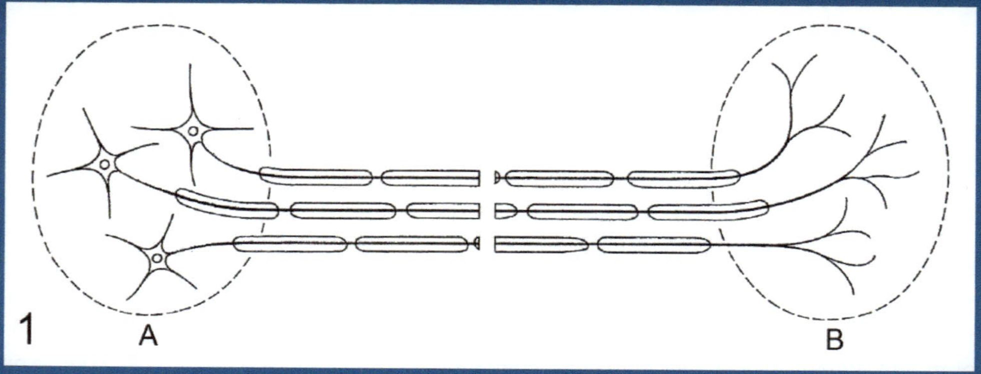

1. Simple fibre tract projecting from nucleus A to nucleus B

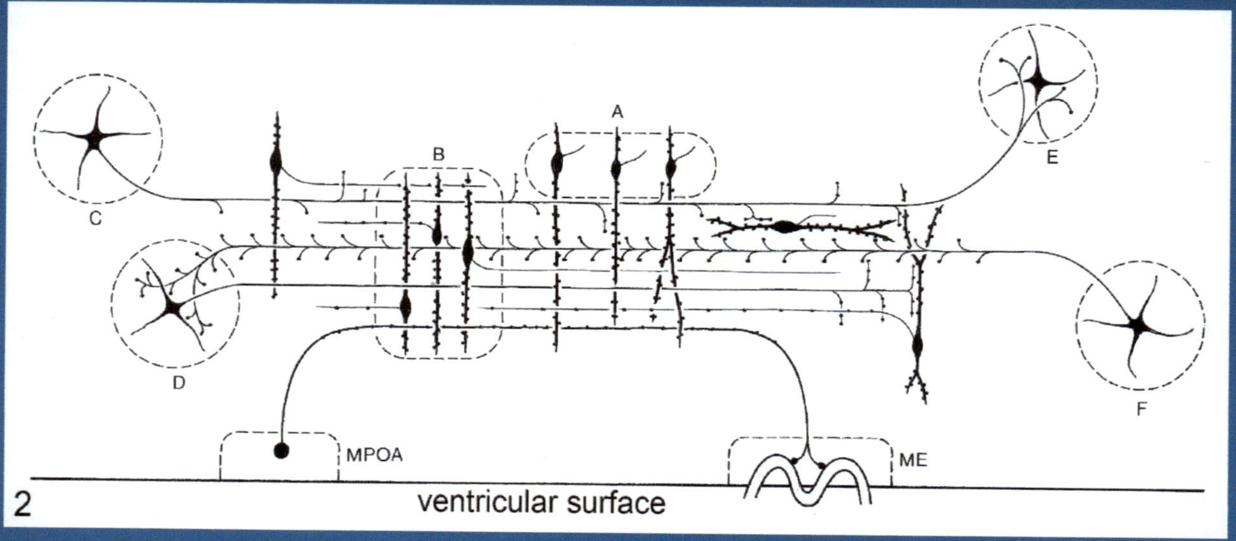

2. Diagrammatic representation of the preoptico-hypothalamic trajectory of the medial forebrain bundle. A-F, Various centres which contribute fibres to the bundle; ME, median eminence; MPOA, medial preoptic area. (After Nieuwenhuys 1998g).

125

Plate 125

Back to our two hodological models, the simple tract shown in Plate 125, Fig. 1, and the complex fibre system shown in Fig. 2.

In classic neurobiology, the overall functional significance of neural centres and fibre systems is crystal clear: the centres process and integrate information, whereas the fibre systems transfer information from centre to centre. The hypothetical fibre system depicted in Fig. 1 perfectly fits in this division of roles. We attribute the following properties to this tract:

1. It originates from a single centre
2. Its fibres emanate from cells of a single type.
3. Its fibres do not branch and do not emit any collaterals.
4. All of its fibres are myelinated.
5. All of its fibres are of the same calibre.
6. All of its fibres course in the same direction.
7. All of its fibres follow the same trajectory throughout.
8. All of its fibres are ‘closed’, i.e. they fransfer information exclusively at their distal ends.
9. All of its fibres terminate in a single centre.
10. The terminal ramifications of all of its fibres are of the same type.
11. The terminal ramifications of all of its fibres make typical synaptic contacts with neurons in its target centre.
12. The synaptic contacts of all of its fibres use the same neuroactive principle.

The data presented earlier in this section have made clear that many fibre systems do not exhibit all of the properties just enumerated. Many fibre systems originate from more than one centre; the fibres of many tracts emanate from several different cell types; the fibres of many tracts do emit collateral branches, and so on.

The complex fibre system portrayed in Plate 125, Fig. 2, which reflects the organization of the mammalan medial forebrain bundle (MFB), is remarkable in that it exhibits *none* of the 12 features attributed to the simple fibre system discussed above. Ultrastructural studies (Sipe and Mooore 1977; Nieuwenhuys et al. 1978) have shown that the MFB of the rat contains small unmyelinated as well as myelinated fibres. The abundant unmyelinated fibres measure 0.1–0.4 μm in diameter. The myelinated fibres, with diameters between 0.5 and 8 μm, are grouped in small bundles. The MFB is known to be composed of numerous (about 50) long and short ascending and descending components (*cf.* Plate 120, Fig. 2), only a few of which have been included in our diagram. Each of these components is presumably characterized by its own neuromediator profile and many of them may contain one or several neuropeptides. [Data collected almost 30 years ago from the literature revealed that, up to that time, at least 15 different neuromediators, among which 10 different neuroeptides, had been localized in MFB fibres (Nieuwenhuys 1985)]. The MFB is surrounded by grisea composed of a diversity of neurons, which extend their dendrites mainly into its trajectory ('fishing out' signals), and some neuronal groups are embedded in the bundle. Some of the components of the MFB project from one centre to another; however, the fibres forming these components have the potential for influencing many other centres adjacent to, or within the bundle. The dendritic trees of the cells in these centres are oriented perpendicular to the bundle and so are optimally placed to sample information from it. Interestingly, several sets of peptidergic fibres originating from the mdial preoptic area pass lateraly, join the MFB, and pass medially again, to terminate in the median eminence, *i.e.* the neuro-vascular interface between hypothalamus and pituitary. It is likely that this peculiar detour enables these fibres to make their specific contribution to the complex multipeptidergic language by which the MFB communicates with the cell groups lying within and around it.

With regard to communication with their surrounding structures, our two hodological models depicted in Plate 125, Figs. 1, 2, represent the extremes of a spectrum, ranging from entirely closed to entirely open fibre systems. Entirely closed systems, answering to all of the 12 criteria enumerated above, do exist but are presumably very rare.

The CNS of anamniote gnathostomes contains numerous pathways consisting mainly of unmyelinated fibres, while in cyclostomes myelinated fibres are entirely lacking. It is generally assumed (e.g. Herrick 1948; Heier 1948) that unmyelinated pathways are in functional contact with all of the grisea that they pass through, via contacts established either by their main axons or their collateral branches; hence the most primitive status represented by open systems.

In the mammalian brain there is a considerable structural and functional difference between the, globally laterally placed cognito-motor domains and the, globally medial greater limbic domains (see for details: Nieuwenhuys 1985, 1996, 2000; Nieuwenhuys et al. 1989). The 'classical' sensory and motor systems within the lateral domains are composed mainly of discrete centres that are interconnected by well-myelinated through-conducting, compact fibre bundles. By contrast, in the medial greater limbic domains, which are involved most directly in behavioural patterns aimed at the survival of the individual organism and of the species, loosely arranged pathways consisting largely of thin, unmyelinated, varicose fibres prevail. The wealth of neuropeptides, present in the fibres forming these pathways suggests that non-synaptic neurotransmission plays a prominent role in these domains.

11 The "wiring" of the brain
11.2 The development of fibre systems

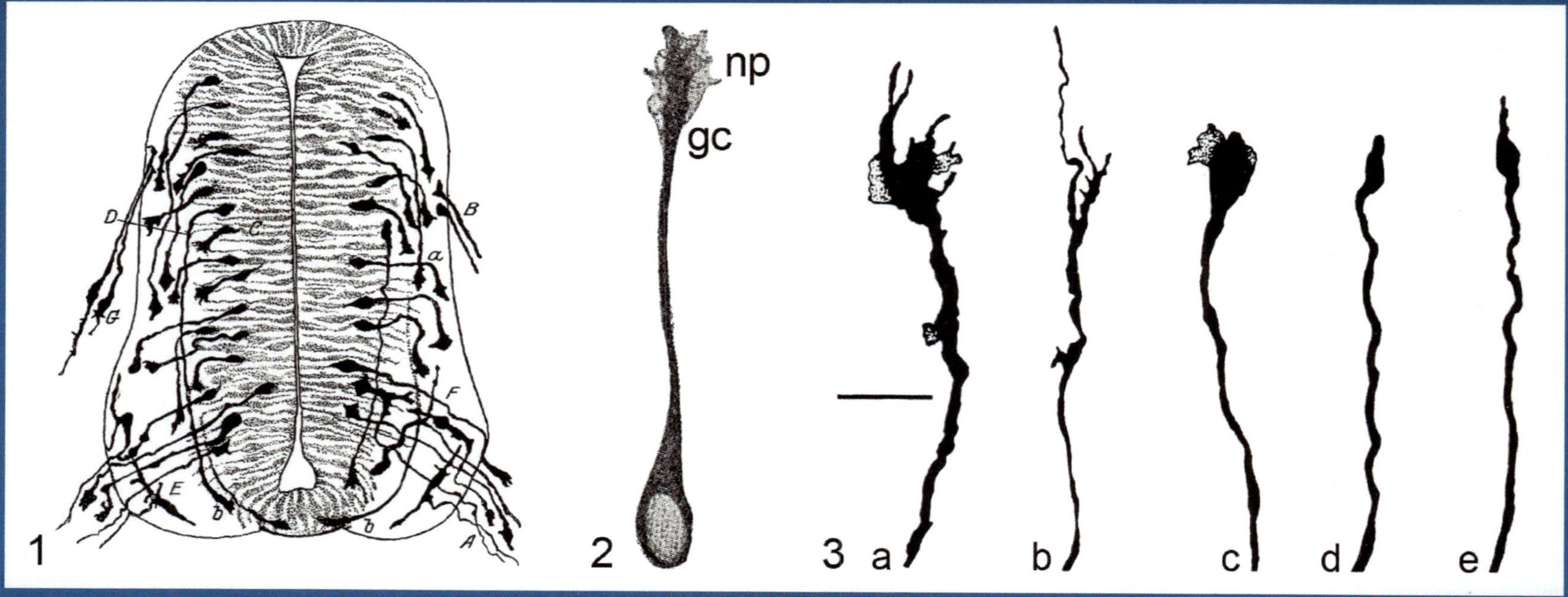

1. Transversal section through the spinal cord of a chick embryo (3 day of incubation) showing numerous neuroblasts and their primordial axons.

2. Neuroblast, the growth cone (gc) of which shows typical neuroplasmatic extensions (np).
(1 and 2 after Ramon y Cajal 1909)

3. Growth cone configurations of embryonic and early larval spinal cords of *Xenopus.*
Examples of five different characteristic shapes (a-e). Scale bar: 10 µm. (After Nordlander 1987)

126

11.2 Development of Fibre Systems

Plate 126

During early development the primordial axons of differentiating neuroblasts pass to the marginal zone of the neural tube, where they assemble in distinct bundles packaged between the peripheral parts of the processes of radial glia cells. During further development, these superficial bundles increase in size and finally merge to become the white matter of the CNS (Keveter and Lasek 1982; Nordlander 1987).

Plate 126, Fig. 1 shows numerous young neuroblasts in a transverse section through the spinal cord of a 4-day-old chick embryo. The primordial axons of some of these elements pass simply peripheralward to the marginal zone, other axons follow a tangential course over some distance, and still others courses all the way down to the floor plate, These so-called arcuate fibres (marked D) are known to decussate to the contralateral side. Figure 1 shows, in addition, that the axons that originate from developing spinal ganglion cells (G) enter the spinal cord, forming bilateral dorsal roots (B), and that the primordial axons of developing motoneurons (E) pass through the external limiting membrane of the cord to assemble in ventral roots (A).

It is of note that the primordial axons of all of the neuroblasts, which are depicted in their entirety in Fig. 1, end in bulbous or irregular expansions, the size of which often exceeds that of their respective parent somata. These structures were first described by Ramón y Cajal (1890) who designated them as:"cono de crecimiento de los axones primordiales", *i.e.* growth cones of the primordial axons. Ramón y Cajal (1909) mentions that these *organs* (his terminology) are composed of two elements, (1) a centrally placed bundle of neurofibrils, and (2) a surrounding neuroplasmatic substance giving rise to appendices and excrescences of variable size and shape (Fig. 2; modernly we speak of *filopodia* and *foliopodia*; filopodia are thin and elongated; foliopodia are ampler, flat, and membranous)). Cajals ideas concerning the putative functions of the axonal growth cones are reflected in the following quotation, derived from Swanson and Swanson's (1995) translation of the French edition (Ramón y Cajal 1909, 1911) of his great work on the Histology of the Nervous System of Vertebrates: "From the functional point of view, we may say that the growth cone is like a club or battering ram endowed with exquisite chemical sensitivity, rapid ameboid movements, and a certain motive force allowing it to circumvent obstacles in its path, thus coursing between various cells until reaching its destination."

Numerous later studies have fully confirmed the observations and speculations of Cajal. It is now well established that in both vertebrates and invertebrates the axons of developing neurons are provided with growth cones and that these structures play a pivotal role in pathfinding and navigating of axons through the developing brain.

Ross Harrison (1910) confirmed with his classic studies on axon outgrowth in tissue cultures that growth cones are active, motile and highly adaptable structures.

Ultrastructural studies, summarized by Strochlic et al. (2007), have shown that the central domain of growth cones contains bundles of microtubules whereas the peripheral domain is rich in actin filaments. The actin filaments may either assemble in small bundles, which form the centres of finger-like processes, known as *filopodia*, or they form networks in thin veil-like extensions, termed *foliopodia* or *lamellipodia*. Where a growth cone has several filopodia, the lamellipodia may form web-like connections between these processes. The tips of the actin-rich filopodia apparently adhere at aleatory points of the surrounding substrate, using a variety of adhesivity components present at the cell memdrane. Thereafter the actin filaments of all the filpodia contract, generating traction on the cone's central domain in mutually competing directions. Eventually one filopodium wins and the cone advances one step in the corresponding direction. New filopodia are next emitted, and so on. Apart of variable adhesion, other effects encountered may be attraction/repellence by diffused molecules, or facilitation/permission of growth, versus inhibition of growth, exerted by ambiental chemoarchitectonic conditions (intercellular matrix, neuronal or glial surfaces, and even bloodvessel surfaces).

The astonishing pluriformity of growth cones is well illustrated by the results of a study of Nordlander (1987). She found that in the spinal cord of embryonc and early larval clawed toads (*Xenopus laevis*) these structures fall into five general categories. Figure 3 shows representative examples of each of these, which include: (a) complex with wide-ranging filopodia and distinct lamellipodia; (b) slender with several long filopodia but no lamellipodia; (c) bulbous with no filpodia and ruffled lamellipodia; (d) clavate or club-shaped, and (e) fusiform.

As regards the formation of fibre systems, in general small numbers of precociously differentiating neurons provide the axons, which pioneer the various pathways. The axonal ends of these cells project a growth cone, which navigates through the developing brain by following a stereotyped and precise route. These primary or pioneer neurons are believed to be responsible for establishing the first functional circuits in the CNS. Other axons, developing from vastly more numerous, later-maturing neurons, join the pioneer axon(s) and, when the wiring is accomplished, the pioneer neurons undergo cell death, leading to degeneration of their axons (Wilson et al. 1990; Knyihar-Csillik et al. 1995). It has been established that the axon growth cones of pioneer fibres are generally larger and more complex than those of 'follower' axons (Reh and Constantine-Paton 1985; Nordlander 1987; LaMantia and Rakic 1990; Williams et al. 1991; Kim et al. 1991).

11 The "wiring" of the brain
11.2 The development of fibre systems

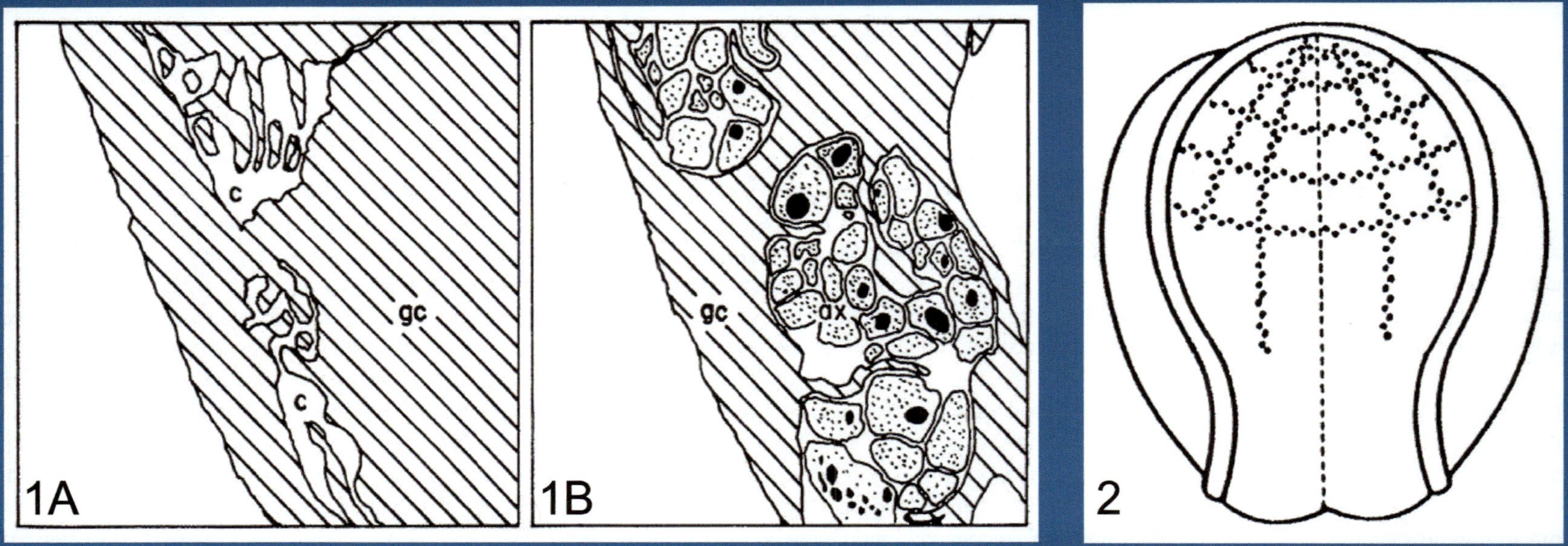

1. Drawings of transversal sections through regenerating newt spinal cords showing intercellular channels (c) between radial processes of glial cells (gc) before axons have reached them (A) and after axons are growing within them (B).
(Based on data of Nordlander in Katz et al. 1980)

2. An approximation of the pattern of the substrate pathway precursor fields projected upon the neural plate of a salamander. (After Lasek et al. 1980)

127

Plate 127

Singer and his associates discovered that in the marginal zone of the spinal cord of the newt *Triturus viridescens* and the clawed frog *Xenopus laevis* (Singer et al. 1979), prior to the appearance of axons, longitudinally oriented, tunnel-like spaces appear between the peripheral processes of the radial glial cells, and that these channels are subsequently invaded by growing axons. Similar channels were observed in the spinal cord in regenerating newts (Nordlander and Singer 1978; Plate 127, Fig. 1) and lizard tails (Egar and Singer 1972). Comparable phenomena have been observed during normal development in the spinal cord of the rhesus monkey (Knyihar-Csillik et al. 1995) and in the optic nerve of the mouse (Silver and Robb 1979).

Singer et al. (1979) proposed that the preformed intercellular channels guide growing axons during development and that the processes of the primitive glia cells contain a developmental genetic *'blueprint'* of the primary nervous pathways.

Katz and Lasek (1978, 1979, 1981) found that in *Xenopus* tadpoles optic nerve and Mauthner cell axons transplanted to unnatural places in the CNS behave as if they are following discrete, reproducible pathways in the substrate. On the basis of these experiments, Katz et al. (1980) and Katz and Lasek (1981) postulated the presence of *substrate pathways* in the developing CNS, which they defined as:"sets of similar guidance cues, aligned in continuous discrete routes" (Katz et al. 1980, p. 825). Their experiments suggested the presence of two stereotyped longitudinal routes, viz. a ventral substrate pathway extending in the ventral marginal zone from the caudal diencephalon to the spinal cord, running the length of the basal plate, and hence called the *basal substrate pathway,* and a dorsal, *alar substrate pathway*, extending in the alar plate of the developing hindbrain and spinal cord.

In order to obtain some idea of how the pattern of substrate pathways is foreshadowed in the neural plate stage, Katz et al. (1980) sketched the routes of a number of major tracts, such as the radix descendens of V, the anterior commissure, and the fasciculus retroflexus (habenulo-interpeduncular tract), on Jacobson's (1959) neural plate fate map of the axolotl *Ambystoma mexicanum*. The resultant simple pattern of 'meridians' and 'parallels' is shown in Plate 127, Fig. 2. They proposed that this pattern is topologically preserved as the neural plate is transformed into the neural tube, and that it manifests itself then as the three-dimensional organization of the early marginal zone with its channels of extracellular space lined by the peripheral processes of radial glial elements.

The preceding paragraphs, dealing with the 'blueprint' theory of Singer et al. (1979) and the related 'substrate pathway' theory of Katz et al. (1980), are largely derived from an earlier treatise (Nieuwenhuys 1998b). We consider it important to reemphasize these two theories here, because they are in our opinion (1) unjustly almost completely forgotten and neglected by now, and (2) fully compatible with the results of current molecular studies on axon pathfinding, to be dicussed below.

11 The "wiring" of the brain

11.3 Molecular studies on the development of fibre systems

1

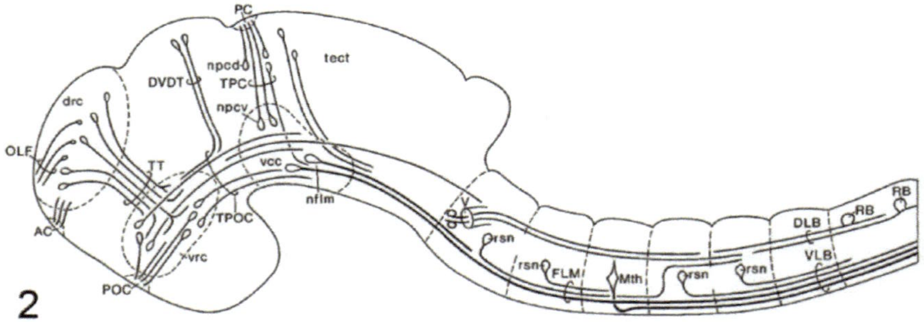

2

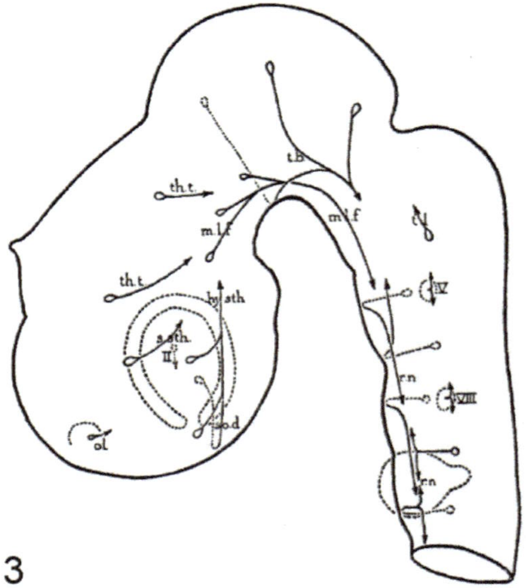

3

1. Developing axon end with long, exploring filopodia (After Weiss 1955)
The original caption of this figure reads: "Five consecutive phases in the advance of an axon tip (semidiagrammatic). Arrows indicate directions of flow, thrust and drain of protoplasm. In e, dichotomous branching of fibre has been initiated. Dotted portions represent the location of earlier protrusions that have been sucked back by the draining force of the axon".

2. Axonal tracts in the brain of a 30-hours embryo of the zebrafish (Compiled from data in Wilson et al. 1990, Chitnis and Kuwada 1990, and Ross et al. 1992)

3. Axonal tracts in the brain of a chick embryo of 60 hours of incubation, as observed in pyridine silver preparations (After Windle and Austin 1935)

128

11.3 Molecular Studies on the Development of Fibre Systems

Plate 128

During the last few decades our understanding of the molecular mechanisms involved in the guidance of growing axons to their destinations has expanded significantly (for reviews, see: Sanes and Jessell 2000; Strochlic et al. 2007; O'Donnell et al. 2009; ten Donkelaar 2014). Four major groups of axonal guidance cues, designated as chemoattractants, contact attractants, chemorepellents and contact repellents, have been identified.

The *chemoattractants* include soluble proteins such as the *netrins*, which stimulate axonal outgrowth as extracellular cues. The *contact attractants* include adhesive cell-surface molecules such as the *cadherins* and the *neural cell adhesive molecule (NCAM)*. The most important members of the *chemorepellent* and *contact repellent* groups include the *semaphorins* and the *ephrins*, respectively.

Axonal growth cones are provided with specializations that represent the molecular identity of the parent cell bodies in terms of receptor complexes, adhesion proteins and various other sets of ligand-related molecules expressed at the cell membrane. They accordingly have different specific sensitivities to variations in the regional environment, and are able to take growth-orienting and terminal or collateral sprouting decisions. Hence, the growth cones, particularly their highly mobile filopodia, represent key structures in sensing, integrating and reacting to the information provided by the various guidance cues in their environment (Plate 128, Fig. 1). The growing tips of the early-developing 'pioneering' axons are set up against the formidable task in making numerous navigational decisions during their long adventurous journeys toward their destinations. It is amazing that within the context of the wiring of the brain, vast numbers of functionally appropriate targets are reached during normal development.

Observations on the developing CNS in different vertebrates, including the zebrafish *Brachydanio rerio* (Wilson et al. 1990; Chitnis and Kuwada 1990; Ross et al. 1992: Plate 128, Fig. 2), the chick (Windle and Austin 1935; Lumsden and Keynes 1989; Figdor and Stern 1993; Plate 128, Fig. 3) and the mouse (Easter et al. 1993; Marin et al. 2002; Puelles et al. 2012), have shown that early differentiating neurons lay down a stereotyped, orthogonal axonal scaffold that closely corresponds to the pattern of molecularly defined histogenetic domains (FMUs). It is noteworthy in this context that (1) the roles of several members of the groups of guidance molecules mentioned above are highly conserved during evolution (Strochlic et al. 2007), and that (2) several protein families, including sonic hedgehog (SHH), bone morphogenetic proteins (BMPs), wingless type proteins (WNTs) and fibroblast growth factors (FGFs), which during early development play a crucial role as diffusible/diffusing morphogens in the DV and AP patterning of neuroepithelial domains (see Plates 21, 33, 35), have also been implicated in axonal growth cone guidance (Butler and Dodd 2003; Charron et al. 2003; Guan and Rao 2003; Yoshikawa and Thomas 2004). It seems that substrate pathways, morphogens and the various attractant and repellent cues discussed all play a role in laying out the plan of the wiring of the CNS.

11 The "wiring" of the brain

11.4 Basic organization plan of the human brain. (Revised from Nieuwenhuys et al. 2008)

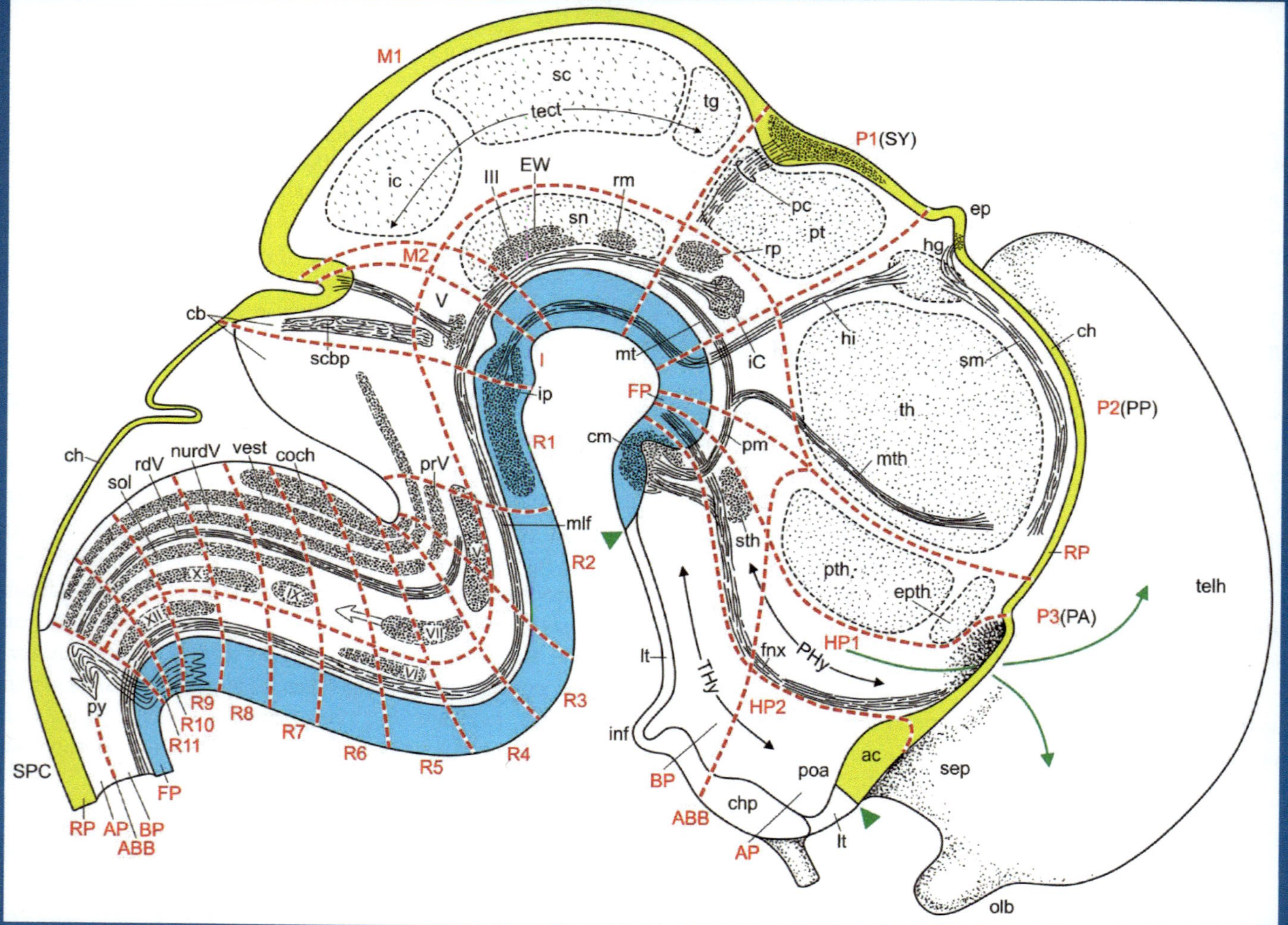

Morphological units and fibre systems mapped in a medial view of a human embryo of about 8 weeks

129

11.4 The Basic Organization Plan of the Human Brain

Plate 129

Plate 129 summarizes the principal aspects of the molecular regionalization and the wiring of the CNS as discussed during this course. It shows a medial view of the bisected brain of a human embryo of about eight weeks of gestation. The ventricular surface of the rhombencephalon has been exaggerated and turned toward the beholder to show the zonal and neuromeric organization of this brain part.

Abbreviations

ABB	Alar-basal boundary
Ac	Anterior commissure
AP	Alar Plate
BP	Basal Plate
Cb	Cerebellum
Ch	Choroidal tela
Chp	Chiasmatic plate
Cm	Corpus mamillare
Coch	Cochlear nuclei
Ep	Epiphysis
Epth	Eminentia prethalamica
EW	Edinger-Westphal nucleus
Fnx	Fornix
FP	Floor plate
Hg	Habenular ganglion
Hi	Habenulo-interpeduncular tract
HP1, 2	Hypothalamo-telencephalic neuromeres
I	Isthmic neuromere
iC	Interstitial nucleus of Cajal
ic	Inferior colliculus
inf	Infundibulum
ip	Interpeduncular nucleus
lt	Lamina terminalis
M1, 2	Mesomeres
mlf	Medial longitudinal fasciculus
mt	Mamillotegmental tract
mth	Mamillothalamic tract
olb	Olfactory bulb
P1, 2, 3	Prosomeres
PA	Parencephalon anterius
Pc	Posterior commissure
PHy	Peduncular hypothalamus
Pm	Principal mamillary fasciculus
Poa	Preoptic area
PP	Parencephalon posterius
prV	Principal trigeminal nucleus of V
pt	Pretectum
pth	Prethalamus
py	Pyramidal tract
R1, 2 etc	Rhombomeres
rdV	Radix descendens of V
rm	Magnocellular red nucleus
rp	Parvocellular red nucleus
RP	Roof plate
sc	Superior colliculus
scbp	Superior cerebellar peduncle
sep	Septum
sm	Stria medullaris
sn	Substantia nigra
sol	Solitary nucleus
SPC	Spinal cord
Sth	Subthalamic nucleus
SY	Synencephalon
Tect	Mesencephalic tectum
Telh	Telencephalic hemisphere
Th	Thalamus
THy	Terminal hypothalamus
Tg	Tectal grey primordium
Vest	Vestibular nuclei

The brain of this embryo is strongly curved, emphasizing once again that the descriptors rostral and caudal (or anterior and posterior), and dorsal and ventral, can only be adequately used in a topological context.

The neural walls of the brain can be divided into a number of basic units or fields, formed by the intersection of molecularly defined transversely oriented neural segments and longitudinal zones. The boundaries between these units are indicated by dashed red curves.

Two longitudinal domains, the basal and alar plates, extend throughout the length of the neuraxis. The border zone between these two domains, i.e. the alar-basal boundary ends just behind the optic chiasma. The basal plates are ventrally interconnected by the floor plate (in blue), whereas the alar plates are interconnected by the roof plate (in yellow).

The neural segments or neuromeres are oriented topologically perpendicular to the longitudinal axis of the brain, the course of which is parallel to the alar-basal boundary. The brain encompasses 19 neuromeres, eleven rhombomeres (R1-R11), one isthmic neuromere (I), two mesomeres (M1, M2), three diencephalic prosomeres (P1-P3), and two hypothalamo-telencephalic neuromeres (HP1, HP2). The neural tube ends rostrally at the terminal wall, also known as the acroterminal domatn of the hypothalamus and preoptic region (Puelles et al. 2012; Puelles and Rubenstein 2015).

The cerebellum develops from the alar domains of the isthmic (I) and the first rhombencephalic neuromere (R1). The alar region of M1, i.e. the mesencephalic tectum, is relatively large at this developmental stage. During further development it will be covered completely by the expanding cerebral and cerebellar hemispheres. The hypothalamo-telencephalic neuromeres (HP1, HP2) participate, as their names indicate, both in the formation of the hypothalamus as well as the telencephalon. The alar region of HP1 extends into the evaginated cerebral hemispheres (green arrows). The telencephalic component of HP2 is confined to the unevaginated preoptic area.

The rostral ends of the floor plate (mamillary region) and of the roof plate (anterior commissure) are marked with large arrowheads in Plate 129. The sizable median forebrain area between these two points represents a singular hypothalamic region (the acroterminal domain) where the alar and basal longitudinal zones of the left and right lateral walls meet front to front across the midline.

The well-delimited primary FMUs, defined by a specific combinatorial molecular identity constitute distnct major progenitor and histogenetic domains. In Plate 129 we have included a number of cellular products of the various FMUs, some represented as individual nuclei (*e.g.* the subthalamic, Edinger–Wesphal and motor V nuclei); others as nuclear complexes (the vestibular and cochlear complexes), and still others as primordial fields (the pretectum, thalamus and prethalamus). In the embryonic CNS, the axons growing out from clusters of early differentiating neurons often fasciculate together and tend to course close to the border zones of the FMUs. Thus, the pioneering tracts form together a distinct and stereotyped scaffold of transversely and longitudinally oriented axonal bundles. Many fibre systems in the adult brain derive directly from this early embryonic axonal scaffold, serving therefore as natural topologic landmarks. This striking correlation between the pattern formed by the FMUs and the course of fibre systems is illustrated in Plate 129. It will be seen that the postcommissural fornix, the superior cerebellar peduncle and the decussation of the pyramidal tract are all transversely oriented in the immediate vicinity of neuromeric borders. The stria medullaris, the initial principal mamillary fasciculus and its direct mamillotegmental continuation, as well as the medial longitudinal fasciculus and the optic tract (not illustrated) show a typical longitudinal course, whereas the routes taken by the mamillothalamic and habenulo-interpeduncular tracts include transverse and longitudinal stretches.

12 General Summary, Conclusions and Perspective

12.1 Antecedents of the New Neuromorphology

Neuromorphological concepts have stood and evolved upon methodological foundations. Early anatomical dissection of brains led to pioneering classifications of grisea and fibre tracts. Subsequent microscopic analysis introduced the cellular theory in neuromorphology, substantially correcting and increasing the number of structural details recognized. The theory of evolution by natural selection (Darwin 1859) set studies on comparative neuroanatomy in a new light Simultaneously, embryology emerged as a significant branch of biology, extending into neuromorphology, and Haeckel's notion of developmental recapitulation of evolution interconnected evolutionary and ontogenetic studies. Experimental embryological studies (e.g., Spemann and Mangold 1924) produced pioneering causal interpretations of forms and structures studied, during a time in which the novel discipline of genetics was not yet molecularly based.

Around the end of the 19th century, accrued developmental data allowed a novel sort of anatomical analysis. For the first time, this was not based on the adult brain, but on the embryonic one, and featured notions about a fundamental organization of the developing neural tube (a Bauplan), found to be shared among all vertebrates. His (1893b) brought into the first Nomina Anatomica his concept of four general longitudinal domains of the neural tube (the floor, basal, alar and roof plates), complemented with a number of transverse vesicular sectors of the tube (Plate 27).

A sound functional implication was that the basal zone contains motor centres, whereas the alar zone receives sensory afferent inputs and contains analytic centres processing such signals. While the His model was widely accepted, theorists of brain organization unfortunately diverged subsequently on the significance given to the transverse units known as neuromeres, or brain segments (His never mentioned them, though he suggested the isthmus as a neuromere-like unit.

Those designated 'neuromeric' or 'segmental' authors (e.g. Orr 1887; von Kupffer 1906; Ziehen 1906) held that such units were as relevant for the fundamental organization of the brain as the four longitudinal zones of His (1893b). This line of thought was developed most extensively by what is called the Swedish School of neuroembryologists (Palmgren 1921; Holmgren 1922, 1925; Rendahl 1924; Bergquist 1932; Bergquist and Källén 1954; Rüdeberg 1961; later also Vaage 1969). These authors examined embryos of diverse vertebrate species in order to map the neuromeric units. They also investigated the relationship of such units with proliferative discontinuities observed at the intersection of the longitudinal and transverse territories (minimal proliferation at the borders), thus reaching the concept of a shared checkerboard pattern of proliferative and histogenetic developmental units termed 'migration areas' (Plate 10; Bergquist and Källén 1954). The adult brain was held to result from the differentiation (neurones, glia, connections), growth, and associated morphogenetic deformation of these embryonic units.

Other authors doubted about the relevance of neuromeric phenomena for understanding adult brains. It was argued that neuromeres could be artefacts resulting from post-mortem delay in fixation. Moreover, the doctrine of sensory and motor nerve components did not attribute any functions to the hindbrain and spinal neuromeric units. Because the characteristic outer constrictions that delimit overt neuromeres disappear as development proceeds, it was concluded by Kuhlenbeck (1935, 1956, review in 1973) that neuromeres are transient early morphological configurations that have no permanent effect on the developing structure of the brain wall ["they show no relationship whatsoever with the definitive Bauplan"; Kuhlenbeck 1956, p. 27].

This conclusion was preceded by the finding that the apparently main histogenetic process in the development of the hindbrain was the differentiation of four columnar territories of neurons within the basal and alar plates. These were related respectively to the somato- and visceromotor, or the viscero- and somatosensory components of the cranial nerves. Function accordingly seemed associated to longitudinal columns of neurons, and not to the seemingly transient orthogonal neuromeres, which were thereafter neglected.

R. Nieuwenhuys and L. Puelles, *Towards a New Neuromorphology*,
DOI 10.1007/978-3-319-25693-1_12

Herrick (1910), working on the adult brain of urodeles, further proposed that the system of four hindbrain columns could be extrapolated to the forebrain, so that diencephalon and telencephalon could also be understood as higher centres involved in the columnar visceral-versus-somatic motor and sensory functions (Plate 51). Herrick's columnar schema received considerable support, particularly from the studies of Kuhlenbeck (see 1973 for review), who purportedly corroborated Herrick's columns in a number of embryonic anamniotes and amniotes, including mammals. The American School of comparative neuroanatomy that emerged around Herrick subsequently firmly established his columnar forebrain model in the literature, where it subsists to the present time, mainly under the variant form proposed by Swanson (1987, 1992, 2003, 2012); this author also lays emphasis on functional neuromorphology rather than a causal form-analytic approach. The post-war neuroanatomic period that developed better methods for studies of connectivity (silver degeneration techniques, axonal transport of dyes), chemoarchitectonic procedures (histochemistry, immunocytochemistry, modern dyes) and electron microscopy was thoroughly dominated by the columnar conception of brain structure, whereas neuromeric contributions were reduced to occasional appearances (see Puelles 1995 for review).

The crucial morphologic difference between the neuromeric and columnar models of the forebrain, apart from the contrasting attitude towards neuromeres, lies in the definition of the *length axis*. Neuromeric studies placed the rostral end of the neuraxis in the hypothalamus, as was first postulated by His (1893a), but investigations based on the longitudinal column concept viewed the long axis as extending into the telencephalon.

12.2 The Genoarchitectonic Basis of the New Neuromorphology

The most recent methodological leap in our field is represented by the advent of *genoarchitecture*, that is, the possibility to study gene expression patterns in embryonic and adult brains. This was made possible by parallel advances in molecular biology and genetics. The deciphering of the genetic molecular code led to the invention of protocols for in situ hybridization of messenger RNA present in the cytoplasm of cells (as a consequence of gene transcription in the cell nucleus) with probes selective for individual genes. Public databases displaying the genomic sequences of *Drosophila, Mus, Homo,* and a diversity of other animals now allow the in silico design and construction not only of optimal in situ probes for any species, but also various transgenic strategies to label, delete or insert genomic sequences in the germ line, with effects to be studied in the progeny (analysis of gene functions). This progress has enormously expanded our descriptive knowledge of which genes are typically expressed in neural structures over time (or associated with given differentiation and functional situations). It has also underlined the role of transcription factors (proteic gene products that return to the cell nucleus and attach to specific DNA sequences) in the progressive regulation of the developmental reading out of genomic information by the developmental units, the molecularly distinct progenitor domains. The experimental loss-of-function and gain-of-function analysis of genetic mechanisms controlling brain development has been also extraordinarily instructive for understanding how these units emerge and develop.

Major results of this overpowering molecular era in our field of study are represented by the following findings:

1. Developmental gene patterns have strongly supported not only the existence of longitudinal zones, but also a standard set of transverse neuromeric subdivisions in all vertebrates, thus unexpectedly confirming the old tenets of the abandoned neuromeric schools.
2. Such neuromeric expression patterns persist well beyond the moment when neuromeric bulges fade out. In fact, molecular fate mapping has shown that the neuromeric boundaries and domains can be easily traced in the adult brain (Plate 92).
3. Importantly, it was found that the axis of the neural tube cannot be arbitrarily defined, because it hangs together with the early inductive effects of the axial mesoderm—i.e. the notochord– upon the floor and basal plates of the neural tube (this process starts at neural plate stages). The resulting 'natural' axis turns out to be identical with the concept defended in neuromeric models.
4. It became manifest that the floor plate induced by the axial mesoderm does not reach the telencephalon and stops at the mamillary hypothalamus, thus supporting His (1893a, b), and falsifying the contrary columnar axial concept.
5. Causal analysis of molecular patterning at neural plate and early neural tube stages greatly helped to explain why the neural primordium of all vertebrates gets patterned into neuromeres and longitudinal zones along the anteroposterior and dorsoventral dimensions, respectively (Plates 34, 35).
6. The longitudinal zones result from the activity of linear organizer (signalling) centres within the neural primordium (the floor and the roof); this activity implies opposed, directional gradiental diffusion of different morphogen molecules, whose dorsoventral gradiental interplay across the lateral wall of the tube has antagonistic effects, leading to a combined differential specification of the basal and alar plates of His. The neuromeres result from analogous patterning mechanisms oriented along the antero-posterior dimension of the neural tube, with primary opposed signals spreading out of the terminal hypothalamic wall and the caudal or nodal end of the CNS. As the neural primordium elongates by longitudinal intercalation of new-born progenitors within clonally expanding early neural cell populations intermediate new transverse secondary organizers emerge at given positions, such as the isthmo-mesencephalic boundary (releasing FGF8), or the zona limitans intrathalamica (releasing SHH and WNTs). The obtaining concentration differences of morphogens at various distances from the sources provide differential positional information in two dimensions (anteroposterior and dorsoventral), which can be read out by the individual progenitor neuroepithelial cells in terms of differential gene expression activity (the genomic gene regulatory regions—so-called enhancers and repressors—are sensitive to the amount of these signals). Some contacting early gene expression domains show strong antagonistic effects (e.g. mutual downregulation) across their boundary, leading to its secondary adjustment into a position where the effects equilibrate. The brain primordium thus becomes regionalized (molecularly differentiated) into a primary intercrossing pattern of longitudinal and transverse fields.
7. This process of molecular regionalization continues beyond the specification of the primary alar and basal longitudinal domains and the transverse neuromeres (Plate 93), with progressive further regionalization into secondary and tertiary partitions, quickly generating well-delimited finer areal subdivisions with differential molecular phenotypes (progenitor domains, subdomains, and final microzones; Plate 95).
8. At some point these dynamics slow down and stabilize (this occurs at different times in different brain parts), producing the apparently final set of fundamental morphogenetic units (FMUs), each characterized by a field of matrix cells (progenitors) showing homogeneously a particular combination of active and inactive genes; this combination defines (codes) the molecular identity or profile of each FMU. The molecular profile controls the future downstream genomic readout by the neural progenitors, and accordingly decides all the future developmental capacities of the FMUs (either shared or differential) in terms of proliferative activity, production of specific cell types and their differentiation traits, axonal navigation properties of locally-originated axons, cell migration properties (ordered radial stratification of early versus late-born cells, and tangential migrations), and, finally, synaptogenesis and trophic properties.

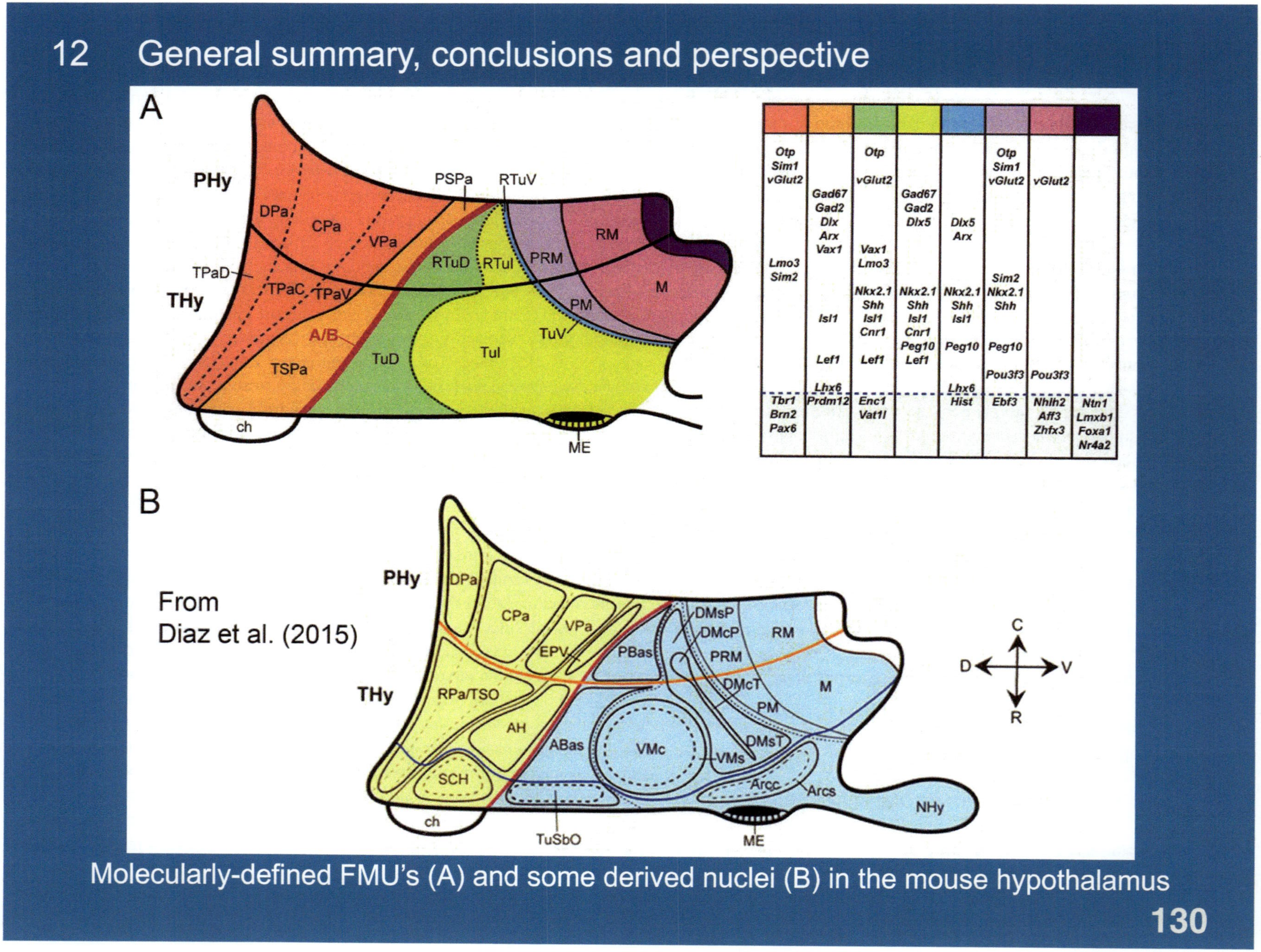
12 General summary, conclusions and perspective
A
PHy
THy
DPa
CPa
VPa
PSPa
RTuV
TPaD
TPaC
TPaV
A/B
TSPa
RTuD
RTuI
PRM
RM
M
PM
TuV
TuD
TuI
ch
ME
Otp Sim1 vGlut2 Lmo3 Sim2 Tbr1 Brn2 Pax6
Gad67 Gad2 Dlx Arx Vax1 Isl1 Lef1 Lhx6 Prdm12
Otp vGlut2 Vax1 Lmo3 Nkx2.1 Shh Isl1 Cnr1 Lef1 Enc1 Vat1l
Gad67 Gad2 Dlx5 Nkx2.1 Shh Isl1 Cnr1 Peg10 Lef1
Dlx5 Arx Nkx2.1 Shh Isl1 Peg10 Lhx6 Hist
Otp Sim1 vGlut2 Sim2 Nkx2.1 Shh Peg10 Pou3f3 Ebf3
vGlut2 Pou3f3 Nhlh2 Aff3 Zhfx3
Ntn1 Lmxb1 Foxa1 Nr4a2
B
From
Diaz et al. (2015)
PHy
THy
DPa
CPa
VPa
EPV
RPa/TSO
AH
SCH
ABas
PBas
DMsP
DMcP
PRM
RM
M
DMcT
PM
DMsT
VMc
VMs
Arcc
Arcs
NHy
TuSbO
ch
ME
C
D
V
R
Molecularly-defined FMU's (A) and some derived nuclei (B) in the mouse hypothalamus
130

Plate 130

9. Note that the molecular parcellations leading to the definition of FMUs precede the differentiation and growth of the neural wall structure, so that by using the new methodology we have gained access to the causally relevant molecular progenitor status that precedes and controls morphologic development. Classic authors just saw an apparently homogeneous neuroepithelium, but we can now see a blueprint of future brain structure in terms of multiple progenitor regions with differential molecular profiles (Plate 130).

12 General summary, conclusions and perspective

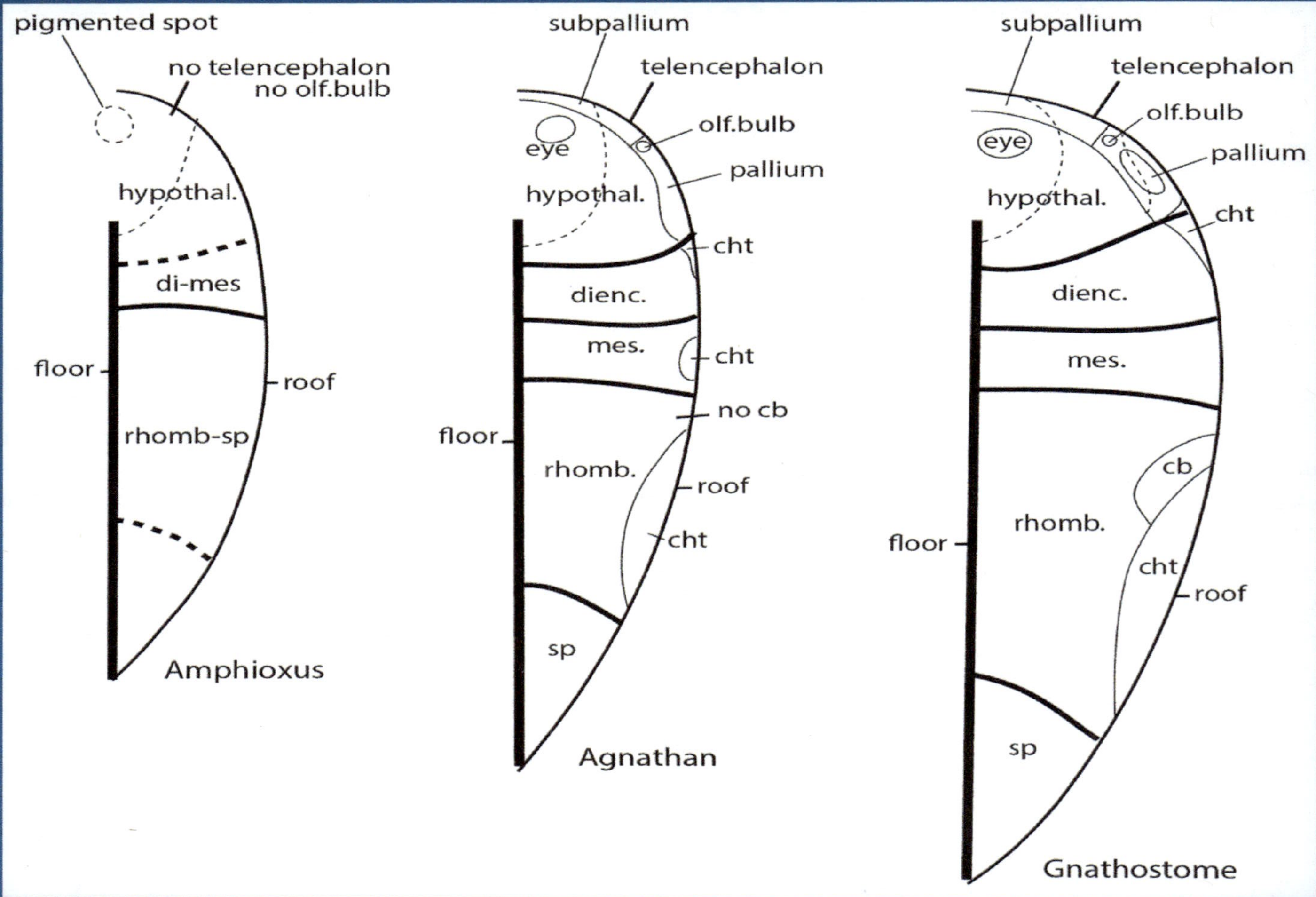

Evolutionary variation associated to complexity of fundamental brain organization at the acraniate/craniate and agnathan/gnathostome transitions.
cb, cerebellum: cht, choriod tela: sp, spinal cord

131

Plate 131

10. Finally, comparative observations across vertebrates, acraniates and invertebrates have already demonstrated that evolution, as would be expected, has widely conserved the fundamental features of the molecular organization of the flat or tubular neural primordium, irrespective of allowing a multiplicity of quantitative variations in the different taxonomic branches. There exist nevertheless distinct steps in molecular Bauplan complexity at the successive appearance of chordates, agnathans and gnathostomes. Notably, acraniate chordates such as amphioxus and the larval tunicate *Ciona* seem to lack many of the secondary organizers, and, as a result, are devoid of several of the standard brain parts (e.g. no evaginated eyes and telencephalic vesicles; no rhombic lip). Agnathans show a more complete Bauplan, having developed evaginated eyes and telencephalic vesicles, but their hemispheres remain small and incompletely patterned; also, they lack a cerebellum. Gnathostomes (starting with cartilaginous fishes) uniformly display the full molecular and structural Bauplan that can be observed in anamniote and amniote vertebrates (Plates 24, 131).

It stands to reason that the new neuromorphology has to be construed on this ample basis, leaving aside the now obsolete columnar forebrain model, and turning to an updated neuromeric model. This new neuromorphology must necessarily attend to the molecular causal mechanistic underpinning of the emerging structural parts and distinct neuronal populations of the brain. Simultaneously, we should profit from the rich comparative possibilities offered by the existence of a common Bauplan across most vertebrates.

12.3 Significance of a Common Bauplan

Mutations and transpositions are continuously introduced in a haphazard way into all genomes, causing their variation, and, as a consequence of accrued variations, the eventual evolution of new species. Natural selection ensures that only those genomes producing individuals fit to the environmental survival conditions can be propagated over long time periods. Selection also operates during development itself, since the latter's progress can be blocked or altered by mutations of genes with crucial developmental functions. There usually is a critical period of development, which is particularly sensitive to mutational changes—the so-called *zootypic stage* –, so that the processes of development occurring at this stage tend to remain invariant in the survivors, while earlier or later periods of development show higher variability (i.e. admit more mutations without loss of survival). These crucial developmental mechanisms are those that establish the basic organization of the organism, ensuring that all needed sorts of tissues and organs are produced in a correct pattern. The composition of such essential body constituents also needs proper mutual neighbourhood relationships between the diverse units, that is, a particular *invariant spatial arrangement* of the crucial parts is needed, to which we refer as the Bauplan. Further development beyond the zootypic stage will variously deform via differential growth or emergent novelty the parts of the embryo until the adult morphology is attained, but the Bauplan relationships will persist, since they are invariant among the survivors of development (there is a death penalty for varying the Bauplan). This conservation of the Bauplan during development is also reflected in evolution, leading to an apparent resistance to variation of the genomic portions that are responsible for the Bauplan in all living species (these genomic sequences actually vary as much as the others, but all versions incompatible with the Bauplan are weeded out by natural selection). This leads to the idea of a common, very slowly changing Bauplan for a large clade of animal forms, such as the vertebrates. Major step-like elaborations of neural Bauplans –e.g. at the invertebrate to vertebrate transition—were preceded by genomic duplications; variant mutation of some of the extra copies of the genes can introduce significant developmental and structural novelty, while survival is maintained by one or more of the available gene copies, which largely remain invariant.

Topology is the branch of geometry that deals with invariant positional relationships, irrespective of deformations. The Bauplan used in morphology can be seen as a topologic map of the essential parts of the body, or of any particular organ, such as the central nervous system. It encompasses those structural aspects and relationships that become established at the zootypic stage of development (normally at the end of gastrulation, that is, the neural plate), or slightly afterwards (the early neural tube). It has been amply documented that all vertebrates share a common organization of the brain primordium at these stages (Plates 34, 35).

As a consequence, the brain Bauplan of vertebrates is the fundamental morphologic reference for comparative analysis of species similarities and differences in the brain of vertebrates, and represents the necessary basis for the establishment of homology relationships (homo-logous entities being those 'thought to be the same' on the grounds of their corresponding developmental topologic positions within the shared Bauplan). Note that the existence of a common ancestor of all vertebrates is implicitly assumed in the foregoing argument. If two animal forms have wholly different Bauplans, they cannot be compared; if they have a similar Bauplan, but it varies in some aspects (e.g. Plate 131), those variant aspects cannot be compared. However, vertebrates can be compared with invertebrates by means of defining, for instance, the more primitive shared Bauplan of *Bilateria*, a group to which many invertebrates as well as the vertebrates belong. This often entails one-many homologies, meaning that one structural component of one species may be thought to correspond topologically to a set of various elements in the other species, due to the different levels of structural complexity that need to be compared (*e.g.* the heads of *Planaria* and vertebrates); we refer to these correspondences as field homolologies.

12 General summary, conclusions and perspective

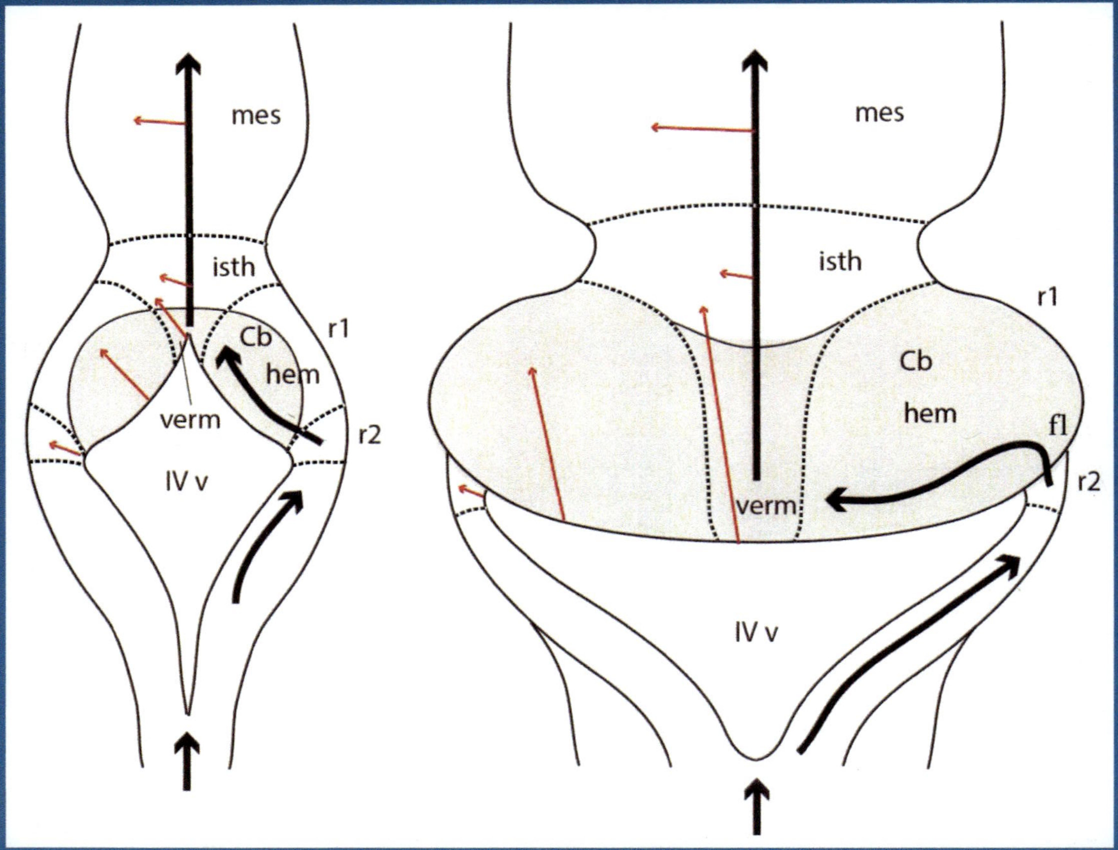

Early (A) and intermediate stages (B) of mammalian cerebellar development in dorsal view, illustrating its fate-mapped dorsal position within isthmus and rhombomere 1 (r1) -gray-shaded area-. Thick black arrows emphasize the deformed longitudinal dimension of the hindbrain (along the rhombic lip), as well as the deformed transversal planes (red arrows). Note that corresponding arrows are shown in both diagrams.

132

12.4 Necessity of a Rigorous Topologic Approach

Plate 132

Once we have a common Bauplan for the brain of vertebrates, and we want to study the structural complexity developing within a particular vertebrate brain, it is logical to classify our observations systematically according to their topologic ascription to the brain parts distinguished in the Bauplan. This insures comparability, and these are, so to speak, the 'natural' subdivisions of the brain (particularly because they are a result of natural selection over millions of years). Any other man-made subdivision pales in comparison.

A practical topological approach using natural coordinates is needed because we know that developing brains get variously deformed as they grow (Plates 64, 65), causing given derivatives to appear quite dislocated from their original positions, particularly when studied in sections. It is therefore convenient to think out at each step of analysis which growth-related deformations alter the apparent positions of the anatomic entities; it is important to refer the latter systematically back to the undeformed topologic Bauplan map in order to check the invariant neighbourhood relationships. This is also relevant in the case of mapping morphogen gradients, or any positionally oriented developmental entity (cell migrations; axonal navigation; topographically-ordered synaptic fields). For instance, just to quote some gross deformations, the posterior amygdaloid pole of the early human telencephalic vesicle gets transformed into the temporal pole of the adult); in the case of the cerebellum, its early longitudinal dimension along the rhombic lip gets stretched medio-laterally in the adult form (the vermis being topologically rostral and the flocculus caudal; verm and fl in Plate 132).

From the point of view of the Bauplan configuration, the adult human brain is a grossly deformed structure. One task of comparative neuromorphology is to keep track of the different sorts of morphogenetic brain deformation observable across the diverse vertebrate lineages, irrespective that they all continue sharing the Bauplan.

This topologic endeavour was helped in classic times, and continues to be now, by the identification of morphologic landmarks already visible at early developmental stages, whose topological position is well-defined and invariant over time. Nerve roots, commissures and decussations, and transverse or longitudinal tracts count among these elements (Plates 91, 129), and the number of the landmarks we can recognize increases with experience.

Nowadays we are further helped in keeping track of relative positions by the existence of a large variety of molecular markers (chemo- or genoarchitectonic), which selectively label given parts or areas of the Bauplan components.

A related practical comment is that topologic morphological analysis also dependss on a careful selection of the section planes used in research. Since the brain axis is bent at several points (particularly at the pontine and cephalic flexures; Plate 47), the topologic (natural) transverse and horizontal planes change from one brain part to another, often quite significantly (note atlases misdirect you into thinking that the straight axis of the microtome is the axis of the brain; any good neuromorphologist should learn to track where atlas 'cross-sections' or 'coronal sections' cease to merit those names). Moreover, the habit to give your brain specimens to a technician for 'cutting', without more precise instructions, insures that you will not obtain optimally sectioned series, increasing the risk of wrong interpretations due to oblique sectioning of given FMUs. The best guide is to first explore axial bending at the area of interest in sagittal sections; these will suggest how best to orient any needed cross-sections or horizontal sections. Keeping this problem in mind is also fundamental for interpreting any morphological or gene patterns, and even more so in the case of mutant phenotypes, where some habitual landmarks may be missing, wrongly dimensioned or misplaced.

12.5 Molecular Causal Underpinning

Neuromorphology does not deal exclusively with describing the brain parts we see. We also inquire into their ontogenetic background, that is, the developmental causal mechanisms that give rise to them (patterning, regionalization, differentiation), and their phylogenetic background, i.e. the tracing of these parts back to their apparent emergence in evolution, by means of comparative analysis.

The appearance of structural changes, or apparently new structures within a conserved Bauplan most frequently results from the accumulation of quantitative changes in morphogenesis—e.g. concerning the range of diffusion of a morphogen, or the timing of cell cycles—which obey subtly altered genomic regulatory equilibria among a host of molecular variables. These changes may be understood as developing cryptic morphologic potencies within the set of all theoretically possible forms (i.e. the morphospace) of a Bauplan, and may also explain the occasional occurrence of independent parallel evolution of some morphological characters in animals lacking a common ancestor. The causal nature of these events will now be summarized.

(A) *Patterning, regionalization and natural subdivisions*. Every structural aspect of a brain has definite causes that can be traced back to neural induction; that any two adjacent brain parts can be distinguished from each other implies a causal boundary between them. It is our role as causal neuromorphologists to investigate such relationships and discover the preceding conditions that led to the observed morphologic detail. Topologic analysis of such boundaries within the Bauplan will no doubt suggest which causal mechanisms oriented in space–anteroposterior, ventrodorsal or radial–, seem significant for the problem at hand. Our knowledge about patterning mechanisms, patterns of regionalization, and histogenetic mechanisms pertaining to the relevant natural brain subdivisions will aid us in generating at least a tentative explanatory hypothesis, which can be further explored experimentally. In older times, such causal interpretations were limited, and often rested on speculative reasoning. Nowadays it is generally possible to reduce the problem to a molecularly fleshed-out hypothetical developmental mechanism, so that neuromorphological details increasingly fall into examinable and explicable categories, though often not all the way down to the networked intricacies of genomic regulatory networks. This greater insight also allows our morphologic knowledge to become more detailed and inclusive.

(B) *Genoarchitectonically delimited FMUs*. We have seen that in a first level of analysis, the longitudinal zones and neuromeric boundaries of the neural tube wall jointly delimit primary FMUs such as, e.g., the pretectum taken as a whole. Mere cytoarchitectonic analysis allowed Rendahl (1924) to propose a subdivision into *commissural* and *precommissural pretectal domains*, whose differential histogenetic development clearly gave rise to distinct nuclear structures (Plate 88). The application of gene marker analysis (genoarchitectonics) to this problem (secondary FMUs; Ferran et al. 2007) implied a higher resolving power, so that an additional intermediate *juxtacommissural domain* was distinguished, and all three anteroposterior domains were observed to become subdivided dorsoventrally into three or more subdomains (tertiary FMUs; Plate 89). This probably would not have been possible without the change in methodology. As a consequence, the analysis of structural development in this more precisely subdivided area allowed the ascription of all previously known pretectal nuclei to one of the tertiary FMUs, and, unexpectedly, 15 new nuclei or distinct cell populations were discovered, which also were precisely ascribed to specific FMUs (Ferran et al. 2009). Though the general descriptive procedure did not change from Rendahl (1924) to Ferran et al. (2009), the molecular dissection tools clearly improved the analysis of FMU-structure within the primary pretectum domain, as well as our understanding of the number of individual nuclei derived from it. This increased knowledge is also advantageous for the comparative study of other species. If secondary and tertiary FMU subdivisions detected in a vertebrate species turn out to be present in all vertebrates, they become incorporated into the Bauplan (or if limited to to a particular group of species, we might distinguish a Bauplan variant characteristic for those species which share such subdivisions).

(C) *FMU histogenesis*. The modern differential molecular definition of tertiary FMUs reduces significantly issues associated with identifying and precisely following through histogenesis their derivatives, thanks to the reduced size of the relevant progenitor cell populations and the consequent reduction of potential derivatives of each FMU. A difficulty is that over time, some FMUs produce several cell types over time, as was recently reported by Diaz et al. (2015) in the hypothalamus, and is known for other places (e.g., cerebellum, superior colliculus). Of course, care has to be exerted to choose markers that are really selective for the studied FMU, and to keep an eye on possible surrounding areas which might exhibit the same marker and potential tangential migrations.

(D) *Ontogeny of grisea beyond FMUs: tangential migrations*. Tangential migrations are a much more frequent phenomenon than was expected a number of years ago (see Chap. 10). There are few brain areas that do not produce or receive one or several of them. Of course, a number of major tangential migrations were discovered in pre-molecular times (e.g., the rhombic lip ones; Plates 104–106), but even these have been studied in more detail by means of molecular markers, particularly as regards their intimate causal mechanisms (effects of attractors or repellents, or of trophic factors; e.g. the marginal pontine migration needs a trophic factor released by the covering pia mater). Accrued data suggest that in general the translocated cells retain at least some of their original FMU-related identity, although they may downregulate some of the initially expressed genes and upregulate new marker genes. They do not seem to acquire a new identity due to re-specification by the territory they invade (however this is not the case with ectopically grafted progenitor cells, which may easily become molecularly re-patterned if exposed to a different patterning field).

12 General summary, conclusions and perspective

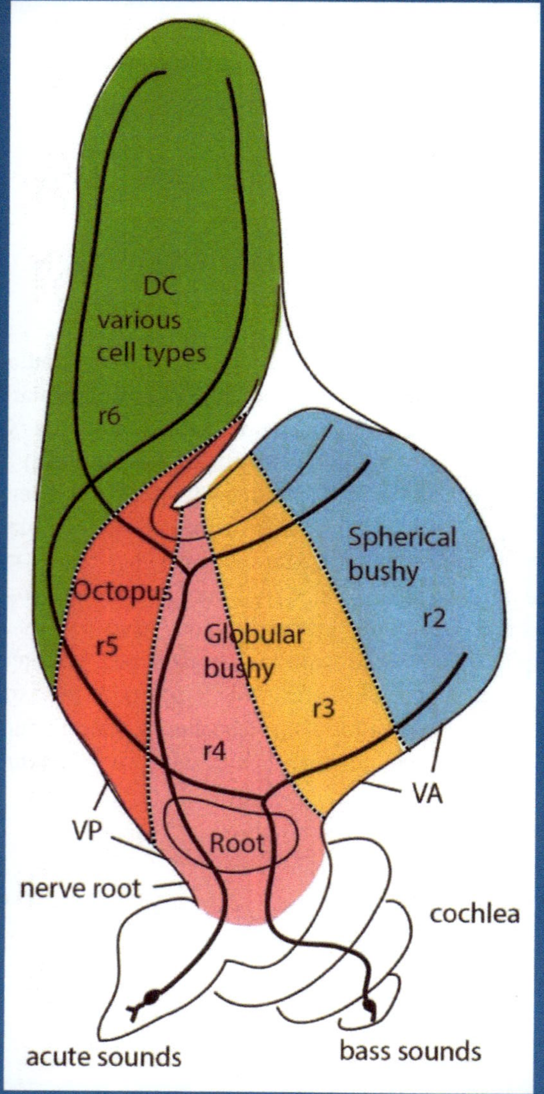

Diagram simplified and modified from Malmierca and Merchán (2004), illustrating a lateral view of the cat cochlear nuclei, with attached cochlear nerve root and sensory organ. Two primary afferents vehiculating extreme acute versus bass sound signals are shown as they bifurcate and distribute longitudinally along isotony lamellae. The VA cochlear nucleus lies in r2 and r3, the VP nucleus occupies r4 and r5, and the DC nucleus appears in r6 molecular fate-mapping of Farago et al. (2006).

This analysis reveals that the complex is actually a distorted column (compare Plate 132).

Some characteristic projection cell types seem to occupy preferentially given neuromeric domains; they are known to project to different targets, and show differential firing algorythms in response to a given sound stimulus.

see more data in Malmierca and Merchán (2004).

133

Plate 133

(E) *Plurisegmental centres*. Plurisegmental centres (*i.e.* nuclei formed across several neuromeres; Plates 99, 100) arise because a number of adjacent neuromeres, irrespective of the unique aspects of their molecular profiles, still share enough dorsoventral patterning phenomena that they produce analogous neuronal structures, which are found placed approximately at corresponding dorsoventral positions along a series of neuromeres. Though the component cells generally reveal segmental peculiarities on close examination, a superficial analysis may suggest that the individual neuromeric components are serially continuous (fused) and structurally homogeneous, so that a unitary complex is present. This is the mechanism that produces the classic hindbrain columns stretching along several rhombomeres (Plate 99; Marín and Puelles 1995; Cambronero and Puelles 2000; Tomás-Roca et al. 2014), or the mesodiencephalic substantia nigra. Detailed connectivity studies usually detect the participation of each neuromeric unit of the columns in differential circuitries. This is well shown in the cochlear column, which extends between r2 and r6 (Farago et al. 2006) and is coposed of serial neuromeric groups of different types of neurons—e.g. spherical and glomerular bushy cells, stellate cells, octopus cells—which were classically distinguished without it having been realized that this was a plurisegmental pattern (Malmierca and Merchán 2004; Plate 133). It may be generalized that the segmental units of all columns represent centres for specialized analysis and relay of signals, due to the differential molecular properties of the constituent neurons, irrespective of superficial similarity (Plates 99, 133). Analogous data have been reported in the vestibular column (e.g., Straka et al. 2001, 2002, 2003; Straka and Baker 2013), in the reticular formation, and in the serotonergic raphe nuclei (Alonso et al. 2012). Such data illustrate that specific cell types, connections and functions are reserved for the neuromeric units of each column. These differential functional aspects are subtler than the column's general function, defined by the nature of its selective afferentiation by a given sensory nerve fibre component, for instance, auditory fibres in the cochlear nuclei (incidentally, this reflects a shared plurisegmental property, namely the capacity of all cochlear neuromeric units to receive selectively synaptic afferents from this specific sort of fibres). Accordinly, at the present state of neuroscience we can ascribe particular functions both to the longitudinal and the transverse constituents of the neural wall.

(F) *A conceptual framework for comparative neuro-ontology. The classification of brain structure into the complex but nevertheless realistic set of developing FMUs underlined by recent genoarchitectonic research introduces a large measure of order for understanding its tri-dimension complexity. Developmental neural wall pigeon-holes (tertiary FMUs and dorsoventral subdomains within FMUs) are distinguished wherein all neural centres acquire a fixed position and a causal background, the whole being susceptible to systematization within the Bauplan.*

The simplest version of the Bauplan schema is two-dimensional (*i.e.* the initial blueprint of FMUs before neurons start to be produced; Plate 35), but more realistic tri-dimensional versions are possible, in which the diverse morphogenetic deformations due to proliferation and histogenesis are included, in order to approximate adult morphology. Further consideration of radial and tangential aspects of the detailed positioning of cell populations in the mantle zone completes a conceptual framework that takes us out of the old 'potatoes in a potato sack' approach, and gives us access to a convenient conceptual representation of the brain parts and their relative positions within a rigorous topological approach (logically consistent positional classifications of grisea; e.g. see Plate 92 and reference atlases developed by L.P. for the Allen Developmental Mouse Brain Atlas; www.developingmouse.brain-map.org; note that this atlas also employs the related Ontogenetic Ontology of neural entities; Puelles et al. 2013). Two-dimensional Bauplan models are nevertheless sufficient for some comparative analyses of homology, as well as for the unification of neuro-ontologies and neuro-terminologies.

12.6 New Neuromorphology and Fibre Systems

The development of fibre connections is an essential aspect of brain structural ontogeny. The selective routes often shown by fibre tracts in the brain are the result of sequential navigational decisions by the axonal growth cones, entailing molecular (and, consequently, mechanical) interactions between the neuron and its complex cellular and intercellular environment. Thus the molecular profile of the neuron relevant to axonal growth and navigation, is determined by which genes are transcribed in the cell nucleus and the subset of those gene products that is transported distally along the axon to the growth cone. Here some of these products are inserted into the cell membrane of the axonal growth cone (or secreted through it into the extracellular matrix), enabling the growth cone to sense and react to the complex chemical composition of the extracellular space plus the neuronal or glial cell membranes contacted by it in its advance. The latter properties will change as the growth cone penetrates different FMUs, forcing the axon to take a series of decisions (stopping, continuing as before, turning, or issuing collaterals; fasciculation into tracts versus de-fasciculation relates to parallel adhesive interactions occurring between contacting fibres). In essence, the growth cone compares its molecular sensitivity and capacity for oriented growth (which primarily depends on its FMU and nucleus of origin) with the environmental molecular profiles it encounters en route, through the series of relevant FMUs. Axonal growth cones are thus highly sophisticated navigating organs, which eventually can reach appropriate targets via alternative routes. For instance, the axons of the large reticular Mauthner neurons of anamniotes, which normally target the contralateral spinal cord, can be deviated experimentally all the way into the telencephalon and may still find their way back to their standard target.

Neuromeric studies have emphasized that most axonal routes seem to react to the orthogonal arrangement of longitudinal and transverse zones (FMUs) in the brain wall, so that changes of direction mostly occur at right angles (Plates 128, 129). Fibres that cross the floor plate sometimes change their properties, so that the route followed after the crossing differs from the preceding route (e.g., crossed arcuate fibres that become ascending or descending fibres at the other side). On the whole, it should be possible to map in natural coordinates the route taken by each set of fibres across a series of FMUs, then study the details of adhesivity proteins, cadherins and attractor/repellent factors, etc. in these areas (see Chap. 11), and finally explore experimentally the causality of all the involved axonal growth decisions. Some of these phenomena seem strongly conserved across vertebrates, leading to a more or less conserved fundamental framework of tracts (sometimes called 'blueprint' of fibre paths; Plate 129). However, some variability also occurs between species, due to the creation of novel collaterals, the deletion of some branches, or, more rarely, the invasion of alternative innervation fields.

12.7 Structure and Function

The approach discussed in this course addresses the study of brain morphology in a form- and causal-analytic background, with emphasis on the advantages brought into this endeavour by the use of molecular markers and the consideration of causal developmental mechanisms. It should also be noted that the development of neuronal synaptic connections establishes functional systems, with the further help of neurochemical maturation of the circuits and some trophic effects. Functions as such are complementary to brain structure, resulting from varied and transient electrochemical state dynamics occurring among neuronal complexes of small, large, or gigantic (billions of interacting elements) proportions to which must be added accompanying glial and vascular phenomena.

From previous experience, it is simplistic to try to attach specific functions to isolated columns, neuromeres or FMUs, since the neuronal derivatives of such units never act alone. Individual derivatives or components of such units normally collaborate with many other formations of a similar or diverse nature in what is known as *functional systems*. These are usually spread out side by side through several brain parts, if not through the whole brain. Calling a column 'vestibular' is a rough functional characterization serving to distinguish it from other 'non-vestibular' sensory columns, but it does not reveal the multiple individual functions that occur within its interior. Elemental functions can often be attributed to individual neuron types, implying either one or several distant effects (e.g., thalamic disinhibition through inhibition of pallidal tonic inhibitory signals by SP-spiny striatal neurons) or a local effect (e.g., lateral inhibition exerted by an interneuron). Small or large groups of neurons within a column, a neuromere or a FMU may integrate into local circuits whose typical function can be defined, at least tentatively (e.g., decoding the relative position of a sound source within the superior olivary complex, which is located wholly within rhombomere 5). However, it is rarely the case that such local circuits are not modulated by various other influences coming from a variety of distant brain areas (e.g., preoculomotor neurons controlling detailed oculomotor activity receive crucial input from the superior colliculus and pretectum, but also receive additional signals from vestibular and cerebellar neurons, as well as from the frontal oculogyric motor area; analysis of the functional meaning of each of these inputs with regard to the shared task of moving the eyes appropriately is obviously possible). In order to understand precisely such individual cell-based functional properties we need numerous subcellular and molecular details, though clearly, when the neuronal complexes involved are large or gigantic we are limited to statistical analysis.

It seems reasonable to think that complex functions slowly evolved out of simpler ones, though the primary global brain function of ensuring organismal survival must have been there from earliest ancestral invertebrate life forms. Early swimming vertebrates such as cyclostomes may have experimented with the ascription of strong functional autonomy to their serially repeated segmental brain units. The simple dorsoventral circuitry generated within segments by sensory cells axons entering the alar plate and relaying sensory signals via intermediate neurons (interneurons) to basal plate motoneurons (already present in amphioxus; Bone 1960) established the basis of reflex responses to a variety of specific sensory signals. Such simple patterns of connections were probably complemented early on by bilateral interconnections and serial activity patterns sparked along the series of metameric units, a configuration needed for swimming. However, some long integrative longitudinal connections between such units must certainly have been advantageous (e.g., reticulospinal contralateral descending activation patterns due to Mauthner-like large reticular neurons are needed to evade predators by flicking the tail appropriately). Moreover, major sensory systems probably always targetted sets of neuromeres rather than isolated ones (e.g., the visual pathway in lampreys, or Rohon-Beard sensory cells). At some points in evolutionary time, individual segments developed more distinctive properties, due to an expanded variation of their molecular profiles. As evolution progressed, the integrative suprasegmental pathways became more sophisticated, potent and pervasive, to the point of distorting or hiding the subjacent segmental circuitry. It can be maintained, nevertheless, that both longitudinal integrative and segmental discriminative functional aspects persist in the brains of all extant vertebrate species.

As yet it is unclear why given derivatives of a developmental unit (be it longitudinal, transverse or areal) participate in particular functional systems or subsystems. For instance, why is the thalamus located in alar prosomere 2, the major diencephalic centre dedicated to the relay of various data streams into specific telencephalic target areas? Did this singular functional property emerge once in evolution by chance, due to an aleatory genomic transformation affecting this neuromere, and was it conserved and developed thereafter as a nearly unique and extremely useful functional recourse? There is of course the more primitive case of the parabrachial complex in alar rhombomere 1, with analogous, but more limited telencephalopetal connective properties.

The family of cadherins (calcium-dependent adhesive proteins) has been invoked as being involved in recruiting series of centres into interconnected functional systems, because homotypic cadherin adhesion phenomena are favoured during axonal navigation and synaptogenesis (Redies 1995; Suzuki et al. 1997; Obst-Pernberg et al. 2001); however, neuroembryologists have not explored in detail yet what decides the expression of homotypic sets of cadherins in selected derivatives of different FMUs. Chance and variation of such patterns may have operated initially in evolution, with subsequent selection associated to functional benefits and resulting increased fitness. It remains to be explored how the characteristic molecular properties of the FMUs and their neuronal derivatives impose restrictions on given connections (e.g., no retinal input for the inferior colliculus), or favor in any way the establishment, diversification and activities of functional systems.

Sensory nerve afferents typically distribute longitudinally across a series of neuromeric domains, establishing a columnar pattern in doing so. The diverse segmental receptive fields so formed must share among their molecular profiles permissive and perhaps attractive axonal guidance properties, which correlate with the navigational molecular properties of particular afferent fibres. This columnar arrangement seems favorable for functional purposes, since it allows individual segmental components of a column to perform variant computations on the complete copies of the afferent data that reach them (see the case of the cochlear nuclei; Plate 133). The consequent variant outputs of the columnar units, also subject to natural selection, diversify the functional impact and advantages of the system with a minimum of genomic instruction. Apart from sensory afferents, most intrinsic longitudinal projection systems show a reiterated distribution of signal copies along a series of neuromeres. For instance, tectospinal descending fibres give out a collateral branch at most of the neuromeric levels they cross in the hindbrain. Likewise, the medial longitudinal fasciculus connects to all oculomotor nuclei, as well as to some reticular and vestibular cell groups (including the prepossitus hypoglossi nucleus) and the cervical and thoracic spinal cord motoneurons and interneurons. Similarly, the ascending spinothalamic tract connects variously along its course with hindbrain, midbrain, pretectal, thalamic and hypothalamic targets. The consequences of such hodologic arrangements, which are clearly responsive to the properties of developmental units, is a higher level of functional integration; each unit receiving a copy of the transmitted signal will perform a locally relevant variant interpretation, thus multiplying the functional benefits of the hodologic design. An extreme case of such multi-target pathways is represented by the dispersed and opportunistic systems projecting, variously, histamine, dopamine, serotonine, noradrenaline, and acetylcholine signals to a large variety of segmental brain units. Their evolutionary origins seem to have been quite primitive.

We feel that the contribution of the new neuromorphology to the eventual understanding of brain functions lies mainly in the insistence on increasingly precise parcellations and maps of each brain part, eschewing any simplistic analysis, since it stands to reason that functional studies can in no way achieve ultimate success by leaving out of consideration elements of the brain machinery.

12.8 Final Conclusion

"Niemals werden wir mit der Erforschung des Lebens endgültig abschliessen, und wenn wir einen vorläufigen Abschluss zeitweise versuchen, so wissen wir sehr wohl, dass auch das Beste, was wir geben können, nicht mehr bedeutet als eine Stufe zu Besserem".

"We will never completely finish the investigation of life, and if we occasionally try out a provisional conclusion, we know very well that the best we can offer signifies no more than a step towards a better concept".

August Weismann (1913)

References

Albuixech-Crespo B, Irimia M, Burguera D, Maeso I, Sánchez-Arrones L, Moreno JA, Somorkai I, Anaya JP, Puelles E, Bovolenta P, Garcia-Fernàndez J, Puelles L, Ferran JL (2015) Genoarchitecture of the larval nervous system of amphioxus suggests a primitive common origin of the thalamus, pretectum and midbrain of vertebrates (submitted)

Alonso A, Merchán P, Sandoval JE, Sánchez-Arrones L, Garcia-Cazorla A, Artuch R, Ferrán JL, Martínez-de-la-Torre M, Puelles L (2012) Development of the serotonergic cells in murine raphe nuclei and their relations with rhombomeric domains. Brain Struct Funct 218:1229–1277

Altman J, Bayer SA (1987) Development of the precerebellar nuclei in the rat. I. The precerebellar neuroepithelium of the rhombencephalon. J Comp Neurol 257:477–489

Altman J, Bayer SA (1987) Development of the precerebellar nuclei in the rat. II. The intramural olivary migratory stream and the neurogenetic organization of the inferior olive. J Comp Neurol 257:490–512

Altman J, Bayer SA (1987) Development of the precerebellar nuclei in the rat. III. The posterior precerebellar extramural migratory stream and the lateral reticular and external cuneate nuclei. J Comp Neurol 257:513–528

Altman J, Bayer SA (1987) Development of the precerebellar nuclei in the rat. IV. The anterior precerebellar extramural migratory stream and the nucleus reticularis tegmenti pontis and the basal pontine gray. J Comp Neurol 257:529–552

Anderson SA, Eisenstat DD, Shi L, Rubenstein JL (1997) Interneuron migration from basal forebrain to neocortex: dependence on Dlx genes. Science 278:474–476

Anthony TE, Klein C, Fischell G, Heintz N (2004) Radial glia serve as neuronal progenitors in all regions of the central nervous system. Neuron 41:881–890

Ariëns Kappers CU (1929) The evolution of the nervous system in invertebrates, vertebrates and man. Bohn, Haarlem

Aroca P, Lorente- Cánovas B, Mateos FR, Puelles L (2006) Locus coeruleus neurons originate in alar rhombomere 1 and migrate into the basal plate: studies in chick and mouse embryos. J Comp Neurol 496:802–818

Bergquist H (1932) Zur Morphologie des Zwischenhirns bei niederen Wirbeltieren. Acta Zool 13:57–303

Bergquist H (1953) On the development of the diencephalic nuclei and certain mesencephalic relations in *Lepidochelys olivacea* and other reptiles. Acta Zool 34:155–190

Bergquist H (1954a) Morphogenesis of diencephalic nuclei in homo. Kgl Fysiogr Sällsk Lund Handl N F 64(20):2–38 + plates I–VII.

Bergquist H (1954) Ontogenesis of diencephalic nuclei in vertebrates. Kgl Fysiogr Sällsk Lund Handl N F 65(6):1–34

Bergquist H, Källén B (1954) Notes on the early histogenesis and morphogenesis of the central nervous system in vertebrates. J Comp Neurol 100:627–659

Berrill NJ (1955) The origin of vertebrates. Clarendon Press, Oxford

Berry M, Rogers AW (1965) The migration of neuroblasts in the developing cerebral cortex. J Anat 99:491–509

Blakemore WF (1969) The ultrastructure of the subependymal plate in the rat. J Anat 104:423–433

Bok ST (1926) Gibt es einen gemeinsamen Bauplan in den Kernen- und Bahnensystemen der verschiedenen Querschnitte des Neuralrohres? Z ges Neurol Psychol 100:678–699

Bone Q (1960) The central nervous system in *amphioxus*. J Comp Neurol 115:27–64

Bone Q (1992) On the locomotion of ascidian tadpole larvae. J Mar Biol Ass U K 72:161–186

Bourrat F, Sotelo C (1990) Early development of the rat precerebellar system: migratory routes, selective aggregation and neuritic differentiation of the inferior olive and lateral reticular nucleus neurons: an overview. Arch Ital Biol 128:151–170

R. Nieuwenhuys and L. Puelles, *Towards a New Neuromorphology*,
DOI 10.1007/978-3-319-25693-1

Brusca RC, Brusca GJ (2003) Invertebrates, 2nd edition. Sunderland Mass.: Sinauer Associates

Buhl EH, Oelschläger HA (1988) Morphogenesis of the Brain in Harbor Porpoise. J Comp Neurol 277: 109–125

Bulfone A, Puelles L, Porteus MH, Frohman MA, Martin GR, Rubenstein JL (1993) Spatially restricted expression of Dix-1, Dix-2 (Tes-1), Gbx-2, and Wnt-3 in the embryonic day 12.5 mouse embryo defines potential transverse and longitudinal boundaries. J Neurosci 13:3155–3172

Bullock TH (1984) Comparative neuroscience holds promise for quiet revolutions. Science 225:473–478

Bullock TH (1993) How are more complex brains different? Brain Behav Evol 41:88–96

Buma P (1988) Synaptic and nonsynaptic release of neuromediators in the central nervous system. Acta Morphol Neerl-Scand 26:81–113

Buma P, Roubos EW, Buys RM (1984) Ultrastructural demonstration of exocytosis of neural, neuroendocrine and endocrine secretions with an in vitro tannic acid (TARI-)metthod. Histochemistry 80:247–256

Buma P, Veening J, Hafmans T, Joosten H (1992) Ultrastructure of the periaqueductal grey matter of the rat: an electron microscopical and HRP study. J Comp Neurol 319:519–535

Bupesh M, Legaz I, Abellán Medina L (2011) Multiple telencephalic and extratelencephalic embryonic domains contribute neurons to the medial extended amygdala. J Comp Neurol 519:1505–1525

Burnie D (2013) The Animal Book; A visual encyclopedia of life on earth. London: Dorling Kindersley Publishing

Busch HFM (1961) An anatomical analysis of the white matter in the brain stem of the cat. Thesis Univ Leiden (the Netherlands).

Butler SJ, Dodd J (2003) A role for BMP heterodimers in roof plate-mediated repulsion of commissural axons. Neuron 38:389–401

Ciceri G, Dehorter N, Sols I, Huang ZJ, Maravall M, Marin O (2013) Lineage-specific laminar organization of cortical GABAergic interneurons. Nat Neurosci 16:1199–1210

Cambronero F, Puelles L (2000) Rostrocaudal nuclear relationships in the avian medulla oblongata: a fate map with quail chick chimeras. J Comp Neurol 427:522–545

Charron F, Stein E, Jeong J, McMahon AP, Tessier-Lavigne M (2003) The morphogen sonic hedgehog is an axonal chemoattractant that collaborates with netrin-1 in midline axon guidance. Cell 113:11–23

Chitnis AB, Kuwada JY (1990) Axonogenesis in the brain of zebrafish embryos. J Neurosci 10:1892–1905

Cooper ERA (1946) The development of the substantia nigra. Brain 69:22–33

Darwin C (1859) On the origin of species by means of natural selection or the preservation of favoured races in the struggle for life. Murray, London

Dastjerdi FV, Consalez GG, Hawkes R (2012) Pattern formation during development of the embryonic cerebellum. Front Neuroanat 6:1–11

Davidson EH (2006) The regulatory genome. Elsevier, Amsterdam

Dawkins R (1992) Progress. In: Keller EF, Lloyd EA (eds) Keywords in evolutionary biology. Harvard University Press, Cambridge, pp 263–277

de Beer GR (1971) Homology, an unsolved problem. Oxford University Press, London

Delsuc F, Brinkmann H, Chourrout S, Philippe H (2006) Tunicates and not cephalochordates are the closest living relatives of the vertebrates. Nature 439:965–968

De Lange D (1936) The head problem in chordates. J Anat 70:515–547

Díaz C, Morales-Delgado N, Puelles L (2015) Ontogenesis of peptidergic neurons within the genoarchitectonic map of the mouse hypothalamus. Front Neuroanat (Special Issue on Development of the Hypothalamus) 8:1–19. doi:10.3389/fnana.2014.00162

Eagleson GW, Harris WA (1990) Mapping of the presumptive brain regions in the neural plate of *Xenopus laevis*. J Neurobiol 28:146–158

Easter SS Jr, Ross LS, Frankfurter A (1993) Initial tract formation in the mouse brain. J Neurosci 13:285–299

Egar M, Singer M (1972) The role of ependyma in spinal cord regeneration in the urodele, *Triturus*. Exp Neurol 37:422–430

Essick CR (1912) The development of the nuclei pontis and the nucleus arcuatus in man. Am J Anat 13:25–54

Farago AF, Awatramani RB, Dymecki SM (2006) Assembly of the brainstem cochlear nuclear complex is revealed by intersectional and subtractive genetic fate maps. Neuron 50:205–218

Ferran JL, Sanchez-Arrones L, Sandoval JE, Puelles L (2007) A model of early molecular regionalization in the chicken embryonic pretectum. J Comp Neurol 5:379–403

Ferran JL, Sanchez-Arrones L, Bardet SM, Sandoval JE, Martinez-de-la-Torre M, Puelles L (2008) Early pretectal gene expression pattern shows a conserved anteroposterior tripartition in mouse and chicken. Brain Res Bull 75:295–298

Ferran J, de Oliveira ED, Merchán P, Sandoval J, Sánchez-Arrones L, Martínez-De-La-Torre M, Puelles L (2009) Genoarchitectonic profile of developing nuclear groups in the chicken pretectum. J Comp Neurol 517:405–451

Ferran JL, Puelles L, Rubenstein JLR (2015) Molecular codes defining rostrocaudal domains in the embryonic mouse hypothalamus. Front Neuroanat (Special Issue on Development of the Hypothalamus). doi:10.3389/fnana.2015.00046

Figdor MC, Stern CD (1993) Segmental organization of embryonic diencephalon. Nature 363:630–634

Fink AJ, Englund C, Daza RAM, Pham D, Lau C, Nivison M, Kowalczyk T, Hevner RF (2006) Development of the deep cerebellar nuclei: transcription factors and cell migration from the rhombic lip. J Neurosci 26:3066–3067

Flames N, Pla R, Gelman DM, Rubenstein JL, Puelles L, Marín O (2007) Delineation of multiple subpallial progenitor domains by the combinatorial expression of transcriptional codes. J Neurosci 27:9582–9695

Flandin Kimura S, Rubenstein JL (2010) The progenitor zone of the ventral medial ganglionic eminence requires Nk2-1 to generate most of the globus pallidus but few neocortical interneurons. J Neurosci 30:2812–2823

Flechsig P (1876) Die Leitungsbahnen in Gehirn und Rückenmark des Menschen auf Grund entwicklungsgeschichtlicher Untersuchungen dargestellt. Engelmann, Leipzig

Flood PR (1966) A peculiar mode of muscular innervation in *amphioxus*. Light and electron microscopic studies of the so-called ventral roots. J Comp Neurol 126:181–218

Forel A (1877) Untersuchungen über die Haubenregion und ihre oberen Verknüpfungen im Gehirne des Menschen und einiger Säugethiere, mit Beiträgen zu den Methoden der Gehirnuntersuchung. Arch Psychiat Nervenkrankh (Berlin) 7:393–495

Foster M, Sherrington CS (1897) A textbook of physiology, part III: the central nervous system. MacMillan, London

Fraser SE (1993) Segmentation moves to the fore. Fate mapping and cell lineage studies reveal neuromeres in the avian forebrain, suggesting that similar cellular and molecular mechanisms operate in both forebrain and hindbrain. Curr Biol 3:787–789

Fujita S (1963) The matrix cell and cytogenesis in the developing central nervous system. J Comp Neurol 120:37–42

Fujita S (1966) Application of light and elecare on microscopic autoradiography in the study of cytogenesis of the forebrain. In: Hassler R, Stephan H (eds) Evolution of the forebrain. Thieme, Stuttgart, pp 180–190

Garcia-López M, Abellán A, Legaz L, Rubenstein JLR, Puelles L, Medine L (2008) Histogenetic compartments of the mouse centromedial and extended amygdala based on gene expressions during development. J Comp Neurol 506:46–74

Garstang W (1928) The morphology of the Tunicata, and its bearing on the phylogeny of the Chordata. Q J Microsc Sci 72:51–187

Gelman D, Marín O (2010) Generation of interneuron diversity in the mouse cerebral cortex. Eur J Neurosci 31:2136–2141

Gelman D, Griveau A, Dehorter N, Teissier A, Varela C et al (2011) A wide diversity of cortical GABAergic interneurons derives from the embryonic preoptic area. J Neurosci 31:16570–16580

Gilland E, Straka H, Wong TW, Baker B, Zottoli SJ (2014) A hindbrain segmental scaffold specifying neuronal location in the adult goldfish, *Carassius auratus*. J Comp Neurol 522:2446–2464

Goulding M (2009) Circuits controlling vertebrate locomotion: moving in a new direction. Nat Rev Neurosci. 10:507–518

Green MJ, Myat AM, Emämenegger BA, Wechsler-Reys RJ et al (2014) Independently specified *Atoh1* domains define novel developmental compartments in rhombomere 1. Development 141:389–398

Grossmann KS, Giraudin A, Britz O, Zhang J, Goulding M (2010) Genetic dissection of rhythmic motor networks in mice. Progr Brain Res 187:19–37

Guan KL, Rao Y (2003) Signalling mechanisms mediating neuronal responses to guidance cues. Nat Rev Neurosci 4:941–956

Guillemot F (2007) Spatial and temporal specification of neural fates by transcription factor codes. Development 134:3771–3780

Häggquist G (1936) Analyse der Faserverteilung in einem Rückenmarksquerschnitt (Th.3). Z Mikrosk Anat Forsch 39:1–34

Hamill GS, Olschowka JA, Lenn NJ, Jacobowitz DM (1984) The subnuclear distribution of substance P, cholecystokinin, vasoactive intestinal peptide, somatostatin, leu-enkephalin, dopamine-ß-hydroxylase and serotonin in the rat interpeduncular nucleus. J Comp Neurol 226:580–596

Hamilton WJ, Boyd JD, Mossman HW (1947) Human embryology. Heffer and Sons, Cambridge

Hansen DV, Lui JH, Parker PRL, Kriegstein AR (2010) Neurogenic radial glia in the outer subventricular zone of human neocortex. Nature 464:554–561

Hansen DV, Lui JH, Flandin P, Yoshikawa K, Rubenstein JR, Alvarez-Buylla A, Kriegstein AR (2013) Non-epithelial stem cells and cortical interneuron production in the human ganglionic eminences. Nat Neurosci 16:1576–1587

Harkmark W (1954) Cell migrations from the rhombic lip to the inferior olive, the nucleus raphe and the pons. A morphological and experimental investigation on chick embryos. J Comp Neurol 100:115–210

Harrison RG (1910) The outgrowth of the nerve fiber as a mode of protoplasmic movement. J Exp Zool 9:787–846

Hartfuss E, Galli R, Heins N, Götz M (2001) Characterization of CNS precursor subtypes and radial glia. Dev Biol 229:15–30

Hatten M (1999) Central nervous system neuronal migration. Annu Rev Neurosci 22:511–539

Hauptmann G, Söll I, Gerster T (2002) The early embryonic zebrafish forebrain is subdivided into molecularly distinct transverse and longitudinal domains. Brain Res Bull 57:371–375

Heier P (1948) Fundamental principles in the structure of the brain. A study of the brain of *Petromyzon*. Acta Anat [Suppl] VI:1–213.

Herrick CJ (1899) The cranial and first spinal nerves of *Menidia*: a contribution upon the nerve components of the bony fishes. J comp Neurol 9:153–455

Herrick CJ (1908) The morphological subdivision of the brain. J Comp Neurol 18:393–408

Herrick CJ (1910) The morphology of the forebrain in amphibia and reptilia. J Comp 20:413–547

Herrick CJ (1913) Anatomy of the brain. In: The reference handbook of the medical sciences, vol 2, 3rd edn. Wood, New York, pp 274–427.

Herrick CJ (1933) Morphogenesis of the brain. J Morphol 54:233–258

Hinds JW, Ruffett TL (1971) Cell proliferation in the neural tube: an electron microscopic and Golgi analysis in the mouse cerebral vesicle. Z Zellforsch 215:226–264

Hirata T, Li P, Lanuza GM, Cocas LA, Huntsman MM, Corbin JG (2009) Identification of distinct progenitor pools for neuronal diversity in the amygdala. Nat. Neurosci 12:141–149

His W (1889) Die Neuroblasten und deren Entstehung im embryonalen Mark. Arch Anat Physiol Anat Abt 1889:249–300

His W (1890) Die Entwicklung des menschlichen Rautenhirns vom Ende des ersten bis zum Beginn des dritten Monats I. Verlängertes Mark. Abh Kön Sächs Ges Wiss Math Phys Kl 17:1–74

His W (1893) Über das frontale Ende des Gehirnrohres. Arch Anat Physiol Anat Abt 1893:157–172

His W (1893) Vorschläge zur Eintheilung des Gehirns. Arch Anat Physiol Anat Abt 1893:172–180

His W (1904) Die Entwickelung des menschlichen Gehirns während der ersten Monate. Hirzel, Leipzig

Hökfelt T, Everitt B, Meister B et al (1986) Neurons with multiple messengers with special reference to neuroendocrine systems. Rec Progr Horm Res 42:1–70

Holland LZ (2007) A chordate with a difference. Nature 447:153–155

Holland LZ (2009) Chordate roots of the vertebrate nervous system: Expanding the molecular toolkit. Nat Rev Neurosci 10:336–746

Holland LZ (2014) Genomics, evolution and development of amphioxus and tunicates: the goldilocks principle. J Exp Zool B Mol Dev Evol, 2014 Mar 24. doi: 10.1002/jez.b.22569 (Epub ahead of print)

Holmgren N (1922) Points of view concerning forebrain morphology in lower vertebrates. J Comp Neurol 34:459–491

Holmgren N (1925) Points of view concerning forebrain morphology in higher vertebrates. Acta Zool 13:413–477

Horie T, Shinki R, Ogura Y, Kusakabe TG, Satoh N, Sasakura Y (2011) Ependymal cells of chordate larvae are stem-like cells that form the adult nervous system. Nature 469:525–528

Imai JH, Meinertzhagen IA (2007) Neurons of the ascidian larval nervous system in *Ciona intestinalis*: I. Central nervous system. J Comp Neurol 501:316–334

Imai JH, Meinertzhagen IA (2007) Neurons of the ascidian larval nervous system in *Ciona intestinalis*: II. Peripheral nervous system. J Comp Neurol 501:335–352

Imai KS, Stolfi A, Levine M, Satou Y (2009) Gene regulatory networks underlying the compartmentalization of the *Ciona* central nervous system. Development 136:285–293

Jacobson CO (1959) The localization of the presumptive cerebral regions in the neural plate of the *axolotl* larva. J Embryol Exp Morphol 7:1–21

Jacobson M (1991) Developmental Neurobiology, 3rd edition. New York: Plenum Press

Johnston JB (1902) The brain of *Petromyzon*. J Comp Neurol 12:1–86

Johnston JB (1902) An attempt to define the primitive functional divisions of the central nervous system. J Comp Neurol 12:87–106

Johnston JB (1905) The cranial nerve components of *Petromyzon*. Morphol Jahrb 34:149–403

Ju MJ, Aroca P, Luo J, Puelles L, Redies C (2004) Molecular profiling indicates avian branchiomotor nuclei invade the hindbrain alar plate. Neuroscience 128:785–796

Kahle W (1951) Studien über die Matrixphasen und die örtlichen Reifungsuntersunterschiede im embryonalen menschlichen Gehirn. Dtsch Z Nervenheilk 166:273–302

Källén B (1951) The nuclear development in the mammalian forebrain with special regard to the subpallium. Kgl Fysiogr Sällsk Lund Handl N F 61:1–43

Källén B (1951) Contributions to the ontogeny and the ventricular sulci in the vertebrate forebrain. Kgl Fysiogr Sällsk Lund Handl N F 63:1–50

Källén B (1951) Embryological studies on the nuclei in their homologization in the vertebrate forebrain. Kgl Fysiogr Sällsk Lund Handl N F 62:1–36

Källén B (1951) On the ontogeny of the reptilian forebrain. J Comp Neurol 95:307–348

Källén B (1962) Embryogenesis of brain nuclei in the chick telencephalon. Ergeb Anat Entw Gesch 36:61–82

Källén B (1965) Early morphogenesis and pattern formation in the central nervous system. In: De Haan RL, Urschprung H (eds) Organogenesis. Holt, Rinehart and Winston, New York, pp 107–128

Katz MJ, Lasek RJ (1978) Eyes transplanted to tadpole tails send axons rostrally in two spinal-cord tracts. Science 199:202–204

Katz MJ, Lasek RJ (1979) Substrate pathways which guide growing axons in Xenopus embryos. J Comp Neurol 183:817–831

Katz MJ, Lasek RJ (1981) Substrate pathways demonstrated by transplanted Mauthner axons. J Comp Neurol 195:627–641

Katz M, Lasek R, Nauta HJW (1980) Ontogeny of substrate pathways and the origin of the neural circuit pattern. Neuroscience 5:821–833

Kepecs A, Fishell G (2014) Interneuron types are fit to function. Nature 505:318–326

Kevetter GA, Lasek RJ (1982) Development of the marginal zone in the rhombencephalon of *Xenopus laevis*. Brain Res 256:195–208

Keyser AJM (1972) The development of the diencephalon of the Chinese hamster. Acta Anat (Basel) [Suppl] 59/1(83)1–181

Kiecker C, Lumsden A (2005) Compartments and their boundaries in vertebrate brain development. Nat Rev Neurosci 6:553–564

Kiecker C, Lumsden A (2012) The role of organizers in patterning the nervous system. Annu Rev Neurosci 35:347–367

Kim GJ, Shatz CJ, McConnell SK (1991) Morphology of pioneer and follower growth cones in the developing cerebral cortex. J Neurobiol 22:629–642

Kingsbury BF (1920) The extent of the floor-plate of His and its significance. J Comp Neurol 32:113–135

Kingsbury BF (1922) The fundamental plan of the vertebrate brain. J Comp Neurol 34:461–491

Knyihar-Csillik E, Csillik B, Rakic P (1995) Structure of the embryonic primate spinal cord at the closure of the first reflex arc. Anat Embryol 191:519–540

Korschelt E (1936) Vergleichende Entwicklungsgeschichte der Tiere, 2 vols. Fischer, Jena

Kovalevski A (1866) Entwicklungsgeschichte der einfachen Ascidien. Mem Acad St Petersburg 10:1–10

Krabbe KH (1942) Studies on the morphogenesis of the brain in lower mammals. Einar Munksgaard, Copenhagen

Krabbe KH (1956) Studies on the morphogenesis of the brain in *Struthio camelus* L. Einar Munksgaard, Copenhagen

Kriegstein A, Alvarez-Buylla A (2009) The glial nature of embryonic and adult neural stem cells. Annu Rev Neurosci 32:149–184

Kuhlenbeck H (1922) Zur morphologie des gymnophionengehirns. Jena Z Naturwiss 58:453–484

Kuhlenbeck H (1935) Über die morphologische Bewertung der sekundären Neuromerie. Anat Anz 81:129–148

Kuhlenbeck H (1956) Die Forbestandteile der Regio paretectalis des Anamniergehirns und ihre Beziehungen zum Hirnbauplan. Folia Anat Jpn 28:23–44

Kuhlenbeck H (1973) The central nervous system of vertebrates. Part II overall morphologic pattern, vol 3. Karger, Basel

Kuhlenbeck H, Malewitz TD, Beasley AB (1956) Further observations on the morphology of the forebrain in gymnophiona, with reference to the topologic vertebrate forebrain pattern. In: Hassler R, Stephan H (eds) Evolution of the forebrain. Thieme, Stuttgart, pp 9–19

Kuypers HGJM (1987) Pyramidal tract. In: Adelman G (ed) Encyclopedia of neuroscience, vol II. Birkhäuser, Boston, pp 1018–1020

Lacalli TC (1996) Frontal eye circuitry, rostral sensory pathways, and brain organization in amphioxus larvae: evidence from 3D reconstructions. Philos Trans R Soc Lond B 351:243–263

Lacalli TC (2005) Protochordate body plan and the evolutionary role of larvae: old controversies controlled? Can J Zool 83:216–224

Lacalli TC, Holland ND, West JE (1994) Landmarks in the anterior central nervous system of amphioxus larvae. Philos Trans R Soc Lond B 344:165–185

LaMantia AS, Rakic P (1990) Axon overproduction and elimination in the corpus callosum of the developing rhesus monkey. J Neurosci 10:2156–2175

Landsberg RL, Awatramani RB, Hunter NL, Farago AF, DiPietrantonio HJ, Rodriguez CI, Dymecki SM (2005) Hindbrain rhombic lip is comprised of discrete progenitor cell populations allocated by Pax6. Neuron 48:933–947

Leber S, Sanes J (1995) Migratory paths of neurons and glia in the embryonic chick spinal cord. J Neurosci 15:1236–1248

Lee S, Hjerring-Leffler J, Zagha E, Fishell G, Rudy B (2010) The largest group of superficial neocortical GABAergic interneurons expresses ionotropic serotonin receptors. J Neurosci 30:16796–16808

Leenen LPH, Meek J, Posthuma PR, Nieuwenhuys R (1985) A detailed morphometrical analysis of the pyramidal tract of the rat. Brain Res 359:65–80

Leto K, Carletti b, Williams IM, Magrassi L, Rossi F (2006) Different types of cerebellar GABAergic interneurons originate from a common pool of multipotent progenitor cells. J Neurosci 26:11682–11694

Leto K, Bartolini A, Yanagawa Y, Obata K, Magrassi L, Schilling K (2009) Laminar fate and phenotype specification of cerebellar GABAergic interneurons. J Neurosci 39:7079–7091

Levi-Montalcini R (1950) The origin and development of the visceral system in the spinal cord of the chick embryo. J Morphol 86:253–273

Lorente-Cánovas B, Marín F, Corral-San-Miguel R, Hidalgo-Sánchez M, Ferrán JL, Puelles L, Aroca P (2012) Multiple origins, migratory paths and molecular profiles of cells populating the avian interpeduncular nucleus. Devel Biol 361:12–26

Lumsden A (1990) The cellular basis of segmentation in the developing hindbrain. Trends Neurosci 13:329–335

Lumsden A, Keynes R (1989) Segmental patterns of neuronal development in the chick hindbrain. Nature 337:424–428

Lund JS (1984) Spiny stellate neurons. In: Peters A, Jones EG (eds) Cerebral cortex, vol 1. Cellular components of the cerebral cortex. Plenum, New York, pp 255–308.

Lund JS (1988) Anatomical organization of macaque monkey striate visual cortex. Annu Rev Neurosci 11:253–288

Lundberg JM, Hökfelt T (1983) Coexistence of peptides and classical neurotransmitters. Trends Neurosci 6:325–333

Ma T, Wang C, Wang L et al (2013) Subcortical origins of human and monkey neocortical interneurons. Nat Neurosci 16:1588–1597

Machold R, Fishell G (2005) Math1 is expressed in temporally discrete pools of cerebellar rhombic-lip neural progenitors. Neuron 48:17–24

Malmierca MS, Merchán MA (2004) Auditory system. In: Paxinos G (ed) The rat nervous system, 3rd edn. Academic Press, San Diego, pp 997–1082.

Margulis L, Schwartz KL (1998) Five kingdoms. An illustrated guide to the phyla of life on earth, 3d edn. Freeman, New York

Marin F, Puelles L (1995) Morphological fate of of rhombomeres in quail/chick chimeras: a segmental analysis of hindbrain nuclei. Eur J Neurosci 7:1714–1738

Marin F, Aroca P, Puelles L (2008) Hox gene co-linear expression in avian medulla oblongata is correlated with pseudorhombomeric domains. Dev Biol 323:230–247

Marin O, Rubenstein JLR (2001) A long, remarkable journey: tangential migration in the telencephalon. Nat Rev Neurosci 2:780–790

Marin-Padilla M (1988) Early ontogenesis of the human cerebral cortex. In: Jones EG, Peters A (eds) Cerebral cortex, vol 7. Development and maturation of cerebral cortex. Plenum, New York, pp 1–34.

Martinez-Ferre A, Navarro-Garberi M, Bueno C, Martinez S (2013) Wnt-signal specifies the intrathalamic limit and its organizer properties by regulating Shh induction in the alar plate. J Neurosci 33:3967–3980

Massaza A (1922) La citoarchitectonica del midollo spinale umano I. Arch Anat Hist Embryol 1:323–410

Massaza A (1923) La citoarchitectonica del midollo spinale umano II. Arch Anat Hist Embryol 2:1–56

Massaza A (1924) La citoarchitectonica del midollo spinale umano iii. Arch Anat Hist Embryol 4:115–189

Maynard Smith J, Szathmáry E (1995) The major transitions in evolution. Freeman, Oxford

Mayr E (1969) Grundgedanken der Evolutionsbiologie. Naturwissenschaften 56:392–397

Medina L (2007) Field homologies. In: Kaas JH, Striedter GF, Rubenstein JLR (eds) Evolution of nervous systems: a comprehensive reference, vol 1: theories, development, invertebrates. Amsterdam. Elsevier/Academic Press, Oxford, pp. 73–87

Medina L, Abellán A, Vicario A, Desfilis E (2014) Evolutionary and developmental contributions for understanding the organization of the basal ganglia. Brain Behav Evol 83:112–125

Meinertzhagen IA, Okamura Y (2001) The larval ascidian nervous system: the chordate brain from its small beginnings. Trends Neurosci 24:401–410

Meves A (1973) Elektronenmikroskopische Untersuchungen über die Zytoarchitektur des Gehirns von *Branchiostoma lanceolatum*. Z Zellforsch 139:511–532

Millhouse OE (1979) A Golgi anatomy of the rodent hypothalamus. In: Morgane PJ, Panksepp J (eds) Handbook of the hypothalamus, vol 1 Anatomy of the Hypothalamus. Dekker, New York/Basel, pp 221–265

Misson J-P, Edwards MA, Yamamoto M, Caviness VS Jr (1988) Mitotic cycling of the fetal murine cerebral wall: a combined autoradiographic and immunohistochemical study. Dev Brain Res 38:183–190

Misson J-P, Austin CP, Takahashi T, Cepko CL, Caviness VS (1991) The alignment of migrating neural cells in relation to the murine neopallial radial glial fiber system. Cereb Cortex 1:221–229

Miyoshi G, Hjerling-Leffler J, Karayannis TR, Sousa VH, Butt SJ, Battiste J, Johnson JE, Machold RP, Fishell G (2010) Genetic fate mapping reveals that the caudal ganglionic eminence produces a large and diverse population of superficial cortical interneurons. J Neurosci 30:1582–1594

Morales-Delgado N, Merchan P, Bardet S.M,. Ferrán JL, Puelles L, Díaz C. (2011) Topography of somatostatin gene expression relative to molecular progenitor domains during ontogeny of the mouse hypothalamus. Front Neuroanat. doi:10.3389/fnana.2011.00010.

Morest DK (1970) A study of neurogenesis in the forebrain of opossum pouch young. Z Anat Entw Gesch 130:265–305

Morona R, Ferran JL, Puelles L, González A (2011) Embryonic genoarchitecture of the pretectum in *Xenopus laevis*: A conserved pattern in tetrapods. J Comp Neurol 519:1024–1050

Nature Encyclopedia (2001) Oxford: Oxford Univ Press.

Neal HV (1918) Neuromeres and metameres. J Morph 31:293–315

Nicholson C (1998) Brain cell microenvironment. In: Adelman G, Smith B (eds) Encyclopedia of neurosciences, 2nd edn. Elsevier, Amsterdam, pp 261–264 (CD ROM and paper)

Nicol D, Meinertzhagen IA (1991) Cell counts and maps in the larval central nervous system of the ascidian *Ciona intestinalis* (L.). J Comp Neurol 309:415–429

Nielsen C, Parker A (2010) Morphological novelties that detonated the Ediacaran-Cambrian "explosion". Evol Dev 12:345–346

Nieuwenhuys R (1965) The forebrain of the crossopterygian *Latimeria chalumnae* Smith. J Morphol 117:1–24

Nieuwenhuys R (1972) Topological analysis of the brain stem of the lamprey *Lampetra fluviatilis*. J Comp Neurol 145:165–178

Nieuwenhuys R (1985) Chemoarchitecture of the brain. Springer, Berlin

Nieuwenhuys R (1996) The greater limbic system, the emotional motor system and the brain. Prog Brain Res 107:551–580

Nieuwenhuys R (1998) Morphogenesis and general structure. In: Nieuwenhuys R, Ten Donkelaar NJ, Nicholson C (eds) The central nervous system of vertebrates, vol I. Springer, Berlin, pp 159–228

Nieuwenhuys R (1998) Histogenesis. In: Nieuwenhuys R, Ten Donkelaar NJ, Nicholson C (eds) The central nervous system of vertebrates, vol I. Springer, Berlin, pp 229–271

Nieuwenhuys R (1998) Amphioxus. In: Nieuwenhuys R, Ten Donkelaar NJ, Nicholson C (eds) The central nervous system of vertebrates, vol I. Springer, Berlin, pp 365–396

Nieuwenhuys R (1998) Brachiopterygian fishes. In: Nieuwenhuys R, Ten Donkelaar NJ, Nicholson C (eds) The central nervous system of vertebrates, vol I. Springer, Berlin, pp 655–699

Nieuwenhuys R (1998e) The coelacanth *Latimeria chalumnae*. In: Nieuwenhuys R, Ten Donkelaar NJ, Nicholson C (eds) The central nervous system of vertebrates, vol II. Springer, Berlin, pp 1007–1043

Nieuwenhuys R (1998) Comparative neuroanatomy: place, principles and programme. In: Nieuwenhuys R, Ten Donkelaar NJ, Nicholson C (eds) The central nervous system of vertebrates, vol I. Springer, Berlin, pp 274–326

Nieuwenhuys R (1998) Structure and organisation of fibre systems. In: Nieuwenhuys R, Ten Donkelaar NJ, Nicholson C (eds) The central nervous system of vertebrates, vol I. Springer, Berlin, pp 113–157

Nieuwenhuys R (2000) Comparative aspects of volume transmission, with sidelight on other forms of intercellular communication. Prog Brain Res 125:49–126

Nieuwenhuys R (2002) Deuterostome brains: synopsis and commentary. Brain Res Bull 57:257–270

Nieuwenhuys R (2009) The Forebrain of actinopterygians revisited. Brain Behav Evol 73:229–252

Nieuwenhuys R (2009) On old and new comparative neurological sinners: the evolutionary importance of the membranous parts of the actinopterygian forebrain and their sites of attachment. J Comp Neurol 516:87–93

Nieuwenhuys R (2009) Analysis of the structure of the brain stem of actinopterygian fishes: synopsis, documentation and commentary. Brain Struct Funct 215:141–157

Nieuwenhuys R (2011) The structural, functional, and molecular organization of the brainstem. Front Neuroanat 5:1–17

Nieuwenhuys R (2013) The myeloarchitectonic studies on the human cerebral cortex of the Vogt-Vogt school, and their significance for the interpretation of functional neuroimaging data. Brain Struct Funct 218:303–352

Nieuwenhuys R, Pouwels E, Veening JG (1978) Structure and composition of the medial forebrain bundle. Neurosci Lett Suppl 1:190

Nieuwenhuys R, Geeraedts LMG, Veening JG (1982) The medial forebrain bundle of the rat. I. General introduction. J Comp Neurol 206:249–281

Nieuwenhuys R, Veening JG, van Domburg P (1989) Core and paracores; some new chemoarchitectural units in the mammalian brain. Acta Morphol Neerl-Scand 26:131–163

Nieuwenhuys R, ten Donkelaar HJ, Nicholson C (1998) The central nervous system of vertebrates, vols I–III. Springer, Berlin

Nieuwenhuys R, Voogd J, van Huijzen C (2008) The human central nervous system, 4th edn. Springer, Heidelberg

Nieuwenhuys R, Broere CAJ, Cerliani L (2015) A new myeloarchitectonic map of the human neocortex based on data from the Vogt-Vogt school. Brain Struct Funct 220:2551–2573

Noble M, Pröschel C, Mayer-Pröschel M (2004) Getting a GR(i)P on oligodendrocyte development. Devel Biol 265:33–52

Nóbrega-Pereira S, Gelman D, Bartolini G, Pla R, Pierani A, Marín O (2010) Origin and molecular specification of globus pallidus neurons. J Neurosci 30:2824–2834

Noctor SC, Flint AC, Weisman TA, Dammerman RS, Kriegstein AR (2001) Neurons derived from radial glial cells establish radial units in neocortex. Nature 409:714–720

Nordlander RH (1987) Axonal growth cones in the developing amphibian spinal cord. J Comp Neurol 263:485–496

Obermüller-Wilén (1984) Neuroendocrine systems in the brain of the lancelet, *Branchiostoma lanceolatum* (cephalochordata). Thesis, Univ Stockholm

Obst-Pernberg K, Medina L, Redies C (2001) Expression of R-cadherin and N-cadherin by cell groups and fiber tracts in the developing mouse forebrain: relation to the formation functional circuits. Neurosci 106:505–533

O'Donnell M, Chance RK, Bashaw GJ (2009) Axon growth and guidance: receptor regulation and signal transduction. Annu Rev Neurosci 32:383–412

O'Rahilly R, Müller F (1999) The embryonic human brain, 2nd edn. Wiley-Liss, New York

Orr HA (1887) Contributions to the embryology of the lizard. J Morphol 1:311–373

Palmgren A (1921) Embryological and morphological studies on the midbrain and cerebellum of vertebrates. Acta Zool 2:1–94

Parnavelas JG (2000) The origin and migration of cortical neurones: new vistas. Trends Neurosci 23:26–31

Phelps PE, Vaughn JE (1995) Commissural fibres may guide cholinergic neuronal migration in developing rat cervical spinal cord. J Comp Neurol 355:38–50

Pilz GA, Shitamukai A, Reillo I, Pacary E, Schwausch J, Stahl R, Ninkovic J, Snippert HJ, Clevers H, Godinho L, Guillemot F, Borrell V, Matsuzaki F, Götz M (2013) Amplification of progenitors in the mammalian telencephalon includes a new radial glial cell type. Nat Commun 4:2125. doi: 10.1038/ncomms3125

Pombal MA, Megias M, Bardet SM, Puelles L (2009) New and old thoughts on the segmental organization of the forebrain in lampreys. Brain Behav Evol 74:7–19

Pose-Méndez S, Candal E, Adrio F, Rodríguez-Moldes I (2014) Development of the cerebellar afferent system in the shark *Scyliorhinus canicula*: insights into the basal organization of precerebellar nuclei in gnathostomes. J Comp Neurol 522:131–168

Puelles L (1995) A segmental morphological paradigm for understanding vertebrate forebrains. Brain Behav Evol 46:319–337

Puelles L (2013) Plan of the developing vertebrate nervous system: relating embryology to the adult nervous system (prosomeric model, overview of brain organization). In: Rubenstein JLR, Rakic P (eds) Comprehensive developmental neuroscience: patterning and cell type specification in the developing CNS and PNS, vol 3. Academic Press, Amsterdam, pp 187–209

Puelles L (2014) Development and evolution of the claustrum. In: Smythies J, Ramachandran VS, Edelstein L (eds) Functional neuroanatomy of the claustrum. Academic Press, New York, pp 119–176

Puelles L, Ferran JL (2012) Neural genoarchitecture: concept and examples. Front Neuroanat 6:1–8

Puelles L, Martinez S (2013) Patterning of the diencephalon In: Rubenstein JLR, Rakic P (eds) Comprehensive developmental neuroscience: patterning and cell type specification in the developing CNS and PNS. Academic Press, Amsterdam, pp 151–172

Puelles L, Medina L (2002) Field homology as a way to reconcile genetic and developmental variability with adult homology. Brain Res Bull 57:243–255

Puelles L, Rubenstein JLR (2002) Forebrain. In: Ramachandran VS (ed) Encyclopedy of the human brain. Academic Press, New York

Puelles L, Rubenstein JLR (2015) A new scenario of hypothalamic organization: rationale of new hypotheses introduced in the updated prosomeric model. Front Neuroanat (special issue on development of the hypothalamus) 9:27. doi:10.3389/fnana.2015.00027

Puelles L, Kuwana E, Puelles E, Bulfone A, Shimamura K, Keleher J Bulfone A, Rubenstein JLR (2000) Pallial and subpallial derivatives in the embryonic chick and mouse telencephalon. J Comp Neurol 424:409–438

Puelles L, Martinez-de-la-Torre M, Paxinos G, Watson C, Martinez S (2007) The chick brain in stereotactic coordinates. Elsevier/Academic Press, Amsterdam

Puelles L, Martinez-de-la-Torre M, Bardet S, Rubenstein JLR (2012) Hypothalamus. In: Watson C, Paxinos G, Puelles L (eds) The mouse nervous system. Academic Press/Elsevier, New York, pp 221–312

Puelles L, Harrison M, Paxinos G, Watson C (2013) A developmental ontology for the mammalian brain based on the prosomeric model. Trends Neurosci 36:570–578

Rakic P (1971) Neuron-glia relationship during granule cell migration in the developing cerebellar cortex. A Golgi and E. M. study in *Macacus rhesus*. J comp Neurol 141:283–312

Rakic P (1972) Mode of cell migration to the superficial layers of fetal monkey neocortex. J Comp Neurol 145:61–83

Rakic P (1974) Neurons in the rhesus monkey visual cortex: systematic relations between time of origin and eventual disposition. Science 183:425–427

Rakic P (1990) Principles of neural cell migrations. Experientia 46:883–891

Rakic P (2009) Evolution of the neocortex: a perspective from developmental biology. Nature Rev Neurosci 10:724–735

Ramón y Cajal S (1890) À quelle époque apparaissent les expansions des cellules nerveuses de la moëlle épinière du poulet? Anat Anz 5:631–639

Ramón y Cajal S (1909) Histologie du Système Nerveux de l'Homme et des Vertébrés, vol I (reprint 1955, CSIC, Madrid)

Ramón y Cajal S (1911) Histologie du Système Nerveux de l'Homme et des Vertébrés, vol II (reprint 1955, CSIC, Madrid)

Ramón y Cajal S (1995) Histology of the nervous system of man and vertebrates. Oxford University Press, New York, Oxford (translation of the French edition by N Swanson and LW Swanson)

Ranson SW (1928) The anatomy of the nervous system. Saunders, Philadelphia

Redies C (1995) Cadherin expression in the developing vertebrate CNS: from neuromeres to brain nuclei and neural circuits. Exp Cell Res 220:243–256

Reh TA, Constantine-Paton M (1985) Growth cone-target interactions in the frog retinotectal pathway. J Neurosci Res 13:89–100

Remane A (1956) Die Grundlagen des natürlichen Systems, der vergleichenden Anatomie und der Phylogenetik. Geest and Portig, Leipzig

Remane A (1966) Phylogenetische Entwicklungsregeln von Organen. In: Hassler R, Stephan H (eds) Evolution of the forebrain. Thieme, Stuttgart, pp 1–6

Rendahl H (1924) Embryologische und morphologische Studien über das Zwischenhirn beim Huhn. Acta Zool 5:241–344

Rexed B (1954) A cytoarchitectonic atlas of the spinal cord of the cat. J Comp Neurol 100:279–379

Richardson WD, Pringle NP, Yu WP, Hall AC (1997) Origins of spinal cord oligodendrocytes: Possible developmental and evolutionary relationships with motor neurons. Devel Neurosci 19:58–68

Romer AS (1972) The vertebrate as a dual animal—somatic and visceral. Evol Biol 6:121–156

Rose MT, Ahmad KA, Thaller C, Zoghbi HY (2009) Excitatory neurons of the proprioceptive, interoceptive, and arousal hindbrain networks share a developmental requirement for *Math1*. Proc Natl Acad Sci 106:22462–22467

Ross LS, Parrett T, Easter SS Jr (1992) Axonogenesis and morphogenesis in the embryonic zebrafish brain. J Neurosci 12:467–482

Rubenstein JL, Shimamura K, Martinez S, Puelles L (1998) Regionalization of the prosencephalic neural plate. Annu Rev Neurosci 21:445–477

Rüdeberg S-I (1981) Morphogenetic studies on the cerebellar nuclei and their homologization in different vertebrates including man. Thesis Univ Lund

Sánchez-Arrones L, Ferrán JL, Rodríguez-Gallardo L, Puelles L (2009) Incipient forebrain boundaries traced by differential gene expression and fate mapping in the chick neural plate. Dev Biol 335:43–65

Sanes JR, Jessell TM (2000) Guidance of axons to their targets. In: Kandel E, Schwartz JH, Jessell TM (eds) Principles of neural science. McGraw-Hill, New York, NY, pp 1063–1086

Sauer FC (1935) Mitosis in the neural tube. J Comp Neurol 62:377–405

Sauer F (1935) The cellular structure of the neural tube. J Comp Neurol 63:13–23

Sauer FC (1936) The interkinetic migration of embryonic epithelial nuclei. J Morphol 60:1–11

Scheibel MB, Scheibel AB (1958) Structural substrates for integrative patterns in the brain stem reticular core. In: Jasper HH, Proctor LD, Knighton RS et al (eds) Reticular formation of the brain. Little Brown, Boston, pp 31–68

Scheibel MB, Scheibel AB (1968) Terminal axonal patterns in cat spinal cord. II. The dorsal horn. Brain Res 9:32–58

Schmechel DE, Rakic P (1979) A Golgi study of radial glial cells in developing monkey telencephalon: morphogenesis and transformation into astrocytes. Anat Embryol 156:115–152

Schulte H, Tilney F (1915) Development of the neuraxis in the domestic cat to the stage of 21 somites. Ann N Y Acad Sci 24:219–346

Shatz CJ (1992) How are specific connections formed between thalamus and cortex? Curr Opin Neurobiol 2:79–82

Shimamura K, Hartigan DJ, Martinez S, Puelles L, Rubenstein JLR (1995) Longitudinal organization of the anterior neural plate and neural tube. Development 121:3923–3933

Silver J, Robb RM (1979) Studies on the development of the eye cup and optic nerve in normal mice and in mutants with congenital optic nerve aplasia. Dev Biol 68:175–190

Singer M, Nordlander RH, Egar M (1979) Axonal guidance during embryogenesis and regeneration in the spinal cord of the newt: the blueprint hypothesis of neuronal pathway patterning. J Comp Neurol 185:1–22

Sipe JC, Moore RY (1977) The lateral hypothalamic area. An ultrastructural analysis. Cell Tissue Res 179:177–196

Smart IHM (1961) The subependymal layer of the mouse brain and its cellular production as shown by autoradiography. J Comp Neurol 116:325–349

Smart IHM (1985) Differential growth of the cell production systems in the lateral wall of the developing mouse telencephalon. J Anat 141:219–229

Smart IHM, Sturrock RR (1979) Ontogeny of the neostriatum. In: Divac I, Oberg RGE (eds) The neostriatum. Pergamon, Oxford, pp 127–146

Smart IHM, Dehay C, Giroud P, Berland M, Kennedy H (2002) Unique features of the proliferative zones and postmitotic compartments of the neural epithelium giving rise to striate and extrastriate cortex in the monkey. Cereb Cortex 12:37–53

Smeets WJAJ (1988) The monoaminergic systems of reptiles investigated with specific antibodies against serotonin, dopamine, and noradrenaline. In: Schwerdtfeger WK, Smeets WJAJ (eds) The forebrain of reptiles: current concepts of structure and function. Karger, Basel, pp 97–109

Spemann H, Mangold H (1924) Ûber Induktion von Embryonalanlagen durch Implantation artfremder Organisatoren. Roux Arch Entw Mech Org 100:599–638

Storm R, Cholewa-Waclaw J, Reuter K, Bröhl D, Sieber M, Treier M, Müller T, Birchmeier C (2009) The bHLH transcription factor Olig3 marks the dorsal neuroepithelium of the hindbrain and is essential for the development of brainstem nuclei. Development 136:295–305

Straka H, Baker R, Gilland E (2001) Rhombomeric organization of vestibular pathways in larval frogs. J Comp Neurol 437:42–55

Straka H, Baker R, Gilland E (2002) The frog as a unique vertebrate model for studying the rhombomeric organization of functionally identified hindbrain neurons. Brain Res Bull 57:301–305

Straka H, Holler S, Goto F, Kolb FP, Gilland E (2003) Differential spatial organization of otolith signals in frog vestibular nuclei. J Neurophysiol 90:3501–3512

Straka H, Baker R (2013) Vestibular blueprint in early vertebrates. Front Neural Circuits. 7:182. doi: 10.3389/fncir.2013.00182.

Striedter GF (1992) Phylogenetic changes in the connections of the lateral preglomerular nucleus in ostariophasan teleosts: a pluralistic view of brain evolution. Brain Behav Evol 39:329–357

Streeter GL (1933) The status of metamerism in the central nervous system of chick embryos. J Comp Neurol 57:455–467

Strochlic L, Weinl M, Piper M, Holt CE (2007) Axon pathfinding. In: Kaas JH (ed) Evolution of nervous systems, vol I. Elsevier/Academic press, Amsterdam, pp 187–209

Sultan KT, Brown KN, Shi S-H (2013) Production and organization of neocortical interneurons. Front Cell Neurosci 7:1–14. doi:10.3389/fncel.2013.00221

Sun T, Hevner RF (2014) Growth and folding of the mammalian cerebral cortex: from molecules to malformations. Nat Rev Neurosci 15:218–232

Suzuki SC, Inou T, Kimura Y, Tanaka T, Takeichi M (1997) Neuronal circuits are subdivided by differential expression of type-II classic cadherins in postnatal mouse brains. Mol Cell Neurosci 9:433–447

Swanson LW (1987) The hypothalamus. In: Björklund A, Hökfelt T, Swanson LW (eds) Handbook of chemical neuroanatomy, vol 5. Integrated systems of the CNS, Part I: hypothalamus, hippocampus, amygdala, retina. Elsevier, Amsterdam, pp 1–124

Swanson LW (1992) Brain maps: structure of the rat brain. Elsevier, Amsterdam

Swanson LW (2003) Brain architecture. Oxford Univ Press, Oxford

Swanson LW (2012) Brain architecture, 2nd edn. Oxford University Press, Oxford

Szentágothai J (1983) The modular architectonic principle of neural centres. Rev Physiol Biochem Pharmacol 98:11–91

Tekki-Kessaris N, Woodruff R, Hall AC, Gaffield W, Kimura S, Stiles CD, Rowitch DH, Richardson WD (2001) Hedgehog-dependent oligodendrocyte lineage specification in the telencephalon. Development 128:2545–2554

ten Donkelaar HJ (2014) Mechanisms of development. In: ten Donkelaar HJ, Lammens M, Hori A (eds) Clinical neuroembryology, 2nd edn. Springer, Heidelberg, pp 53–104

Thompson DAW (1942) On growth and form. Cambridge University Press, Cambridge

Tomás-Roca L, Corral-San-Miguel R, Aroca P, Puelles L, Marín F (2014) Crypto-rhombomeres of the mouse medulla oblongata, defined by molecular and morphological features. Brain Struct Funct (Epub ahead of print)

Vaage S (1969) The segmentation of the primitive neural tube in chick embryos. Ergeb Anat Entw Gesch 41:1–88

Van Beneden PJ, Julin C (1886) Morphologie des tuniciers. Arch Biol 6:237–474

Verhaart WJC (1970) Comparative aspects of the mammalian brainstem and the spinal cord, vol I, II. van Gorcum, Assen

Verney C, Zecevic N, Puelles L (2001) Structure of longitudinal brain zones that provide the origin for the substantia nigra and ventral tegmental area in human embryos, as revealed by cytoarchitecture and tyrosine hydroxylase, calretinin, calbindin, and GABA immunoreactions. J Comp Neurol 429:22–44

Vogt C, Vogt O, Allgemeine Ergebnisse unserer Hirnforschung. J Psychol Neurol 25:279–468

Von Baer KE (1828) Über Entwicklungsgeschichte der Thiere, Beobachtung und Reflexion. Bornträger, Kõnigsberg

Von Kupffer C (1906) Die Morphogenie des Centralnervensystems. In: Hertwig O (ed) Handbuch der vergleichenden und experimentellen Entwicklungslehre der Wirbeltiere, vol 2, part 3. Fischer, Jena, pp 1–272

Wang VY, Rose MF, Zoghby HY (2005) *Math1* expression redefines the rhombic lip derivatives and reveals novel lineages within the brainstem and cerebellum. Neuron 48:31–43

Wedeen VJ, Rosene DL, Wang R, Dai G, Mortazavi F, Hagmann P, Kaas JH, Tseng W-YI (2012) The geometric structure of the brain fiber pathways. Science 335:1628–1634

Weisman T, Noctor SC, Clinton BK, Honig LS, Kriegstein AR (2003) Neurogenic radial glial cells in reptile, rodent and human from mitosis to migration. Cereb Cortex 13:550–559

Weismann A (1913) Vorträge über Descendenztheorie. Gustav Fischer, Jena

Weiss P (1955) Nervous system. In: Willier BH, Weiss PA, Hamburger VB (eds) Analysis of Development. Saunders, Philadelphia, pp 346–401

Werner HB (2013) Do we have to reconsider the evolutionary origin of myelin? Front Cell Neurosci 7:1–3

Wicht H, Lacalli TC (2005) The nervous system of amphioxus: structure, development, and evolutionary significance. Can J Zool 83:122–150

Williams RW, Borodkin M, Rakic P (1991) Growth cone distribution patterns in the optic nerve of fetal monkeys: implications for mechanisms of axon guidance. J Neurosci 11:1081–1094

Wilson SW, Ross LS, Parrett T, Easter SS Jr (1990) The development of a simple scaffold of axon tracts in the brain of the embryonic zebrafish, *Brachydanio rerio*. Development 108:121–145

Windle WF (1971) Origin and early development of neural elements in the human brain. In: Tobach E (ed) The biopsychology of development. Academic Press, New York, pp 129–169

Windle WF, Austin MF (1936) Neurofibrillar development in the central nervous system of chick embryos up to 5 days incubation. J Comp Neurol 63:431–463

Winkler C, Potter A (1914) An anatomical guide to experimental researches on the cat's brain. Versluijs, Amsterdam

Wonders CP, Taylor L, Welagen J, Mbata IC, Xiang JZ, Anderson SA (2008) A spatial bias for the origin of interneuron subgroups within the medial ganglionic eminence. Dev Biol 314:127–136

Wullimann MF, Mueller T, Distel M, Babaryka A, Grothe B, Köster RW (2011) The long adventurous journey of rhombic lip cells in jawed vertebrates: a comparative developmental analysis. Front Neuroanat 5:1–16

Xu Q, Tam M, Anderson SA (2008) Fate mapping Nkx2.1-lineage cells in the mouse telencephalon. J Comp Neurol 506:16–29

Yeung J, Ha TJ, Swanson J, Choi K, Tong Y, Goldowitz D (2014) *Wls* provides a new compartmental view of the rhombic lip in mouse cerebellar development. J Neurosci 34:12527–12537

Yip JW, Yip YPL, Nakajima K, Capriotti C (2000) Reelin controls position of autonomic neurons in the spinal cord. Proc Nat Acad Scien 97:8612–8616

Yoshikawa S, Thomas JB (2004) Secreted cell-signaling molecules in axon guidance. Curr Opin Neurobiol 14:45–50

Yun K, Garel S, Fischman S, Rubenstein JL (2003) Patterning of the lateral ganglionic eminence by the Gsh1 and Gsh2 homeobox genes regulates striatal and olfactory bulb histogenesis and the growth of axons through the basal ganglia. J Comp Neurol 461:151–165

Zhu PC, Thureson-Klein A, Klein RL (1986) Exocytosis from large dense-core vesicles outside the active synaptic zones of terminals within the trigeminal subnucleus caudalis: a possible mechanism for neuropeptide release. Neuroscience 19:43–54

Ziehen T (1906) Die Morphogenie des Centralnervensystems der Säugetiere. In: Hertwig O (ed) Handbuch der vergleichenden und experimentellen Entwicklungslehre der Wirbeltiere, vol 2, part 3. Fischer, Jena, pp 273–394

Index

Boldface page numbers indicate extensive coverage of the subject

R. Nieuwenhuys and L. Puelles, *Towards a New Neuromorphology*,
DOI 10.1007/978-3-319-25693-1